Ngoc Thanh Nguyen and Edward Szczerbicki (Eds.)

Intelligent Systems for Knowledge Management

Studies in Computational Intelligence, Volume 252

Editor-in-Chief
Prof. Janusz Kacprzyk
Systems Research Institute
Polish Academy of Sciences
ul. Newelska 6
01-447 Warsaw
Poland
E-mail: kacprzyk@ibspan.waw.pl

Further volumes of this series can be found on our homepage: springer.com

Vol. 232. Feng-Hsing Wang, Jeng-Shyang Pan, and Lakhmi C. Jain
Innovations in Digital Watermarking Techniques, 2009
ISBN 978-3-642-03186-1

Vol. 233. Takayuki Ito, Minjie Zhang, Valentin Robu, Shaheen Fatima, and Tokuro Matsuo (Eds.)
Advances in Agent-Based Complex Automated Negotiations, 2009
ISBN 978-3-642-03189-2

Vol. 234. Aruna Chakraborty and Amit Konar
Emotional Intelligence, 2009
ISBN 978-3-540-68606-4

Vol. 235. Reiner Onken and Axel Schulte
System-Ergonomic Design of Cognitive Automation, 2009
ISBN 978-3-642-03134-2

Vol. 236. Natalio Krasnogor, Belén Melián-Batista, José A. Moreno-Pérez, J. Marcos Moreno-Vega, and David Pelta (Eds.)
Nature Inspired Cooperative Strategies for Optimization (NICSO 2008), 2009
ISBN 978-3-642-03210-3

Vol. 237. George A. Papadopoulos and Costin Badica (Eds.)
Intelligent Distributed Computing III, 2009
ISBN 978-3-642-03213-4

Vol. 238. Li Niu, Jie Lu, and Guangquan Zhang
Cognition-Driven Decision Support for Business Intelligence, 2009
ISBN 978-3-642-03207-3

Vol. 239. Zong Woo Geem (Ed.)
Harmony Search Algorithms for Structural Design Optimization, 2009
ISBN 978-3-642-03449-7

Vol. 240. Dimitri Plemenos and Georgios Miaoulis (Eds.)
Intelligent Computer Graphics 2009, 2009
ISBN 978-3-642-03451-0

Vol. 241. János Fodor and Janusz Kacprzyk (Eds.)
Aspects of Soft Computing, Intelligent Robotics and Control, 2009
ISBN 978-3-642-03632-3

Vol. 242. Carlos Artemio Coello Coello, Satchidananda Dehuri, and Susmita Ghosh (Eds.)
Swarm Intelligence for Multi-objective Problems in Data Mining, 2009
ISBN 978-3-642-03624-8

Vol. 243. Imre J. Rudas, János Fodor, and Janusz Kacprzyk (Eds.)
Towards Intelligent Engineering and Information Technology, 2009
ISBN 978-3-642-03736-8

Vol. 244. Ngoc Thanh Nguyen, Rados law Piotr Katarzyniak, and Adam Janiak (Eds.)
New Challenges in Computational Collective Intelligence, 2009
ISBN 978-3-642-03957-7

Vol. 245. Oleg Okun and Giorgio Valentini (Eds.)
Applications of Supervised and Unsupervised Ensemble Methods, 2009
ISBN 978-3-642-03998-0

Vol. 246. Thanasis Daradoumis, Santi Caballé, Joan Manuel Marquès, and Fatos Xhafa (Eds.)
Intelligent Collaborative e-Learning Systems and Applications, 2009
ISBN 978-3-642-04000-9

Vol. 247. Monica Bianchini, Marco Maggini, Franco Scarselli, and Lakhmi C. Jain (Eds.)
Innovations in Neural Information Paradigms and Applications, 2009
ISBN 978-3-642-04002-3

Vol. 248. Chee Peng Lim, Lakhmi C. Jain, and Satchidananda Dehuri (Eds.)
Innovations in Swarm Intelligence, 2009
ISBN 978-3-642-04224-9

Vol. 249. Wesam Ashour Barbakh, Ying Wu, and Colin Fyfe
Non-Standard Parameter Adaptation for Exploratory Data Analysis, 2009
ISBN 978-3-642-04004-7

Vol. 250. Raymond Chiong and Sandeep Dhakal (Eds.)
Natural Intelligence for Scheduling, Planning and Packing Problems, 2009
ISBN 978-3-642-04038-2

Vol. 251. Zbigniew W. Ras and William Ribarsky (Eds.)
Advances in Information and Intelligent Systems, 2009
ISBN 978-3-642-04140-2

Vol. 252. Ngoc Thanh Nguyen and Edward Szczerbicki (Eds.)
Intelligent Systems for Knowledge Management, 2009
ISBN 978-3-642-04169-3

Ngoc Thanh Nguyen and Edward Szczerbicki (Eds.)

Intelligent Systems for Knowledge Management

Springer

Prof. Ngoc Thanh Nguyen
Institute of Informatics
Wroclaw University of Technology
Str. Wyb. Wypsianskiego 27
50-370 Wroclaw
Poland
E-mail: thanh@pwr.wroc.pl

Prof. Edward Szczerbicki
Faculty of Management and Economics
Gdansk University of Technology,
Str. Narutowicza 11/12
80-233 Gdansk,
Poland
E-mail: Edward.Szczerbicki@zie.pg.gda.pl

ISBN 978-3-642-04169-3 e-ISBN 978-3-642-04170-9

DOI 10.1007/978-3-642-04170-9

Studies in Computational Intelligence ISSN 1860-949X

Library of Congress Control Number: 2009934306

Typeset & Cover Design: Scientific Publishing Services Pvt. Ltd., Chennai, India.

Printed in acid-free paper

9 8 7 6 5 4 3 2 1

springer.com

Editorial

If two individuals get together and exchange a dollar, they each walk away with a dollar. If the same individuals get together and exchange an idea, they both walk away with two ideas.

~Thomas Jefferson

The quote by Thomas Jefferson gets at the very heart of what this edited book is all about. This is a platform to share ideas. Our contributors, the authors of carefully selected and peer revived Chapters, are academics, educators, and practitioners who are pursuing a course of study, research and implementation of Artificial Intelligence (AI) and Knowledge Management (KM). They are at the forefront of exploring ways to integrate AI and KM. This book serves as a way for them to connect with others by sharing strategies, ideas, models, approaches that they have successfully implemented in their own environments and case studies.

The Editors of this book, acting over the last decade as Editors and Associate Editors for a number of international journals and book series in the general area of complex intelligent systems and knowledge management, as well as Chairs and Co-Chairs of several conferences in this field, have had the exciting opportunity to closely witness and to be actively engaged in the emerging research area of Knowledge Management (KM). Today, an enterprise at its core can be considered "intelligent enterprise" converting information and knowledge resources into some useful form of the end product. This is a new paradigm which creates enormous challenges for researchers in the new millennium. The purpose of this book is to address some of these challenges, especially those related to smart management of knowledge.

A desire to come to grips with KM has become an area of interest both commercially and academically in recent years primarily as a consequence of the technological revolution. Our ability as a society to manage our existing knowledge base has become the benchmark by which we are able to move forward and expand our knowledge horizons. The expression “avoid re-inventing the wheel” has become the catch cry justification for KM principals. Knowledge builds on knowledge, and managing existing knowledge availability feeds the growth of future knowledge. Acquisition of knowledge, through smart transformation of information, can make the difference between the success and failure of a company in the competitive environment of global economy and knowledge society. Recent studies have established that the aims of KM research have to look for: (a) reducing the gap between theory and practice, and thus providing well-established knowledge strategies, tools and procedures for decision makers, and (b) achieving holistic research which integrates a number of the diverse perspectives of researchers and practitioners. The Chapters of our book pursue these aims with excellent precision – they bridge the gap between academic propositions and real life requirements by integrating different perspectives of theoretical probing and actuality. The book adds significantly to the development and application of intelligent knowledge management techniques and technologies for a complex systems environment in which industries have to efficiently perform their day-to-day operations. These are frontier technologies which have enormous potential to transfer global industries to knowledge based philosophy as required in the age of information society and global economy. The ability to transform information into knowledge, and then to use it in the decision making processes, contributes in a substantial way to intelligence understood as the ability to act appropriately in uncertain environments. This book focuses on the area of smart information use in studying and understanding of knowledge generation, maintenance, and usage.

The first Chapter of the book titled *Learning Automata-based Solutions to Stochastic Nonlinear Resource Allocation Problems* and authored by Ole-Christoffer Granmo and B. John Oommen, deals with computational intelligence aspect of solving hard and complex problems for which finding optimal solution through traditional approaches becomes infeasible. The Authors focus their deliberations on the family of Knapsack Problems (KP), the sub-set of NP-hard problems and show how Learning Automata approach can be used to produce fast and accurate solutions to randomized versions of the KP. Thee presented approach is illustrated with comprehensive experimental results demonstrating its superiority over the previous state-of-the-art schemes. The next Chapter titled *Neural Networks in Model Predictive Control* is authored by Maciej Lawrynczuk and addresses the issues of real-life efficiency and applications of advanced control technique called MPC (Model Predictive Control) for which neural models are used on-line. The Author provides very comprehensive discussion of a number of MPC related issues including model structure selection, training, and stability. The paper recommends computationally efficient algorithms which use on-line linearization of the neural model and demonstrates that they are favorable regarding the outcomes in comparison to those obtained when nonlinear optimisation is used on-line in MPC. The ideas presented are supported by illustrative examples considering the control of a chemical process. The paper clearly sets the direction for future developments in the area of smart advanced control and prediction that have an excellent potential of abundance of real life applications. The following Chapter is titled *Application of a Multi-domain Knowledge Structure: The Decisional DNA* and is authored by Cesar Sanín, Leonardo Mancilla-Amaya, Edward Szczerbicki and Paul CayfordHowell. This Chapter presents three case studies of the implementation of knowledge engineering platform developed by the Authors. It explains, in a comprehensive and at the same time very practical way, cases of collecting formal decision events, that is, decisional experience from: geothermal energy development process, renewable energy case, and net income evaluation issue and uses this information for the construction of decisional chromosomes viewed as building blocks of Decisional DNA. Decisional DNA and the Set of Experience were applied as a novel and innovative knowledge representation structure for gathering experiences for each case studied in the Chapter. The recorded experiences were then implemented in an ontology structure with reflexivity characteristics in order to use it as a tool for decision making processes that can substantially enhance different real life systems with prediction capabilities and facilitate knowledge engineering processes embedded in decision making. In conclusion of the Chapter the Authors discuss the future research directions and challenges related to the idea of Decisional DNA further development and potential applications. The subsequent Chapter *Intelligent-based Mobility Models for Data Management in Wireless Sensor Networks* is authored by Samer Hanoun and Saeid Nahavandi. The Authors propose Wireless Sensor Networks (WSNs) as powerful and efficient means for fine grained monitoring in different classes of real life applications that can be run over extended periods of time at a very low cost. Intelligent-based models taking into consideration the network runtime conditions are proposed by the Authors to overcome the problem the data latency in the network the sensors' energy consumption levels. All existing and proposed models are thoroughly discussed providing the reader with the most comprehensive insight into the possible solutions available in this domain. The models are additionally compared considering various metrics and design goals. The Chapter concludes with directions of further research in this dynamically evolving area of data management. In the following Chapter titled *Abductive Inference Based Approach to DSS Designing for Project Portfolio Planning* the Authors Grzegorz Bocewicz and Zbigniew A. Banaszak propose their reference model for smart project planning. This innovative model employs Constraint Programming (*CP*) paradigm describing both an enterprise and a set of project-like production orders. In addition, the model includes consumer orders requirements and available production capabilities providing a powerful formal framework allowing the user to develop a class of decision support systems aimed at interactive production process planning cases that can be subjected to multi-project environment constraints. The proposed concept of a context dedicated constraint satisfaction problem is a new, cutting edge formal framework for developing smart task oriented Decision Support Tool for Project Portfolio Prototyping setting the direction for further research in the emerging area of development of designing of dedicated and interactive decision support systems. The next Chapter titled *CSP Techniques for Solving Combinatorial Queries within Relational Databases by* Malek Mouhoub and Chang Feng addresses the area of relational databases that became the most significant practical platform for data management implementations and query searching. In particular the Authors present a new modeling solution to overcome the problem of conventional relational database systems often failing to obtain answers to the combinatorial query efficiently. The model proposed by the Authors of this Chapter integrates the Constraint Satisfaction Problem (CSP) framework into the database systems. The Authors compare the response time performance of their presented CSP-based model with the traditional way of handling combinatorial queries by conducting several experiments on large size databases. The presented results are very promising and show the superiority of the approach presented in this Chapter comparing to the traditional one. In conclusion the Authors define a very clear path for future research in this area with the aim of further reduction of processing time. The Chapter that follows is titled *Predictive Capabilities of Adaptive and Evolutionary Fuzzy Cognitive Maps -A Comparative Study* and authored by Wojciech Froelich and Przemyslaw Juszczuk. The Authors introduce Fuzzy Cognitive Maps (FCMs) a a new, efficient, and user friendly graph based representation of decision processes. After comprehensive introduction and state-of-the-art literature review in the field, the focus of the Chapter is directed towards comparative study of adaptive and evolutionary FCMs based on type of learning that is applied. Then the Authors propose an approach where diverse learning methods are applied to the same prediction problem and use the effectiveness of prediction as a quality measure to evaluate the trained FCMs. The ultimate goal of this research is to develop real life application recommendations regarding the particular method that should be used for a given prediction problem. The presented approach is illustrated with an example of weather conditions prediction problem. The next Chapter is titled *An Architecture for the Semantic Enhancement of Virtual Engineering Applications* and is authored by Carlos Toro, Manuel Graña, Jorge Posada, Cesar Sanín and Edward Szczerbicki. This Chapter introduces the exciting field of Virtual Engineering (VE) defined as the integration of geometric models and related engineering tools within a computerized environment that facilitates multidisciplinary and collaborative product development. The Authors propose their own novel architecture for the semantic enhancement of Virtual Engineering Applications (VEA) through the embedded Virtual Engineering Tools that they developed. The proposed architecture follows a User-Intention-Domain schema and makes use of state of the art technologies like the Set of Experience Knowledge Structure and the Reflexive Ontologies in order to enhance user experience with VEA. Conclusions of this Chapter include future directions of this innovative research area. The following Chapter titled *Effective Backbone Techniques for Ontology Integration* is authored by Trong Hai Duong, Ngoc Thanh Nguyen, and Geun Sik Jo. The Authors of this Chapter introduce a novel approach to ontology integration that reduces computational complexity related to traditional approaches based on similarity measurements between ontological entities. This new approach starts with introduction of an enriched ontology model for formal ontological analysis that facilitates development of ontologies with a clean and untangled taxonomic structure. The Authors propose *importance* and *types of concepts* for priority matching and direct matching between concepts. *Identity-based similarity* replaces comparisons of all properties related to each concept, while matching between concepts, and thus reduces the computational complexity. The proposed approach reduces also number of conflicts in ontology integration making it an important development in the area of knowledge engineering. The Chapter that follows is titled *Cluster Analysis in Personalized E-learning Systems* and is authored by Danuta Zakrzewska. To enhance the performance of an e-learning system, the Author focuses on ways of improving adaptation of such systems. The Chapter introduces an approach in which cluster analysis is applied to arrive at student groupings according to their dominant learning styles and usability preferences. The Author develops cluster-based smart student model in which personalized teaching paths as well as information content organization are adjusted according to different requirements. Presented in the Chapter research efforts focus on developing clustering technique that guarantees clusters of good quality and fulfils tutor requirements at the same time. Such technique was developed and proposed by the Author. The technique is based on novel approach to clustering thresholds which decide about cluster qualities. The approach is illustrated with informative examples and Chapter conclusions define comprehensive path of further research in this area of intelligent tools development to support e-learning processes. The next Chapter titled *Agent-based Modelling and Analysis of Air Traffic Organisations* is authored by Alexei Sharpanskykh. It introduces a comprehensive concept of developing a smart formal agent-based organisational model of an Air Traffic Organisation (ATO) accounting for typical strong structural and behavioural complexities that present a challenge both for modelling and analysis of such an organisation. The Author develops and justifies an automated approach for modelling and analysis of complex ATOs. The presented approach is illustrated by a case study in which ATO's tasks related to movement of aircraft on the ground and to safety occurrence reporting are considered. Conclusions of the Chapter outline further research needs in this area. The Chapter that follows is titled *Identification of Contact Conditions by Active Force Sensing* and is authored by Takayoshi Yamada, Tetsuya Mouri, Nobuharu Mimura, Yasuyuki Funahashi, and Hidehiko Yamamoto. The Authors introduce and discuss the problem of parameter identification of contact conditions between an object grasped by a robot and its environment. They propose an algorithm distinguishing four contact types and verify its effectiveness through numerical examples. The advantages and the novelty of the proposed approach are discussed and it is shown that the method worked out by the Authors is applicable to the identification of the contact conditions in a variety of tasks, including the contact between grasped object and an external environment, the contact between grasped object and each finger of a robot hand, the contact between a manipulated object and each arm of a humanoid robot, and the contact between a floor surface and each foot of a legged robot. The last Chapter of this selection is titled *Modelling User-Centric Pervasive Adaptive Systems – The REFLECT Ontology* and is authored by Gerd Kock, Marko Ribarić, and Nikola Šerbedžija. The Authors aim is to supplement current smart systems with supportive behaviour in terms of doing what the users want, i.e. feel and desire in a seamless and personalized way. They propose an innovative and novel approach to reach this goal by introducing reflective technology that strives for simplicity offering a generic software/hardware solution for a wide range of application domains, ranging from vehicular, home ambient up to the embedded real-time systems. Implementation and application issues are thoroughly discussed by the Authors who finish the Chapter by prpviding directions for further research in this exciting area of intelligent systems development and implementation.

The briefly introduced above Chapters of this book represent a sample of an effort to provide guidelines to develop tools for smart processing of knowledge and information that is available to decision makers acting in information rich environments of our knowledge based society. The guide does not presume to give ultimate answers but it poses models, approaches, and case studies to explore, explain and address the complexities and challenges of modern KM issues.

Ngoc Thanh Nguyen
Wroclaw University of Technology, Wroclaw, Poland

Edward Szczerbicki
Gdansk University of Technology, Gdansk, Poland

Table of Contents

Learning Automata-Based Solutions to Stochastic Nonlinear Resource Allocation Problems

Ole-Christoffer Granmo[1] and B. John Oommen[2,⋆,⋆⋆]

[1] Dept. of ICT, University of Agder, Grimstad, Norway
[2] School of Computer Science, Carleton University, Ottawa, Canada

Abstract. "Computational Intelligence" is an extremely wide-ranging and all-encompassing area. However, it is fair to say that the strength of a system that possesses "Computational Intelligence" can be quantified by its ability to solve problems that are intrinsically hard. One such class of NP-Hard problems concerns the so-called family of *Knapsack* Problems, and in this Chapter, we shall explain how a sub-field of Artificial Intelligence, namely that which involves "Learning Automata", can be used to produce fast and accurate solutions to "difficult" and randomized versions of the *Knapsack* problem (KP).

Various increasingly complex versions of the KP have been discussed in the literature. The most complex of these is probably the *stochastic* Nonlinear Fractional Knapsack Problem (NFKP), which is a randomized version of the Fractional Knapsack Problem (FKP). The FKP concerns filling a knapsack of fixed volume with a mixture of n materials so as to attain a maximal value, where each material is characterized by its *value per unit volume.* Although the original problem can be posed in such abstract terms, we shall, at the outset, argue that it can be used to model a host of real-life problems.

This Chapter introduces two novel solutions to the stochastic NFKP which involve a *Team* of deterministic Learning Automata (LA). The first solution is referred to as the decentralized Learning Automata Knapsack Game (LAKG). Using the general LA paradigm, this scheme improves a current solution in an online manner, through a series of informed guesses which move towards the optimal solution. The second solution is a completely new on-line LA system, namely, the *Hierarchy of Twofold Resource Allocation Automata* (H-TRAA) whose primitive component is a *Twofold Resource Allocation Automaton* (TRAA). The Chapter formally states the convergence properties of the TRAA and the H-TRAA, and also presents empirical results demonstrating "computationally intelligent" properties which are superior to those possessed by the traditional estimation-based methods.

Keywords: *Nonlinear Knapsack Problems, Hierarchical Learning, Learning Automata, Stochastic Optimization, Resource Allocation.*

⋆ The second author was partially supported by NSERC, the Natural Sciences and Engineering Research Council of Canada.

⋆⋆ *Chancellor's Professor*; *Fellow: IEEE* and *Fellow: IAPR*. The Author also holds an *Adjunct Professorship* with the Dept. of ICT, University of Agder, Norway.

N.T. Nguyen and E. Szczerbicki (Eds.): Intel. Sys. for Know. Management, SCI 252, pp. 1–30.
springerlink.com

1 Introduction

1.1 Chapter Overview

There are numerous problems that are intrinsically hard. There are also problems for which the solution space is combinatorially explosive, and so searching for an optimal solution by a brute-force search becomes infeasible. It is exactly here that a researcher or developer seeks an "intelligent" solution, whereby one can attain to a good or almost-optimal solution without having to execute an unreasonably large number of computations.

A question which has been posed rhetorically since the earliest days of Artificial Intelligence (AI) is that of inferring whether a proposed solution or system can be truthfully called "intelligent". AI researchers and practitioners are willing to assert that a system or algorithm possesses "Computational Intelligence" if it can solve such intrinsically hard problems and "quickly" come up with an acceptable solution. One such class of NP-Hard problems concerns the so-called family of *Knapsack* Problems, and these have been studied by scientists in Computer Science and *Operations Research* for almost half a century.

In this Chapter, we shall explain how a sub-field of AI, namely that which involves Learning Automata (LA), can be used to produce fast and accurate solutions to "difficult" and randomized versions of the *Knapsack* problem (KP). To be more specific, we are concerned with a generalization of the Fractional Knapsack Problem[1] (FKP) which involves filling a knapsack of fixed volume with a mixture of n materials so as to attain a maximal value, where each material is characterized by its *own value per unit volume.* In particular, we shall concentrate on the *stochastic* Nonlinear Fractional Knapsack Problem (NFKP), which is a randomized version of the FKP.

The goal of this Chapter is to present accurate and efficient LA-based solutions[2] to solve the NFKP.

1.2 Motivation of Application-Domain Problems

In a multitude of real-world situations, resources must be allocated based on incomplete and noisy information. However, in many cases, such incomplete and noisy information render traditional resource allocation techniques ineffective. As mentioned, we consider problems that can be modeled in terms of the *stochastic* NFKP, informally described above.

With regard to applications, we argue that a solution to the stochastic NFKP can be effectively applied to (at least) two resource allocation problems dealing with the world-wide web. The first real-life problem relates to resource allocation

[1] All of these problems will be formally described in a subsequent section. Also, although the original problem can be posed in such abstract terms, we shall, at the outset, argue that it can be used to model a host of real-life problems including those related to the world-wide web.

[2] Due to space limitations, the details of the proofs of the properties reported here will be omitted.

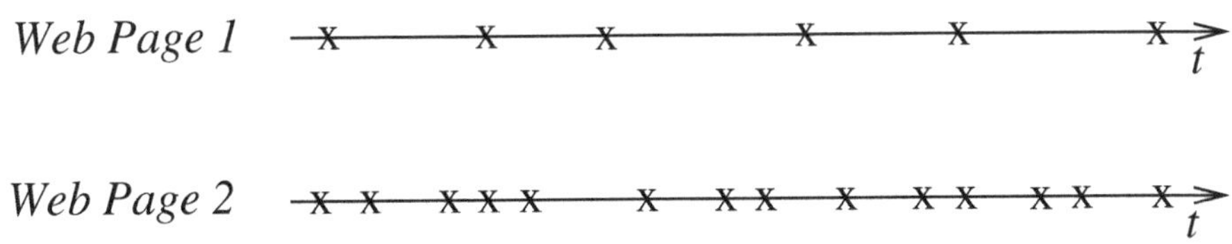

Fig. 1. Web page changes occurring over time. An 'x' on the time-lines denotes that the respective web page has changed. Observe that the occurrence of this event is not observable to the outside world unless the web page is polled.

in web monitoring so as to *optimize* information discovery when the polling capacity is constrained. The second problem concerns allocating limited sampling resources in a "real-time" manner with the purpose of estimating multiple binomial proportions. This is the scenario encountered when the user has to evaluate multiple web sites by accessing a limited number of web pages, and the proportions of interest are the fraction of each web site that is successfully validated by an HTML validator.

The world wide web is an extremely vast resource-thirsty field, which probably consumes a major portion of the computing resources available today. Searching, updating and examining web-pages is, undoubtedly, one of the primary tasks currently done by both individuals and companies. This, in turn, leads to numerous extremely interesting real-life resource allocation and scheduling problems, such as the two problems mentioned above, i.e., the "Web polling" problem, and the optimal sampling problem.

Web Polling. Web page monitoring consists of repeatedly polling a selection of web pages so that the user can detect changes that occur over time. Clearly, as this task can be prohibitively expensive, in practical applications, the system imposes a constraint on the *maximum* number of web pages that can be polled per unit of time. This bound is dictated by the governing communication bandwidth, and by the speed limitations associated with the processing. Since only a fraction of the web pages can be polled within a given unit of time, the problem which the system's analyst encounters is one of determining which web pages are to be polled. In such cases, a reasonable choice of action is to choose web pages in a manner that maximizes the number of changes detected, and the optimal allocation of the resources involves trial-and-error. As illustrated in Fig. 1, web pages may change with varying frequencies (that are unknown to the decision maker), and changes appear more or less randomly. Furthermore, as argued elsewhere, [1,2], the probability that an individual web page poll uncovers a change on its own decreases monotonically with the polling frequency used for that web page.

Optimal Sampling. The second problem which we study in this chapter, is the fascinating problem of learning distributions when the number of "Classes" is

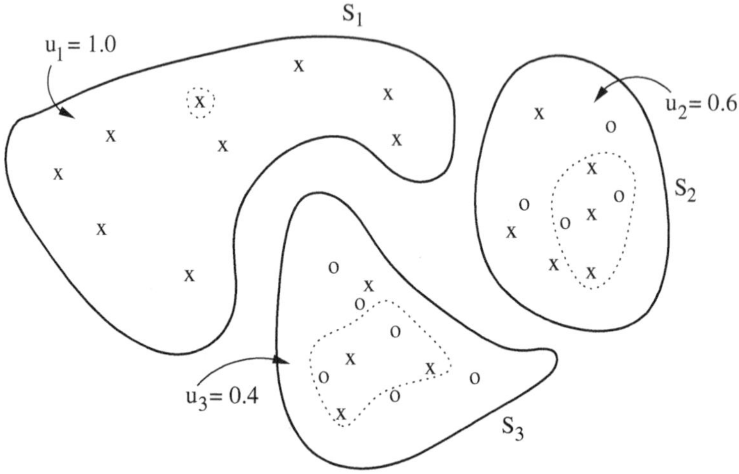

Fig. 2. The figure shows three sets S_1, S_2 and S_3 containg elements of two classes 'x' and 'o'. Since S_1 is homogenous and contains elements of only one class ('x'), a *single* sample is sufficient to determine this. Sets S_2 and S_3 are non-homogenous, although the proportions of the classes in each is different. The problem involves determining how the sampling is to be done when the proportions are unknown.

large, and the number of elements (i.e., data points) per class is also large, but when the resources available are limited. In particular, we concentrate on methods which are forced to resort to sampling techniques because of the cardinalities of the set of classes, and of the set of data points per class.

The problem studied can be stated as follows: We consider the case when the user is presented with 'n' sets of data points, $S_1, S_2, \ldots, S_n$ (see Fig. 2). The set S_i has N_i points drawn from two classes $\{\omega_1, \omega_2\}$. A random sample in set S_i belongs to ω_1 with probability u_i and it belongs to ω_2 with probability $1-u_i$. The problem we study involves estimating $\{u_i\}$ for $i = 1, 2, \ldots n$. However, to render the problem both meaningful and non-trivial, we assume that we are allowed to sample *a maximum of only* c points from the sets, $\{S_i\}$, and thus we have to determine how samples must be drawn (with replacement) from each set so that we can obtain both accurate and efficient estimates of $\{u_i\}$. The purpose is to make the estimates as accurate as possible, and so, we will here pursue the goal of minimizing the variance of the proportion estimates when the total number of samples available for estimating the proportions is restricted to c.

The problem we study is exactly the one encountered when, for instance, the task at hand is to determine the proportion of a web site that is successfully validated by an HTML validator [3] and/or a WCAG accessibility validator [4], and that n web sites are to be evaluated by only accessing c web pages.

Again, the problem becomes intriguing when the underlying distributions are unknown, which often is the case when dealing with the world wide web. Indeed, dealing with the web introduces further complexity to the problem in focus because web data is becoming increasingly dynamic, and is possibly relevant for only a limited period of time as it either rapidly changes, gets replaced, or becomes part of the deep web[3].

Although this problem is quite general, we shall proceed to solve it by suggesting that it falls within the model of knapsack-based problems.

1.3 Formal Formulation of Knapsack-Based Problems

We first formulate, in a fairly general setting, a set of knapsack-based problems that are, in actuality, related to the web-polling and optimal sampling problems. Indeed, we address one such model which can be translated into a family of problems as below:

> *Imagine that you have to allocate a limited amount of time among n different activities. The problem is such that spending a time instant on an activity randomly produces one of two possible outcomes — the time instant is either spent "fruitfully" or "unfruitfully". In this generic setting, your goal is to maximize the expected amount of fruitfully spent time. Unfortunately, you are only given the following information regarding the activities:*
>
> 1. *Each instant of time spent on an activity has a certain probability of being fruitful, and*
> 2. *This probability decreases with the amount of time spent on the activity.*
>
> *To render the problem even more realistic, you do not have access to the probabilities themselves. Instead, you must rely on solving the problem by means of trial-and-error, i.e., by attempting different allocations, and observing the resulting random outcomes.*

Instances of the above problem can be formulated as instantiations of the *Stochastic Non-linear Fractional Equality Knapsack (NEFK) Problem* as clarified earlier [5, 1, 2], and as further explained below.

In order to appreciate the qualities of the Stochastic NEFK Problem, it is beneficial to view the problem in the light of the classical *linear* Fractional Knapsack (FK) Problem, which is what we shall do now.

1.3.1 Related Foundational Problems

The Linear Fractional Knapsack (FK) Problem: The linear FK problem is a classical continuous optimization problem which also has applications within the field of resource allocation. The problem involves n materials of different value v_i per unit volume, $1 \leq i \leq n$, where each material is available in a certain amount $x_i \leq b_i$. Let $f_i(x_i)$ denote the value of the amount x_i of material i, i.e., $f_i(x_i) = v_i x_i$. The problem is to fill a knapsack of fixed volume c with the material mix $\boldsymbol{x} = [x_1, \ldots, x_n]$ to yield a maximal value for $\sum_1^n f_i(x_i)$ [6].

[3] Data from the deep web is dynamically produced in response to a direct request, usually involving a searchable database.

The Nonlinear Equality FK (NEFK) Problem: One important extension of the above classical problem is the *Nonlinear Equality* FK problem with a separable and concave objective function. The problem can be stated as follows [7]:

$$\begin{aligned}&\text{maximize } f(\boldsymbol{x}) = \textstyle\sum_1^n f_i(x_i)\\&\text{subject to } \textstyle\sum_1^n x_i = c \text{ and } \forall i \in \{1, \ldots, n\}, x_i \geq 0.\end{aligned}$$

Since the objective function is considered to be concave, the value function $f_i(x_i)$ of each material is also concave. This means that the derivatives of the material value functions $f_i(x_i)$ with respect to x_i, (hereafter denoted f_i'), are non-increasing. In other words, the material value *per unit volume* is no longer constant as in the linear case, but decreases with the material amount, and so the optimization problem becomes:

$$\begin{aligned}&\text{maximize } f(\boldsymbol{x}) = \textstyle\sum_1^n f_i(x_i), \text{where } f_i(x_i) = \int_0^{x_i} f_i'(x_i)dx_i\\&\text{subject to } \textstyle\sum_1^n x_i = c \text{ and } \forall i \in \{1, \ldots, n\}, x_i \geq 0.\end{aligned}$$

Efficient solutions to the latter problem, based on the principle of Lagrange multipliers, have been devised. In short, the optimal value occurs when the derivatives f_i' of the material value functions are equal, subject to the knapsack constraints [8,9]:

$$\begin{aligned}&f_1'(x_1) = \cdots = f_n'(x_n)\\&\textstyle\sum_1^n x_i = c \text{ and } \forall i \in \{1, \ldots, n\}, x_i \geq 0.\end{aligned}$$

The Stochastic NEFK Problem: In this chapter we generalize the above nonlinear equality knapsack problem. First of all, we let the material value per unit volume for any x_i be a *probability* function $p_i(x_i)$. Furthermore, we consider the distribution of $p_i(x_i)$ to be *unknown*. That is, each time an amount x_i of material i is placed in the knapsack, we are only allowed to observe an instantiation of $p_i(x_i)$ at x_i, and not $p_i(x_i)$ itself. Given this stochastic environment, we intend to devise an on-line incremental scheme that learns the mix of materials of maximal *expected* value, through a series of informed guesses. Thus, to clarify issues, we are provided with a knapsack of fixed volume c, which is to be filled with a mix of n different materials. However, unlike the NEFK, in the Stochastic NEFK Problem the unit volume value of a material i, $1 \leq i \leq n$, is a random quantity — it takes the value 1 with probability $p_i(x_i)$ and the value 0 with probability $1 - p_i(x_i)$, respectively. As an additional complication, $p_i(x_i)$ is nonlinear in the sense that it decreases monotonically with x_i, i.e., $x_{i_1} \leq x_{i_2} \Leftrightarrow p_i(x_{i_1}) \geq p_i(x_{i_2})$.

Since unit volume values are random, we operate with the expected unit volume values rather than the actual unit volume values themselves. With this understanding, and the above perspective in mind, the expected value of the amount x_i of material i, $1 \leq i \leq n$, becomes $f_i(x_i) = \int_0^{x_i} p_i(u)du$. Accordingly, the expected value per unit volume[4] of material i becomes $f_i'(x_i) = p_i(x_i)$.

[4] We hereafter use $f_i'(x_i)$ to denote the derivative of the expected value function $f_i(x_i)$ with respect to x_i.

In this stochastic and non-linear version of the FK problem, the goal is to fill the knapsack so that the expected value $f(\boldsymbol{x}) = \sum_1^n f_i(x_i)$ of the material mix contained in the knapsack is maximized as below:

$$\text{maximize } f(\boldsymbol{x}) = \sum_1^n f_i(x_i), \text{where } f_i(x_i) = \int_0^{x_i} p_i(u)du, \text{ and } p_i(x_i) = f_i'(x_i),$$
$$\text{subject to } \sum_1^n x_i = c \text{ and } \forall i \in \{1, \ldots, n\}, x_i \geq 0.$$

A fascinating property of the above problem is that the amount of information available to the decision maker is limited — the decision maker is only allowed to observe the current unit value of each material (either 0 or 1). That is, each time a material mix is placed in the knapsack, the unit value of each material is provided to the decision maker. The actual outcome probabilities $p_i(x_i), 1 \leq i \leq n$, however, remain *unknown*. As a result of the latter, the expected value of the material mix must be maximized by means of trial-and-error, i.e., by experimenting with different material mixes and by observing the resulting random unit value outcomes.

1.3.2 State-of-the-Art

To the best of our knowledge, prior to our work reported in [2], the stochastic NEFK problem was not addressed in the literature before. However, several studies on related problems have been reported. For example, the works of [10, 11] consider solution policies for stochastic generalizations of the so-called NP-hard *linear* integer knapsack problem. In these papers, value distributions were considered known and constant, making dynamic programming a viable solution. Another variant of the knapsack problem is found in [12] where a deterministic knapsack is used, however, with objects arriving to and departing from the knapsack at random times. The optimization problem considered was to accept/block arriving objects so that the average value of the knapsack is maximized.

The first reported generic treatment of the stochastic NEFK problem itself can be found in [2]. Various instantiations of the problem have, however, appeared sporadically, particularly within the web monitoring domain. In these latter instantiations, the unknown parameters are *estimated* by means of a tracking phase where web pages are polled mainly for estimation purposes [13, 14]. One major disadvantage of such an approach is that the parameter estimation phase significantly delays the implementation of an optimal solution. This disadvantage is further aggravated in *dynamic* environments where the optimal solution changes over time, introducing the need for parameter re-estimation [5].

With regard to the particular application domain, recent approaches to resource allocation in web monitoring attempt to *optimize* the performance of the system when the monitoring capacity is restricted [13, 14]. The principle cited in the literature essentially invokes Lagrange multipliers to solve a *nonlinear equality* knapsack problem with a separable and concave objective function [7]. Thus, for example, a basic web monitoring resource allocation problem may involve n web pages that are updated periodically, although with different periods. Clearly, each web page can be polled with a maximum frequency - which would result in a sluggish system. The problem which we study involves determining the web page polling frequencies (i.e., how often each web page is accessed by the monitoring system) so as to maximize the number of web page updates detected. Observe that this must be achieved without exceeding the available monitoring capacity — e.g., the maximum number of web pages that can be accessed per unit of time as dictated by the governing communication bandwidth and processing speed limitations.

In contrast to the above approaches, we base our work on the principles of Learning Automata (LA) [15, 16]. LA have been used to model biological systems [17], and have attracted considerable interest in the last decade because they can learn the optimal actions when operating in (or interacting with) unknown stochastic environments. Furthermore, they combine rapid and accurate convergence with low computational complexity.

1.4 Contributions of This Chapter

The contributions of this chapter are the following:

1. We present the first reported solution to the Stochastic NEFK Problem in which the material values per unit volume are *stochastic* with *unknown* distributions.
2. We present the Learning Automata Knapsack Game (LAKG), which is the first reported solution to the Stochastic NEFK Problem which uses a team of LA operating in a multi-dimensional Environment.
3. We present the first reported solutions to the Polling Frequency Determination problem and to the so-called Sample Size Determination problem when modelled in unknown stochastic settings.
4. We report the first *analytical* results for schemes that solve the Stochastic NEFK Problem using a formal asymptotically optimal solution to the Stochastic NEFK Problem.
5. We propose a novel scheme for the *two-material* resource allocation problem, namely, the *Twofold Resource Allocation Automaton (TRAA)*. From the perspective of LA, the TRAA, in itself, is the first reported LA which is *artificially* rendered ergodic.
6. We report the first *hierarchical* solution to the Stochastic NEFK Problem, based on a hierarchy of TRAAs, namely, the H-TRAA.
7. We verify empirically that the H-TRAA provides orders of magnitude faster convergence when compared to the LAKG.

As a result of the above contributions, we believe that the LAKG and H-TRAA are the first reported viable and realistic strategies for solving the optimal web-polling problem. Indeed, they can also be used for other problems involving the optimal allocation of sampling resources in large scale web accessibility assessment [3].

2 Overview of Non-LA Solutions

In order to put our work in the right perspective, we first provide a brief review of the concepts and the solution found in [2] - which are also relevant for more "primitive" variants of the knapsack problem.

As indicated in the introduction, solving the classical linear FK problem involves finding the most valuable mix $\boldsymbol{x}^* = [x_1^*, \ldots, x_n^*]$ of n materials that fits within a knapsack of fixed capacity c. The material value per unit volume for each material i is given as a constant v_i, and each material is available in a certain amount $x_i \leq b_i$, $1 \leq i \leq n$. Accordingly, the value of the amount x_i of material i, $f_i(x_i) = v_i x_i$, is linear with respect to x_i. In other words, the derivative of $f_i(x_i)$ — i.e., the material value per unit volume — is fixed: $f_i'(x_i) = v_i$. Because a fraction of each material can be placed in the knapsack, the following greedy algorithm from [6] finds the most valuable mix: *Take as much as possible of the material that is most valuable per unit volume. If there is still room, take as much as possible of the next most valuable material. Continue until the knapsack is full.*

Let us now generalize this and assume that the material unit volume values are *random* variables with *constant* and *known* distributions. Furthermore, for the sake of conceptual clarity, let us only consider binary variables that *either* instantiate to the values of 0 or 1. Since the unit volume values are random, let p_i denote the probability of the unit volume value $v_i = 1$ for material i, $1 \leq i \leq n$, which means that the probability of the unit volume value $v_i = 0$ becomes $1 - p_i$. With some insight, it becomes evident that under such conditions, the above greedy strategy can again be used to maximize the *expected* value of the knapsack, simply by selecting the material based on the *expected* unit volume values, $E[v_i] = 0 \times (1 - p_i) + 1 \times p_i$, rather than actual unit volume values.

The above indicated solution is, of course, inadequate when the p_i's are unknown. Furthermore, the problem becomes even more challenging when the p_i's are no longer constant, but rather depend on their respective material amounts x_i, $1 \leq i \leq n$. Let $p_i(x_i)$ denote the probability that the current unit volume value of material i is $v_i = 1$, given that the amount x_i has already been placed in the knapsack. Then, the expected value per unit volume of material i, $1 \leq i \leq n$, becomes $E[v_i] = 0 \times [1 - p_i(x_i)] + 1 \times p_i(x_i) = p_i(x_i)$, and accordingly, the expected value of the amount x_i becomes $f_i(x_i) = \int_0^{x_i} p_i(u)du$.

Our aim, then, is to find a scheme that moves towards optimizing the following NEFK problem on-line:

$$\begin{array}{l} \text{maximize } f(\boldsymbol{x}) = \sum_1^n f_i(x_i), \text{where} f_i(x_i) = \int_0^{x_i} p_i(u)du, \text{ and } p_i(x_i) = f_i'(x_i), \\ \text{subject to } \sum_1^n x_i = c \text{ and } \forall i \in \{1, \ldots, n\}, x_i \geq 0. \end{array}$$

Note that we allow only instantiations of the material values per unit volume to be observed. That is, each time an amount x_i of material i is placed in the knapsack, an instantiation v_i at x_i is observed.

Because of the above intricacies, we approach the problem by relying on informed material mix *guesses*, i.e., by experimenting with different material mixes and learning from the resulting random unit volume value outcomes. We shall assume that x_i is any number in the interval $(0, 1)$. The question of generalizing this will be considered later. The crucial issue that we have to address, then, is that of determining how to change our current guesses on x_i, $1 \leq i \leq n$. . We shall attempt to do this in a discretized manner by subdividing the unit interval into N points $\{\frac{1^\lambda}{N^\lambda}, \frac{2^\lambda}{N^\lambda}, \ldots, \frac{(N-1)^\lambda}{N^\lambda}, 1\}$, where N is the resolution of the learning scheme and $\lambda > 0$ determines the linearity of the discretized solution space[5]. We will see that a larger value of N will ultimately imply a more accurate solution to the knapsack problem.

3 A Nonlinear Knapsack Game of Finite Automata

3.1 Details of the LAKG Solution

At the heart of our scheme is a *game* between n finite automata that interact with a Scheduler and a stochastic Environment. Fig. 3 provides an overview of this interaction, and is explained below.

1. **Stochastic Environment:** The Environment consists of a set of stochastic material unit volume value functions $\mathcal{F}' = \{F_1'(x_1), F_2'(x_2), \ldots, F_n'(x_n)\}$. If amount x_i of material i is suggested to the Environment, $F_i'(x_i)$ takes the unit volume value $v_i' = 0$ with probability $p_i^0(x_i)$ and the unit volume value $v_i' = 1$ with probability $p_i^1(x_i)$. In addition, the Environment provides a signal ϕ indicating whether the knapsack is full:
$$\phi = \begin{cases} \textit{True} \text{ If } \sum_{i=1}^n x_i \geq c \\ \textit{False} \text{ Otherwise} \end{cases}$$
2. **Scheduler:** The Scheduler takes material amounts $\boldsymbol{x} = [x_1, \ldots, x_n]$ as its input. The purpose of the Scheduler is:
   (a) To render accesses to the stochastic Environment sequential, and
   (b) To make sure that the unit volume value functions $F_1'(x_1), F_2'(x_2), \ldots,$ $F_n'(x_n)$ are accessed with frequencies proportional to $\boldsymbol{x}$.
   The reader should note that our scheme does not rely on accessing the unit volume value functions sequentially with frequencies proportional to $\boldsymbol{x}$ for solving the knapsack problem. However, this restriction is obviously essential for solving the problem *incrementally* and *on-line* (or rather in a "real-time" manner). For the sake of simplicity, we choose to access the functions randomly by sampling them from a probability distribution proportional to $\boldsymbol{x}$.
3. **The Team of Learning Automata:** Each material i is assigned a finite fixed structure automaton LA$_i$ with the states $1, \ldots, N$. Let the current state of the automaton LA$_i$ be $s_i(t)$. When the automaton acts, it suggests the amount $x_i(t) = \frac{s_i(t)^\lambda}{N^\lambda}$ of material i to the Scheduler, which, in turn, interacts with the stochastic Environment. Assume that $v_i'(t)$ and $\phi(t)$ are the resulting feedback from the stochastic Environment. Then the state of the automaton is updated as follows:
$$\begin{array}{ll} s_i(t+1) := s_i(t) + 1 \textbf{ If } v_i'(t) = 1 \textbf{ and } 1 \leq s_i(t) < N \textbf{ and not } \phi(t) \\ s_i(t+1) := s_i(t) - 1 \textbf{ If } v_i'(t) = 0 \textbf{ and } 1 < s_i(t) \leq N \textbf{ and } \phi(t) \\ s_i(t+1) := s_i(t) \qquad \textbf{Otherwise.} \end{array}$$

[5] The importance of this parameter will become evident in Sect. 5 and Sect. 6 where empirical results are presented.

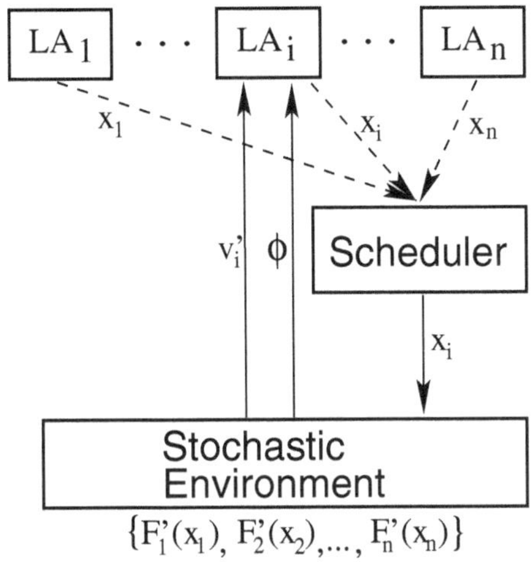

Fig. 3. The *Team* of Learning Automata interacting with a Scheduler and an unknown stochastic Environment.

> Notice that although the above state updating rules are deterministic, because the knapsack is stochastic, the state transitions will also be stochastic.

The purpose of the above LA game is to form a stochastic competition between the n automata so that the competition directs the automata towards the optimal solution. In essence, the competition is governed by two situations:

1. There is still room for more material in the knapsack, and
2. The knapsack is full.

In the first situation, the competition consists of increasing the material amount guesses as quickly as possible. However, a guess x_i can only be increased when the corresponding unit volume value function $F_i'(x_i)$ instantiates to 1. Accordingly, if $p_i^1(x_i) < p_j^1(x_j)$ for material i and j, then automaton j will have an edge in the competition. The second situation is the opposite of the first one. In the second situation, each automaton i *decreases* its material amount guess x_i whenever the corresponding unit volume value function $F_i'(x_i)$ instantiates to 0. Accordingly, if $p_i^1(x_i) < p_j^1(x_j)$ for material i and j, then automaton i will have an edge, rather than automaton j. Operating simultaneously, the two situations of the competition are designed to stochastically move the current allocation of materials towards the knapsack capacity, with the aim of approaching the optimal expected knapsack value.

Note that because each automaton LA_i acts (e.g., polls a web page) with an average frequency proportional to $x_i(t)$, the F_i's are accessed with a frequency corresponding to the current knapsack solution. In other words, our scheme actually applies the *current solution* when seeking its improvement. As we will see in Sect. 5 and Sect. 6 the latter property permits us to use the scheme incrementally and on-line.

4 A Hierarchy of Twofold Resource Allocation LA (H-TRAA)

Before we describe the system which incorporates a hierarchy of LA, we shall first explain the structure and operation of the primitive unit, namely, the Twofold Resource Allocation Automaton (TRAA), which is specifically designed for the two-material problem. We shall then show how a collection of TRAAs can be networked using a tree-topology, to yield a solution for the mutli-material case.

4.1 Details of the TRAA Solution

We first present our LA based solution to *two-material* Stochastic NEFK Problems. The two-material solution forms a critical part of the hierarchical scheme for multiple materials that is presented subsequently. As illustrated in Fig. 4, our solution to two-material problems constitutes of three modules:

1. A Stochastic Environment
2. The TRAA itself, and
3. An Earliest Deadline First (EDF) Scheduler.

We first detail each of the three modules, before we analyze the overall *feedback connection* between them. Finally, we state that the TRAA that we have developed in this section is asymptotically optimal for two-material Stochastic NEFK Problems.

Stochastic Environment: The *Stochastic Environment* for the two-material case is characterized by (a) The capacity c of the knapsack, and (b) Two material unit volume value probability functions $p_1(x_1)$ and $p_2(x_2)$.

In brief, if the amount x_i of material i is suggested to the Stochastic Environment, the Environment replies with a unit volume value $v_i = 1$ with probability

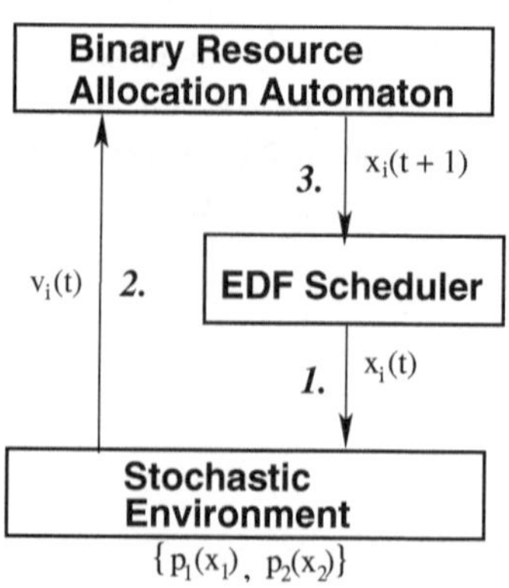

Fig. 4. The Twofold Resource Allocation Automaton (TRAA) interacting with a Scheduler and an unknown Stochastic Environment.

$p_i(x_i)$ and a unit volume value $v_i = 0$ with probability $1 - p_i(x_i)$, $i \in \{1, 2\}$. It should be emphasized that to render the problem both interesting and non-trivial, we assume that $p_i(x_i)$ is unknown to the TRAA.

Twofold Resource Allocation Automaton (TRAA): The scheme which attempts to learn the optimal allocation $\boldsymbol{x}^* = [x_1^*, x_2^*]$ can be described as follows. A finite fixed structure automaton with the states $s(t) \in \{1, 2, \ldots, N\}$ is used to decide the allocation of resources among the two materials. Let the current state of the automaton be $s(t)$. Furthermore, let $q_{s(t)}$ refer to the fraction $\frac{s(t)}{N+1}$, and let $r_{s(t)}$ refer to the fraction: $1 - q_{s(t)}$. Then the automaton's current guess is $\boldsymbol{x} = [q_{s(t)}, r_{s(t)}]$.

If the Stochastic Environment tells the automaton that the unit volume value of material i is $v_i(t)$ at time t, the automaton updates its state as follows:

$$\begin{array}{lll} s(t+1) := s(t) + 1 & \textbf{If} & \textbf{rand}() \le r_{s(t)} \textbf{ and } v_i(t) = 1 \quad (1) \\ & & \textbf{and } 1 \le s(t) < N \textbf{ and } i = 1 \\ s(t+1) := s(t) - 1 & \textbf{If} & \textbf{rand}() \le q_{s(t)} \textbf{ and } v_i(t) = 1 \quad (2) \\ & & \textbf{and } 1 < s(t) \le N \textbf{ and } i = 2 \\ s(t+1) := s(t) & \textbf{Otherwise} & \quad (3). \end{array}$$

Fig. 5 shows the resulting stochastic transition graphs for resolution $N = 5$. The upper graph shows the transitions for feedback from the Stochastic Environment on material 1, and the graph below shows the transitions for feedback on material 2. Notice how the stochastic state transitions are designed to offset the learning bias introduced by accessing the materials with frequencies proportional to $\boldsymbol{x} = [q_{s(t)}, r_{s(t)}]$. Also observe that the overall learning scheme does not produce any absorbing states, and is accordingly ergodic, thus supporting dynamic environments. The effect of these properties is analysed in the next subsection.

Finally, after the automaton has had the opportunity to change its state, it provides output to the EDF Scheduler. That is, it outputs the material amounts $\boldsymbol{x} = [q_{s(t+1)}, r_{s(t+1)}]$ that have been changed.

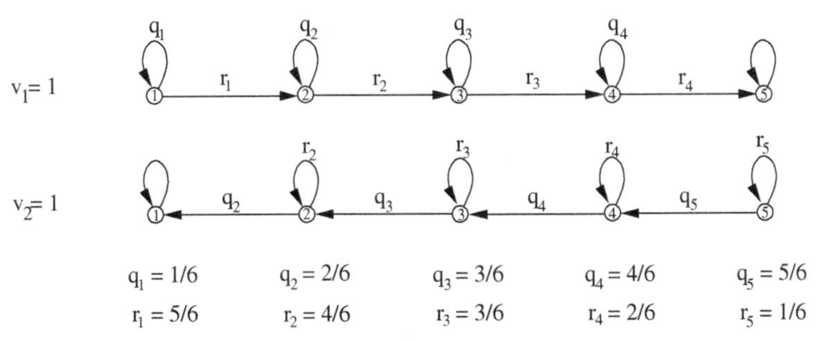

Fig. 5. The stochastic transition graphs of a TRAA with resolution $N = 5$.

Earliest Deadline First (EDF) Scheduler: The Scheduler takes material amounts $\boldsymbol{x} = [x_1, \ldots, x_n]$ as its input (for the two-material case the input is $\boldsymbol{x} = [x_1, x_2]$). The purpose of the Scheduler is (a) To provide accesses to the Stochastic Environment in a sequential manner, and (b) To make sure that the unit volume value functions are accessed with frequencies proportional to $\boldsymbol{x}$. The reader should note that our scheme does not rely on accessing the unit volume value functions sequentially with frequencies proportional to $\boldsymbol{x}$ for solving the knapsack problem. However, this restriction is obviously essential for solving the problem *incrementally* and *on-line* (or rather in a "real-time" manner). Note that since it, in some cases, may be essential to access each unit volume value function with a constant period and not randomly (for example, in the earlier-alluded-to problem which analyzes web page polling), we use the Earliest Deadline First (EDF) Scheduling to access the functions according to $\boldsymbol{x}$.

The analysis of the TRAA follows.

Stochastic NEFK Problem Solution and Analysis of the TRAA Solution. We now characterize the optimal solution to a Stochastic NEFK Problem. Thereafter, we analyze the feedback connection of the TRAA and the Stochastic Environment — we claim that the TRAA is asymptotically optimal in the sense that it can find material allocations arbitrarily close to the solution of the Stochastic NEFK Problem.

Lemma 1. *The material mix $\boldsymbol{x} = [x_1, \ldots, x_n]$ is a solution to a given Stochastic NEFK Problem if (1) the derivatives of the expected material amount values are all equal at $\boldsymbol{x}$, (2) the mix fills the knapsack, and (3) every material amount is positive, i.e.:*

$$f_1'(x_1) = \cdots = f_n'(x_n)$$
$$\textstyle\sum_1^n x_i = c \text{ and } \forall i \in \{1, \ldots, n\}, x_i \geq 0.$$

□

The above lemma is based on the well-known principle of Lagrange Multipliers [8, 9], and its proof is therefore omitted here for the sake of brevity. Instead, we will start by analyzing the *two-material* problem and the TRAA. Multiple TRAAs will then be organized in a hierarchy with the aim of tackling *n-material* problems.

For the two-material problem, let $\boldsymbol{x}^* = [x_1^*, x_2^*]$ denote a solution, as defined above. Note that since x_2^* can be obtained from x_1^*, we will concentrate on finding x_1^*.

Theorem 1. *The TRAA solution scheme specified by (1)–(3) is asymptotically optimal.*

Proof. The proof of the result is quite intricate and involves a detailed analysis of the underlying Markov chain. It is omitted in the interest of brevity, but can be found in [18]. □

4.2 Details of the H-TRAA Solution

In this section we propose a hierarchical scheme for solving n-material problems. The scheme takes advantage of the TRAA's ability to solve two-material problems asymptotically, by organizing them hierarchically.

Construction of Hierarchy. The hierarchy of TRAAs, which we hereafter will refer to as the H-TRAA, is constructed as follows[6]. First of all, the hierarchy is organized as a balanced binary tree with depth $D = \log_2(n)$. Each node in the hierarchy can be related to three entities: (1) a set of materials, (2) a partitioning of the material set into two subsets of equal size, and (3) a dedicated TRAA that allocates a given amount of resources among the two subsets.

Root Node: The hierarchy root (at depth 1) is assigned the complete set of materials $S_{1,1} = \{1, \ldots, n\}$. These n materials are partitioned into two disjoint and exhaustive subsets of equal size: $S_{2,1}$ and $S_{2,2}$. An associated TRAA, $T_{1,1}$, decides how to divide the full knapsack capacity c (which, for the sake of notational correctness will be referred to as $c_{1,1}$) among the two subsets. That is, subset $S_{2,1}$ receives the capacity $c_{2,1}$ and subset $S_{2,2}$ receives the capacity $c_{2,2}$, with $c_{2,1} + c_{2,2} = c_{1,1}$. Accordingly, *this* TRAA is given the power to prioritize one subset of the materials at the expense of the other.

Nodes at Depth d: Node $j \in \{1, \ldots, 2^{d-1}\}$ at depth d (where $1 < d \leq D$) refers to: (1) the material subset $S_{d,j}$, (2) a partitioning of $S_{d,j}$ into the subsets $S_{d+1,2j-1}$ and $S_{d+1,2j}$, and (3) a dedicated TRAA, $T_{d,j}$. Observe that since level $D+1$ of the H-TRAA is non-existent, we use the convention that $S_{D+1,2j-1}$ and $S_{D+1,2j}$ refer to the primitive materials being processed by the leaf TRAA, $T_{D,j}$. Assume that the materials in $S_{d,j}$ has, as a set, been assigned the capacity $c_{d,j}$. The dedicated TRAA, then, decides how to allocate the assigned capacity $c_{d,j}$ among the subsets $S_{d+1,2j-1}$ and $S_{d+1,2j}$. That is, subset $S_{d+1,2j-1}$ receives the capacity $c_{d+1,2j-1}$ and $S_{d+1,2j}$ receives the capacity $c_{d+1,2j}$, with $c_{d+,2j-1} + c_{d+1,2j} = c_{d,j}$.

At depth D, then, each individual material can be separately assigned a fraction of the overall capacity by way of recursion, using the above allocation scheme.

Interaction of the H-TRAA with EDF Scheduler and Environment. As in the single TRAA case, the H-TRAA interacts with an EDF Scheduler, which suggests which unit volume value function $p_i(x_i)$ to access next. A response is then generated from the Stochastic Environment using $p_i(x_i)$. This response is given to all the TRAAs that were involved in determining the material amount x_i, that is, the TRAAs in the hierarchy that have allocated capacacity to a material subset that contains material i. Finally, a new candidate material mix $\boldsymbol{x} = [x_1, \ldots, x_n]$ is suggested by the H-TRAA to the EDF Scheduler.

[6] We assume that $n = 2^\gamma, \gamma \in \mathbb{N}^+$, for the sake of clarity. If the number of materials is less than this, we can assume the existence of additional materials whose values are "zero", and who thus are not able to contribute to the final optimal solution.

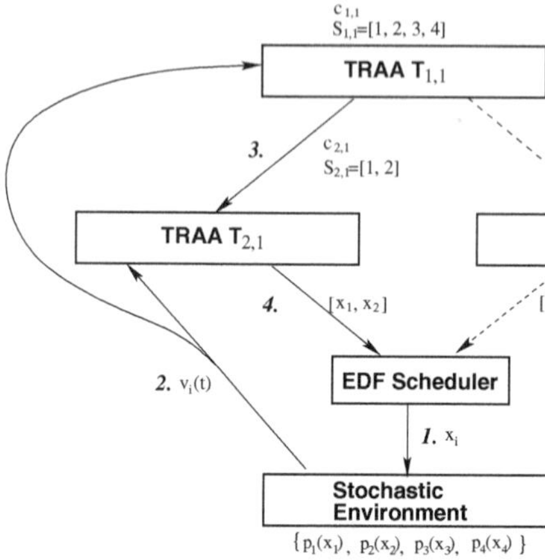

Fig. 6. A Hierarchy of Twofold Resource Allocation Automata (H-TRAA) interacting with a Scheduler and an unknown Stochastic Environment as explained in Example I.

Example I. Consider a 4-material problem. Fig. 6 shows the associated hierarchy, constructed as described above. At the root level the TRAA $T_{1,1}$ divides the knapsack capacity among the two material subsets $\{1, 2\}$ and $\{3, 4\}$, respectively related to the TRAA $T_{2,1}$ and $T_{2,2}$. At the level below, then, the TRAA $T_{2,1}$ allocates its share of the capacity among material 1 and material 2, while TRAA $T_{2,2}$ assigns *its* share of the capacity to material 3 and material 4. Based on the present assignment at time t, the EDF Scheduler selects material i, suggesting the amount $x_i(t)$ to the Stochastic Environment. The Stochastic Environment, in turn, responds with a randomly drawn material unit volume value, $v_i(t)$, using the probability value function $p_i(x_i)$. By way of example, if $i = 3$, the latter feedback is given to TRAAs $T_{1,1}$ and $T_{2,1}$, which update their states accordingly, and the feedback loop continues.

Analysis of the H-TRAA Solution. In the previous section we asserted that an individual TRAA is asymptotically optimal. We will now consider the H-TRAA itself and claim its asymptotic optimality. More specifically, we shall show that if each individual TRAA in the hierarchy has solved its own two-material problem, a solution to the complete n-material Knapsack Problem has also been produced.

Theorem 2. *Let $T_{d,j}$ be an arbitrary TRAA at level d of the H-TRAA associated with the node whose index is j. Then, if every single TRAA, $T_{d,j}$, in the H-TRAA has found a local solution with proportions $c_{d+1,2j-1}$ and $c_{d+1,2j}$ satisfying*

$$f'_{d+1,2j-1}(c_{d+1,2j-1}) = f'_{d+1,2j}(c_{d+1,2j}),$$

the overall Knapsack Problem involving n materials that are hierarchically placed in $\log_2 n$ levels of TRAAs, also attains the global optimum solution.

Proof. The proof of the result is also quite deep. It involves a recursive analysis of the convergence along the various levels of the tree, and some intricate probability arguments which maintain the consistency. It is omitted in the interest of brevity, but can be found in [2]. □

Remarks: Theorem 2 has some very interesting consequences listed below:

1. The proof of Theorem 2 has tacitly assumed that all the automata have converged before the global convergence can be asserted. This implies that the TRAA $T_{d,j}$ is aware of its capacity, and that this is a known quantity to the TRAAs $T_{d+1,2j-1}$ and $T_{d+1,2j}$. In other words, if all the individual TRAAs converge to their local optimum, Theorem 2 states that the global optimum is attained. Conceptually, this can pose a small implementation-related problem. The fact is that the TRAAs of the lower level are converging even while the TRAA at the higher level is attempting to find *its* capacity. Therefore, essentially, the lower level TRAAs are working in a non-stationary environment. The strategy by which we can resolve this is to ensure that the higher level automata converge at a slower rate than the lower ones (thus guaranteeing a certain level of stationarity). In practice, however, we have observed that if the resolution parameter N is large enough (in the order of hundreds) the time varying phenomenon is marginal, and the TRAAs at all the levels tend to converge simultaneously.
2. Theorem 2 claims that the solution obtained by the convergence of the individual TRAAs leads to the global convergence of the overall optimization problem. But this claim means that the ordering of the materials at the leaf nodes does not carry any significance. This is, indeed, true! It turns out that if the nodes at the leaves are ordered in such a way that "more precious materials" lie in the same sub-tree, the weight associated with the sub-tree of the composite material containing these "more precious materials" will have a much larger weight, and the weight of the other sub-trees will be much smaller. As opposed to this, if the "more precious materials" lie in distinct sub-trees, the weights associated with the respective sub-trees will be correspondingly compensated for.

5 Application I: Determination of Optimal Polling Frequency

Having obtained a formal solution to the model in which we set the NEFK, we shall now demonstrate how we can utilize this solution for the first application problem being studied, namely, the optimal web-polling problem.

Although several nonlinear criterion functions for measuring web monitoring performance have been proposed in the literature (e.g., see [13, 14]), from a broader viewpoint they are mainly built around the basic concept of the *update detection probability*, i.e., the probability that polling a web page results in new information being discovered. Therefore, for the purpose of clarification and for the sake of conceptual clarity, we will use the update detection probability as the token of interest in this chapter. To further define our notion of web monitoring performance, we consider that time is discrete with the time interval length T to be the atomic unit of decision making. In each time interval every single web page i has a constant probability u_i of remaining *unchanged* in static environments, and changing in dynamic environments. Furthermore, when a web page is updated/changed, the update is available for detection only until the web page is updated again. After that, the original update is considered lost. For instance, each time a newspaper web page is updated, previous news items are replaced by the most recent ones.

In the following, we will denote the update detection probability of a web page i as d_i. Under the above conditions, d_i depends on the frequency, x_i, that the page is polled with, and is modeled using:

$$d_i(x_i) = 1 - {u_i}^{\frac{1}{x_i}}.$$

Thus, consider the scenario that a web page remains unchanged in any single time step with probability 0.5. Then polling the web page uncovers new information with probability $1 - 0.5^3 = 0.875$ if the web page is polled every 3rd time step (i.e., with frequency $\frac{1}{3}$) and $1 - 0.5^2 = 0.75$ if the web page is polled every 2nd time step. Increasing the polling frequency reduces the probability of discovering new polling information.

Given the above considerations, our aim is to find the page polling frequencies $\boldsymbol{x}$ that maximize the expected number of pollings uncovering new information per time step:

$$\begin{array}{l}\text{maximize } \sum_1^n x_i \times d_i(x_i)\\ \text{subject to } \sum_1^n x_i = c \text{ and } \forall i = 1, \ldots, n, x_i \geq 0.\end{array}$$

Note that in the general web monitoring case, we are not able to observe $d_i(x_i)$ or u_i directly — polling a web page only reveals whether the page has been updated *at least once* since our last poll[7]. As such, web monitoring forms a proof-of-concept application for resource allocation in unknown stochastic environments.

5.1 The H-TRAA Solution

In order to find a H-TRAA Solution to the above problem we must define the Stochastic Environment that the LA are to interact with. As seen in Sect. 4, the Stochastic Environment consists of the unit volume value functions $\mathcal{F}' = \{f_1'(x_1), f_2'(x_2), \ldots, f_n'(x_n)\}$, which are unknown to the H-TRAA. We identify the nature of these functions by applying the principle of Lagrange multipliers to the above maximization problem. In short, after some simplification, it can be seen that the following characterize the optimal solution:

$$\begin{array}{l}d_1(x_1) = d_2(x_2) = \cdots = d_n(x_n), \text{ with}\\ \sum_1^n x_i = c \text{ and } \forall i = 1, \ldots, n, x_i \geq 0.\end{array}$$

[7] Some web pages are also annotated with the time of last update. However, this information is not generally available/reliable [19], and is therefore ignored in our scheme.

Since we are not able to observe $d_i(x_i)$ or u_i directly, we base our definition of $\mathcal{F}'$ on the result of polling web pages. Briefly stated, we want $f_i'(x_i)$ to instantiate to the value 0 with probability $1 - d_i(x_i)$ and to the value 1 with probability $d_i(x_i)$. Accordingly, if the web page i is polled and i has been updated since our last polling, then we consider $f_i'(x_i)$ to have been instantiated to unity. Accordingly, if the web page i is unchanged, we consider $f_i'(x_i)$ to have been instantiated to 0.

5.2 Empirical Results

In this section we evaluate our learning scheme by comparing it with four classical policies using synthetic data. We have implemented the following classical policies:

Uniform: The uniform policy allocates monitoring resources uniformly across all web pages. This is the only classical policy of the four that can be applied directly in an unknown environment.

Proportional: In the proportional policy, the allocation of monitoring resources to web pages is proportional to the update frequencies of the web pages. Accordingly, this policy requires that the web page update frequencies are known.

Estimator: The estimator policy handles unknown web update frequencies by polling web pages *uniformly* in a parameter estimation phase, with the purpose of estimating update frequencies. After the parameter estimation phase, the proportional policy is applied. However, the latter is based on the estimated update frequencies rather than the true ones.

Optimal: The optimal policy requires that web page update frequencies are known, and finds the optimal solution based on the principle of Lagrange multipliers [13,14].

To evaluate web resource allocation policies, recent research advocates Zipf-like distributions [20] to generate realistic web page update frequencies [13,14]. The Zipf distribution can be stated as follows [21]:

$$Z(k; s, N) = \frac{1/k^s}{\sum_{n=1}^{N} 1/n^s},$$

where N is the number of elements, k is their rank, and s is a parameter that governs the skewed-ness of the distribution (e.g., for $s = 0$ the distribution is uniform).

For our experiments, web pages are considered ranked according to their update frequencies, and the update probability of a web page is calculated from its rank. We use the following function to determine the update probability of each web page:

$$u_k(\alpha, \beta) = \frac{\alpha}{k^\beta}.$$

In this case, k refers to the web page of rank k and the parameter β determines the skewed-ness of the distribution, while $\alpha \in [0.0, 1.0]$ represents the magnitude of the update probabilities (i.e., the web page of rank 1 is updated with probability α each time step).

Without loss of generality, we normalize the web page polling capacity in our experiments to a unit poll per time step, and accordingly, we vary the average total number of web page updates per time step instead.

As we will see in the following, it turns that one of the strengths of the H-TRAA is its ability to take advantage of so-called spatial dependencies among materials. As mentioned earlier, in the above experimental setup, materials are spatially related in the sense that the updating probabilities decreases with the rank-index k. In order to starve the H-TRAA from this information, we opted to perturb this spatial structure. Each perturbation swapped the updating probabilities of a randomly selected material and the material succeeding it in the ranking. We conducted our experiments with 10^3, 10^4, 10^5 and 10^6 perturbations.

The results of our experiments are truly conclusive and confirm the power of the H-TRAA. Although several experiments were conducted using various values of α, β, and number of automata, we report for the sake of brevity, mainly the results for 512 web pages (the main case from [13] which we have considered as a benchmark) within the following environments:

- $\alpha = 0.3$; $\beta = 1.5$, where the average number of updates per time step is 0.76, and accordingly, below the web page polling capacity. The web page update distribution is highly skewed, as explored in [13].
- $\alpha = 0.3$; $\beta = 1.0$, where the average number of updates per time step is increased to 2.0 (twice the polling capacity) by making the page update distribution less skewed (the normal Zipf distribution).
- $\alpha = 0.9$; $\beta = 1.5$, where the average number of updates is set to 2.3 by increasing the web page update probability. Because of the high values of both α and β, this environment turns out to be the most challenging one, discriminating clearly between the optimal policy and the proportional policy.

For these values, an ensemble of several independent replications with different random number streams was performed to minimize the variance of the reported results.

Configuring the H-TRAA. The H-TRAA can be configured by various means. First of all, the material amount space $(0, 1)$ need not be discretized uniformly. Instead, a nonlinear material amount space can be formed, as done for the LAKG in [2]. Furthermore, the discretization resolution N must also be set for each TRAA, possibly varying from TRAA to TRAA in the hierarchy. In short, the performance achieved for a particular problem can be optimized using these different means of configuring the H-TRAA. In this section, however, our goal is to evaluate the overall performance of the H-TRAA, without fine tuning. Therefore, we will only use a linear material amount space, as specified in Sect. 4. Furthermore, we will use the same resolution $N = 500$ for all the TRAAs in

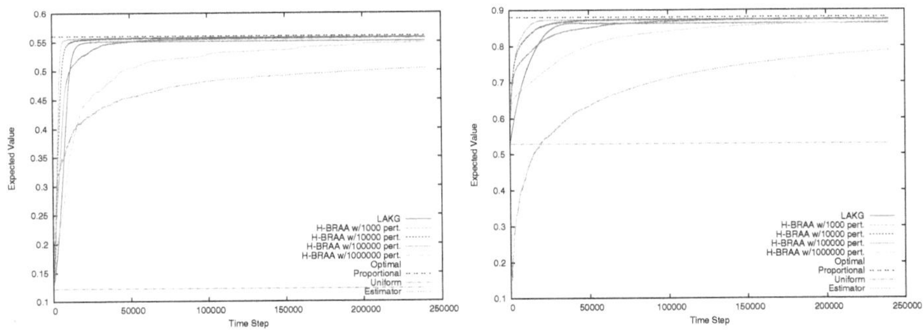

Fig. 7. From the figure on the left we see that in the ($\alpha = 0.3, \beta = 1.5$)-environment, the H-TRAA scheme is superior to the LAKG scheme and the estimator scheme. However, for highly unstructured environments, the LAKG provides better performance. Further, from the figure on the right, for the ($\alpha = 0.3, \beta = 1.0$)-environment, a less skewed web page update distribution makes the uniform policy as well as the estimator policy more successful, mainly because of more widely and abundant updating of the web pages.

the hierarchy, independent of the specific knapsack problem at hand. Thus, our aim is to ensure a fair comparison with the present state of the art, namely, the LAKG scheme.

Static Environments. We see from the figure on the left of Fig. 7 that the proportional policy and the optimal policy provide more-or-less the same solution — a solution superior to the uniform policy solution. We also observe that the performance of the estimator scheme increases steadily with the length of the parameter estimation phase. The figure also shows the performance of the H-TRAA increases significantly quicker than the LAKG and the Estimator schemes. However, when increasing the number of perturbations, the performance of the H-TRAA is reduced. Indeed, with $1,000,000$ perturbations, the LAKG turns out to converge both more quickly and more accurately than the H-TRAA. Note that even with $1,000,000$ perturbations, the H-TRAA provides performance equal to the LAKG if each TRAA in the hierarchy is given a resolution N that is twice as large as the resolution applied by any of its children. However, the performance advantage of the H-TRAA is lost for the less perturbed cases. In this sense, the H-TRAA is more flexible than the LAKG, performing either better or similarly when the H-TRAA configuration is optimized for the problem at hand. Note that in contrast to the Estimator scheme, the performance of both the H-TRAA and the LAKG are improved online (in a real-time manner) *without* invoking any parameter estimation phase.

As seen from the plot in the right of Fig. 7 a less skewed web page update distribution function makes the uniform policy more successful, mainly because a larger number of web pages will have a significant probability of being updated. For the same reason, the estimator scheme is able to lead to an improved performance quicker. In spite of this, the H-TRAA yields a superior performance.

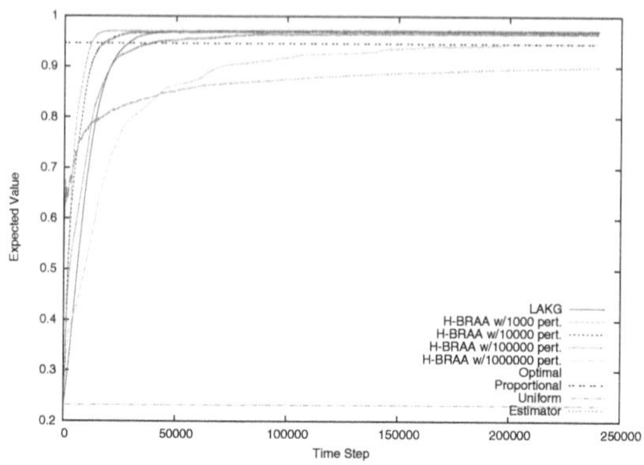

Fig. 8. This figure shows that in the ($\alpha = 0.9, \beta = 1.5$)-environment, the H-TRAA scheme breaches the performance boundary set by the proportional policy, converging towards near-optimal solutions.

The most difficult class of environments we simulate is an environment with a highly skewed web page update distribution ($\beta = 1.5$) combined with a high update probability ($\alpha = 0.9$). In such an environment, the optimal policy performs significantly better than the proportional policy, and so any scheme that converges towards a proportional policy solution will not reach optimal performance. As seen in Fig. 8, both the LAKG and the H-TRAA breach the performance boundary set by the proportional policy, and converges towards near-optimal solutions. The H-TRAA converges slightly quicker compared to the LAKG.

Dynamic Environments. A dynamically changing environment is particularly challenging because the optimal solution is time dependent. In such cases, the current resource allocation solution should be modified according to the environmental changes. When, additionally, the environment and its characteristics are unknown, any changes must first be learned before any meaningful modification can take place.

In order to simulate a dynamic environment, we change the ranking of the web pages at every r^{th} web page poll — a single web page is selected by sampling from the current Zipf-distribution, and this web page switches rank with the succeeding web page in the ranking. As a result, the Zipf-distribution also changes. This means that the web monitor is allowed to conduct r web page polls before the environment changes. The leftmost graph of Fig. 9 demonstrates the ability of our scheme to re-learn in a switching environment for $r = 80,000$. Observe that in the previous dynamic environment, the H-TRAA was able to fully recover to a near-optimal solution because of the low frequency of environmental changes. The graph on the right of Fig. 9 demonstrates the behavior of the automata in a case when this frequency is increased to $r = 1,000$. As seen in the figure, the H-TRAA quickly recovers after the environment has changed, and then moves towards a new near optimal solution. Also, the H-TRAA clearly outperforms the LAKG. Observe that the automata still quickly and steadily improve the

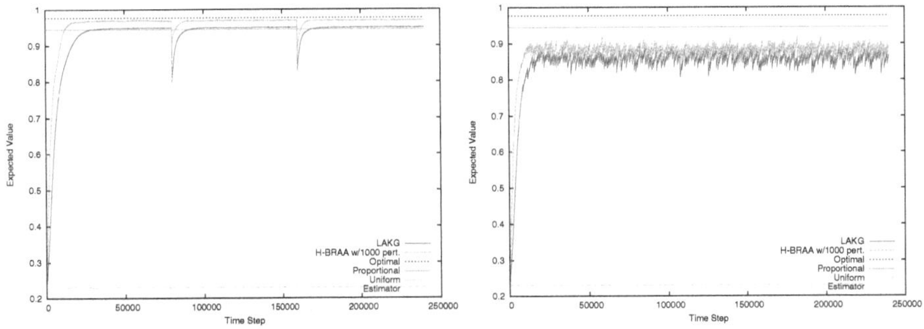

Fig. 9. The figure on the left shows the performance of the schemes for the ($\alpha = 0.9, \beta = 1.5$)-environment where the web page ranking changes at every $80,000^{th}$ web page poll. The H-TRAA finds near-optimal solutions initially, and recovers quickly after the respective environmental changes. Further, consider the figure on the right for the ($\alpha = 0.9, \beta = 1.5$)-environment in which the web page ranking changes every $1,000^{th}$ poll. Observe here that the H-TRAA is able to steadily improve the initial solution, but is never allowed to reach an optimal solution due to the rapid nature of the switching.

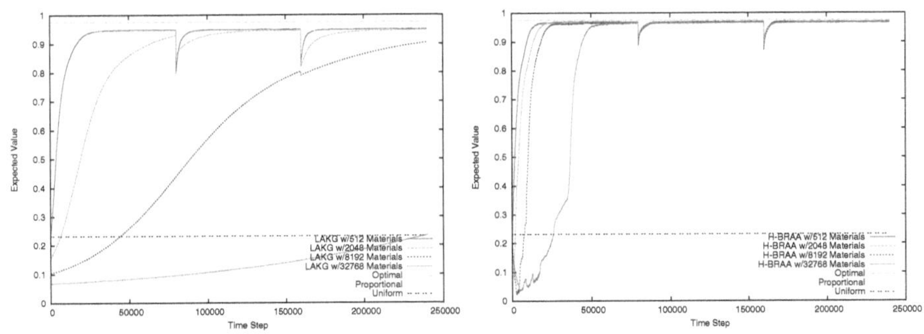

Fig. 10. From the figure on the left we see that extending the number of materials significantly increases the convergence and adaption time of the LAKG. Further, from the figure on the right we see that the H-TRAA scales sub-linearly with the number of materials, and that the impact on the adaptation speed is negligible.

initial solution, but are obviously never allowed to reach an optimal solution. However, the reader should note how the rapidly-changing environment is not able to hinder the automata stabilizing on a solution superior to the solutions found by the estimator scheme. Again, the H-TRAA performs better than the LAKG. Clearly, these results demonstrate how the H-TRAA can perform when the environment is switching with a fixed period (in this case $r = 80,000$ and $r = 1,000$). We believe that similar results will be obtained if r is not fixed but changing, as long as the scheme has enough time to learn the parameters of the updated environment.

Scalability. One of the motivations for designing the H-TRAA was the improved scalability by means of hierarchical learning. As seen in Fig. 10, extending the number of materials significantly increases the convergence time of the LAKG. An increased initial learning phase may be unproblematic in cases where the system will run correspondingly longer, adapting to less dramatic changes as they occur. However, as also seen from the figure, the adaptation speed increases with the number of materials too, when the LAKG is used. The H-TRAA, however, is far less affected by the number of materials. Further, as also seen from Fig. 10 we observe that the initial learning phase is orders of magnitude faster than what can be achieved with the LAKG. Furthermore, the impact on adaptation speed is negligible!

6 Application II: Optimal Sample Size Determination

In this section we consider the problem of estimating the proportion of a population having some specific characteristic. Specifically, we assume that n populations are to be evaluated, and that each population i is characterized by an independent unknown binomial proportion u_i. We will here pursue the goal of minimizing the variance of the proportion estimates when the total number of samples available for estimating the proportions is restricted to c. The purpose is to make the estimates as accurate as possible. As mentioned earlier, for instance, the task at hand could be to determine the proportion of a web site that is successfully validated by an HTML validator [3] and/or a WCAG accessibility validator [4], and that n web sites are to be evaluated by only accessing c web pages.

6.1 Problem Specification

Let x_i be the number of elements sampled randomly from population i and let the count Y_i be the number of the sampled elements that possess a chosen characteristic. For large x_i and when u_i is not too near 0 or 1, the estimator $\hat{u_i} = \frac{Y_i}{x_i}$ is approximately normal with mean u_i and standard deviation $s_i = \sqrt{\frac{u_i(1-u_i)}{x_i}}$ [22]. This standard deviation can be reduced (and the estimate accuracy increased) by increasing the number of samples x_i. In the problem targeted in this section, n different populations can be sampled c times and the goal is to distribute the samples among the populations to minimize the aggregated variance of the estimates. The problem can be reformulated as follows:

$$\begin{array}{ll} \text{maximize} & \sum_{i=1}^{n} -\frac{u_i(1-u_i)}{x_i} \\ \text{subject to} & \sum x_i = c \\ 0 \leq x_i,\ i = 1, \ldots, n. & \end{array}$$

The above optimization problem is an NEFK problem with concave and separable objective function. Since the u_i's are assumed unknown, we apply our H-TRAA to find a near-optimal solution incrementally and online.

6.2 The H-TRAA Solution

We must first define the Stochastic Environment that H-TRAA is to interact with. That is, we must define the stochastic functions $\mathcal{F}' = \{f_1'(x_1), f_2'(x_2), \ldots, f_n'(x_n)\}$. By applying the principles of Lagrange multipliers we find the following conditions that characterize the optimal solution:

$$\frac{u_1(1-u_1)}{{x_1}^2} = \cdots = \frac{u_n(1-u_n)}{{x_n}^2}$$
$$\sum x_i = c$$
$$0 \leq x_i, \; i = 1, \ldots, n$$

Accordingly, we define $f_i'(x_i)$ as follows. First of all, each time $f_i'(x_i)$ is accessed by the H-TRAA, population i is sampled once and the proportion estimate $\hat{u_i}$ is updated accordingly[8]. After $\hat{u_i}$ has been updated, we instantiate $f_i'(x_i)$ by a random draw — $f_i'(x_i)$ is instantiated to the value 0 with probability $1 - \frac{\hat{u_i}(1-\hat{u_i})}{{x_i}^2}$ and to the value 1 with probability $\frac{\hat{u_i}(1-\hat{u_i})}{{x_i}^2}$. In other words, we keep running estimates of the u_i's in order to calculate the outcome probabilities of the $f_i'(x_i)$'s[9].

Configuring the H-TRAA. The H-TRAA can be configured by various means. First of all, the material amount space $(0, 1)$ need not be discretized uniformly. Instead, a nonlinear material amount space can be formed, as done for the LAKG in [2]. Furthermore, the discretization resolution N must also be set for each TRAA, possibly varying from TRAA to TRAA in the hierarchy. In short, the performance achieved for a particular problem can be optimized using these different means of configuring the H-TRAA. In this section, however, our goal is to evaluate the overall performance of the H-TRAA, without fine tuning. Therefore, we will only use a linear material amount space, as specified in Sect. 4. Furthermore, we will use the same resolution $N = 5,000$ for all the TRAAs in the hierarchy, independent of the specific knapsack problem at hand. Thus, our aim is to ensure a fair comparison with the present state of the art, namely, the LAKG scheme.

6.3 Empirical Results

Experimental Set-up. In this sub-section we evaluate our learning scheme by comparing it with the optimal and uniform policies using synthetic data. The reader should appreciate that, in practice, we can only apply the uniform policy, because the optimal policy requires that the u_i's are known.

The data used in the experiment is summarized in Table 1. The table shows the true population proportions used, and the number of populations associated

[8] For a dynamic environment we would utilize a "window-based" strategy and only use the last c samples to estimate the u_i's. However, we are currently studying how recently proposed weak estimators can be used in this setting [23].

[9] Because the outcome probabilities are always available for the populations, we can normalize the outcome probabilities to speed up convergence.

Table 1. The true population proportions used in the experiment, and the number of populations associated with each proportion.

True Proportion	Populations
0.5	6
0.750 / 0.250	5
0.900 / 0.100	41
0.990 / 0.010	51
0.999 / 0.001	409

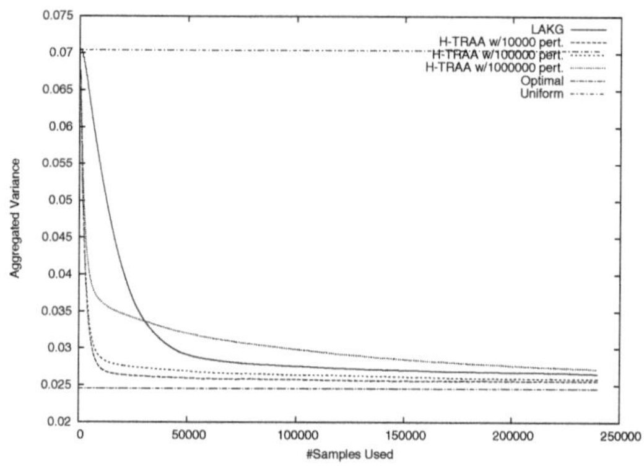

Fig. 11. The H-TRAA steadily reduces the total variance of the initial solution (the uniform policy) as it moves towards near-optimal solutions.

with each proportion. The experiment encompasses 512 populations, and the corresponding proportions are to be estimated by allocating 50,000 samples (window based).

As we will see in the following, it turns that one of the strengths of the H-TRAA is its ability to take advantage of so-called spatial dependencies among materials. To be more specific, as seen in Table 1, the materials are spatially ordered in the sense that the proportion of each material decreases/increases with *its* row-index in the table. In order to starve the H-TRAA from this information, we have rather opted to perturb this spatial structure, and present the *perturbed* information to the learning algorithm. We emphasize though that the algorithm is unaware of the ordering, or the fact that such a perturbation has taken place in the background! Each perturbation swapped the proportions of a randomly selected material and the corresponding proportion of a material succeeding it in the ordering. To enable us to conduct experiments with increasing orders of complexity, we have done our experiments with 10^3, 10^4, 10^5 and 10^6 perturbations. For each of these values, an ensemble of several independent replications with different random number streams was performed so as to minimize the variance of the reported results.

The results of our experiments are truly conclusive and confirm the power of the H-TRAA. Although several experiments were conducted using various setting for various numbers of automata, we report, in the interest of brevity, a brief overview of the results obtained.

Fig. 11 plots the variance of the current solution (as a function of time) each time a unit volume value function $f_i'(x_i)$ has been sampled. The graphs show the results of applying the H-TRAA with 5,000 states and the LAKG with 12,500 states (where the amount of the material added on a transition in the latter is not fixed but varying in a nonlinear manner[10]).

As seen in the figure, the H-TRAA steadily reduces the variance of the initial solution in which the populations are sampled uniformly. Indeed, even by the first 50,000 samples, one can observe a very *significant* reduction. The reader should notice that the H-TRAA converges to a near optimal allocation more expediently and far quicker than the LAKG-scheme, expect for the case with 1,000,000 perturbations where the H-TRAA initally converges faster but subsequently in a more conservative manner.

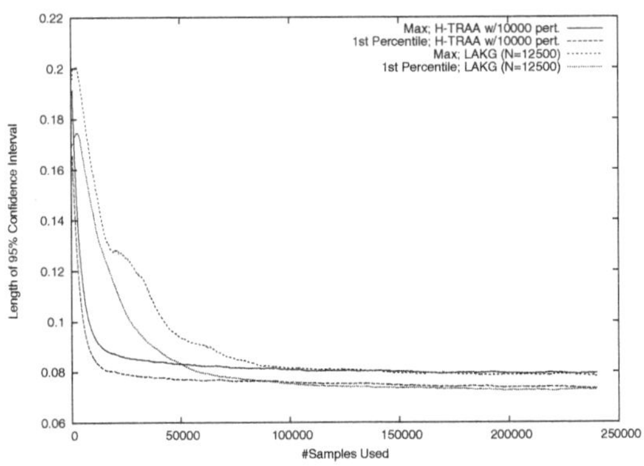

Fig. 12. The confidence interval of each estimated proportion is reduced as the total variance is minimized.

Fig. 12 plots the length of the widest 95% confidence interval among the n estimates after each sampling. We also plot the length of the 5th widest interval (1st percentile), whence we see that the confidence interval of each estimated proportion is reduced by minimizing the total variance.

To conclude, our experimental results demonstrate that the H-TRAA is superior to LAKG in spatially structured environments.

Scalability. One of the motivations for designing the H-TRAA was the improved scalability by means of hierarchical learning. As seen in Fig. 13, extending the number of materials significantly increases the convergence time of the LAKG. From this figure we further observe that while the LAKG does not even

[10] The details of this are omitted. They can be found in [2].

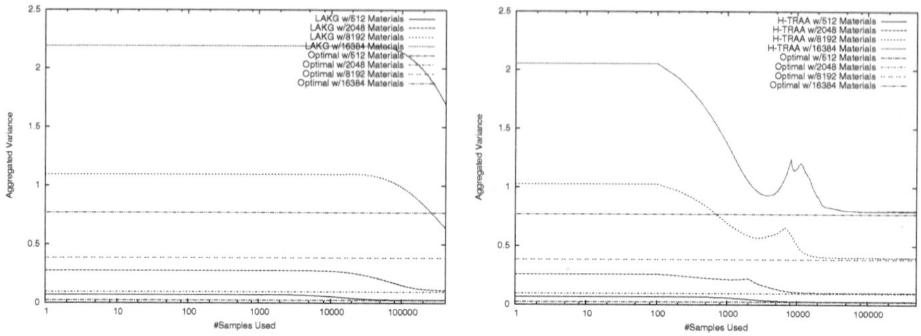

Fig. 13. From the figure on the left we see that extending the number of sets significantly increases the convergence and adaption time of the LAKG. Further, from the figure on the right we see that the H-TRAA scales sub-linearly with the number of materials. Observe the ability of the H-TRAA to emerge out of local optima.

converge, (see the figure on the left), the H-TRAA scales sub-linearly *in every case* with the number of materials (the figure on the right). However, the most interesting phenomenon that we observe from Fig. 13 is the ability of the H-TRAA to emerge out of local optima. The H-TRAA first decreases to a minimum, but when it "discovers" that there is a better solution (which in this cases implies a superior partitioning of the nodes in the tree to their left and right subtrees), it is capable of unlearning the inferior configuration and converging to a superior solution. This, we believe, is quite remarkable, especially because the size of the underlying tree is very large, implying that the number of possible binary trees (which grows exponentially with the size), is even larger. However, by arriving at the global optimum, we see that the H-TRAA has succeeded in learning the best tree structure to resolve the sampling proportions!

7 Conclusions

In this chapter we have considered the fractional knapsack problem and extended the non-LA state-of-the-art in two ways. First of all, we have treated the unit volume values of each material as a *stochastic* variable whose distribution is *unknown*. Secondly, we have worked with the model that the expected value of a material may decrease after each addition to the knapsack. The first learning scheme we proposed for solving this knapsack problem was based on a team of LA that performed a controlled random walk on a discretized fraction space. The second learning scheme was based on a hierarchy of so-called Twofold Resource Allocation Automata (TRAA). Each TRAA works with two materials and moves along the probability space discretized by a resolution parameter, N, with a random walk whose coefficients are not constant. The asymptotic optimality of the TRAA has been proven. Both the TRAA and the H-TRAA have been proven to be asymptotically optimal.

Comprehensive experimental results have demonstrated that performances of both LAKG and H-TRAA are superior to the previous state-of-the-art schemes, with H-TRAA in turn, being superior to LAKG. Additionally, for a given precision, our scheme determines the material fractions of maximal expected value by invoking on-line interactions with the knapsack. We have also demonstrated that both the LAKG scheme and the H-TRAA scheme adapt to switching web distribution functions, allowing us to operate in dynamic environments. Finally, we also provided empirical evidence to show that the H-TRAAs possess a sub-linear scaling property.

References

1. Granmo, O.-C., Oommen, B.J., Myrer, S.A., Olsen, M.G.: Determining Optimal Polling Frequency Using a Learning Automata-based Solution to the Fractional Knapsack Problem. In: Proceedings of the 2006 IEEE International Conferences on Cybernetics & Intelligent Systems (CIS) and Robotics, Automation & Mechatronics (RAM), pp. 73–79. IEEE, Los Alamitos (2006)
2. Granmo, O.-C., Oommen, B.J., Myrer, S.A., Olsen, M.G.: Learning Automata-based Solutions to the Nonlinear Fractional Knapsack Problem with Applications to Optimal Resource Allocation. IEEE Transactions on Systems, Man, and Cybernetics, Part B 37(1), 166–175 (2007)
3. Snaprud, M., Ulltveit-Moe, N., Granmo, O.C., Rafoshei-Klev, M., Wiklund, A., Sawicka, A.: Quantitative Assessment of Public Web Sites Accessibility - Some Early Results. In: The Accessibility for All Conference (2003)
4. Xiaoming, Z.: Evaluation and Enhancement of Web Content Accessibility for Persons with Disabilities. PhD thesis, University of Pittsburgh (2004)
5. Granmo, O.-C., Oommen, B.J.: On Allocating Limited Sampling Resources Using a Learning Automata-based Solution to the Fractional Knapsack Problem. In: Proceedings of the 2006 International Intelligent Information Processing and Web Mining Conference (IIS:IIPW 2006). Advances in Soft Computing, pp. 263–272. Springer, Heidelberg (2006)
6. Black, P.E.: Fractional knapsack problem. Dictionary of Algorithms and Data Structures (2004)
7. Kellerer, H., Pferschy, U., Pisinger, D.: Knapsack Problems. Springer, Heidelberg (2004)
8. Bretthauer, K.M., Shetty, B.: The Nonlinear Knapsack Problem — Algorithms and Applications. European Journal of Operational Research 138, 459–472 (2002)
9. Fox, B.: Discrete optimization via marginal analysis. Management Sciences 13(3), 211–216 (1966)
10. Dean, B.C., Goemans, M.X., Vondrdk, J.: Approximating the Stochastic Knapsack Problem: the benefit of adaptivity. In: 45th Annual IEEE Symposium on Foundations of Computer Science, pp. 208–217. IEEE, Los Alamitos (2004)
11. Steinberg, E., Parks, M.S.: A Preference Order Dynamic Program for a Knapsack Problem with Stochastic Rewards. The Journal of the Operational Research Society 30(2), 141–147 (1979)
12. Ross, K.W., Tsang, D.: The stochastic knapsack problem. IEEE Transactions on Communications 37(7), 740–747 (1989)

13. Pandey, S., Ramamritham, K., Chakrabarti, S.: Monitoring the Dynamic Web to Respond to Continuous Queries. In: 12th International World Wide Web Conference, pp. 659–668. ACM Press, New York (2003)
14. Wolf, J.L., Squillante, M.S., Sethuraman, J., Ozsen, K.: Optimal Crawling Strategies for Web Search Engines. In: 11th International World Wide Web Conference, pp. 136–147. ACM Press, New York (2002)
15. Narendra, K.S., Thathachar, M.A.L.: Learning Automata: An Introduction. Prentice-Hall, Englewood Cliffs (1989)
16. Thathachar, M.A.L., Sastry, P.S.: Networks of Learning Automata: Techniques for Online Stochastic Optimization. Kluwer Academic Publishers, Dordrecht (2004)
17. Tsetlin, M.L.: Automaton Theory and Modeling of Biological Systems. Academic Press, London (1973)
18. Granmo, O.C., Oommen, B.J.: Solving stochastic nonlinear resource allocation problems using a hierarchy of twofold resource allocation automata (submitted for publication, 2009)
19. Castillo, C.: Effective Web Crawling. PhD thesis, University of Chile (2004)
20. Zipf, G.: Human Behavior and the Principle of Least Effort. Addison-Wesley, Reading (1949)
21. Wikipedia: Zipf's law — wikipedia, the free encyclopedia (accessed December 5, 2006)
22. Bhattacharyya, G.K., Johnson, R.A.: Statistical Concepts and Methods. John Wiley & Sons, Chichester (1977)
23. Oommen, B.J., Rueda, L.: Stochastic Learning-based Weak Estimation of Multinomial Random Variables and its Applications to Pattern Recognition in Nonstationary Environments. Pattern Recognition 39, 328–341 (2006)

Neural Networks in Model Predictive Control

Maciej Ławryńczuk

Institute of Control and Computation Engineering, Warsaw University of Technology
ul. Nowowiejska 15/19, 00-665 Warsaw, Poland
Tel.: +48 22 234-76-73
M.Lawrynczuk@ia.pw.edu.pl

Abstract. This work is concerned with Model Predictive Control (MPC) algorithms in which neural models are used on-line. Model structure selection, training and stability issues are thoroughly discussed. Computationally efficient algorithms are recommended which use on-line linearisation of the neural model and need solving on-line quadratic optimisation tasks. It is demonstrated that they give very good results, comparable to those obtained when nonlinear optimisation is used on-line in MPC. In order to illustrate the effectiveness of discussed approaches, a chemical process is considered. The development of appropriate models for MPC is discussed, the control accuracy and the computational complexity of recommended MPC are shown.

Keywords: Process control, Model Predictive Control, neural networks, model identification, optimisation, quadratic programming, linearisation.

1 Introduction

Model Predictive Control (MPC) is recognised as the only advanced control technique which has been very successful in practice [29, 43, 44, 47]. It is mainly because MPC algorithms can take into account constraints imposed on both process inputs (manipulated variables) and outputs (controlled variables), which usually decide on quality, economic efficiency and safety. Moreover, MPC techniques are very efficient in multivariable process control (i.e. for processes with many inputs and outputs) and for processes with difficult dynamic properties, e.g. with significant time-delays. Different versions of MPC algorithms are nowadays used in numerous fields, not only in chemical, food and motor industries, but also in medicine and aerospace [43].

MPC techniques based on easy to obtain linear models are frequently used in practice [43]. In many cases their control accuracy is sufficient, much better than that of the classical PID approach. Nevertheless, in the last two decades numerous MPC algorithms based on nonlinear models have been developed and have gained in popularity [12, 35, 43, 47, 48]. When applied to significantly nonlinear processes, they significantly improve the control accuracy in comparison with MPC approaches which use linear models.

In MPC a dynamic model of the process is used to predict its behaviour over some time horizon and to determine the optimal future control policy. Hence, the

N.T. Nguyen and E. Szczerbicki (Eds.): Intel. Sys. for Know. Management, SCI 252, pp. 31–63.
springerlink.com

choice of the model structure which is used in nonlinear MPC is extremely important. Main measures of model utility are: approximation accuracy, suitability for control and easiness of development [39]. Fundamental (first-principle) models [19, 30], although potentially very precise, are usually not suitable for on-line control. Such models are comprised of systems of nonlinear differential and algebraic equations which have to be solved on-line in MPC at each sampling instant. It is usually computationally demanding as fundamental models can be very complex and may lead to numerical problems (e.g. stiffness, ill-conditioning). Moreover, in many cases the development and validation of fundamental models is difficult, it needs technological knowledge.

This work presents MPC algorithms based on neural models [11]. In spite of the fact that a number of different nonlinear black-box model types are available (e.g. polynomial models, fuzzy models, Volterra series models [39]), neural networks have some unique features thanks to which they can be very efficiently used on-line in MPC. More specifically, neural networks are universal approximators [13], have relatively a small number of parameters and a simple structure. Moreover, they directly describe relations between inputs and outputs of the process, which means that during on-line control it is not necessary to solve complicated systems of nonlinear differential equations at each sampling instant as it is necessary when fundamental models are used. In consequence, neural models can be effectively used in MPC as models of technological processes. In particular, thanks to a simple, regular structure of neural models, the implementation of described algorithms is relatively easy. Neural models are trained using recorded data sets, no technological knowledge is necessary.

Although the literature concerned with MPC algorithms based on neural models is quite rich [1, 2, 3, 4, 5, 8, 10, 12, 14, 16, 18, 20, 21, 22, 23, 24, 25, 26, 27, 28, 33, 35, 37, 40, 41, 42, 47, 48, 49, 50, 51, 52], there are a few issues worth exploring. First of all, the way the nonlinear model is used on-line in MPC is crucial in light of the computational complexity and reliability of the whole control system. In theory, neural models can be used directly without any simplifications but it means that the optimal control policy at each sampling instant must be calculated from a nonlinear optimisation problem. It is not only computationally very demanding but also the optimisation routine is likely to terminate at shallow local minima. In order to reduce the computational burden and increase reliability, in this work suboptimal approaches are recommended. The nonlinear neural model is linearised on-line and next the obtained local approximation is used in MPC algorithms [10,20,21,22,23,24,25,26,27,28,37,47,48]. Thanks to using for control a local linearisation of the original neural model, the necessity of on-line nonlinear optimisation is avoided, it is replaced by an easy to solve quadratic programming problem. It is demonstrated in this work that suboptimal MPC algorithms with on-line linearisation are very precise, the control accuracy is comparable to that obtained when a nonlinear optimisation routine is used on-line at each sampling instant.

The second important issue discussed in this work is the choice of the model structure and training. It is emphasised that MPC algorithms are very model-based, the possible control performance is determined by the accuracy of predictions calculated by means of a dynamic model. The role of the model in MPC cannot be ignored during model structure selection and training. The model has to be able to make good predictions of future behaviour of the process over the whole prediction horizon. Usually, Multi Layer Perceptron (MLP) and Radial Basis Function (RBF) neural networks are used in MPC. In some cases block-oriented nonlinear models which are composed of linear dynamic systems and nonlinear steady-state (static) elements can be efficiently used (neural Hammerstein and Wiener models). In this work two classes of specialised neural models designed with the specific aim of using them in MPC are also discussed: multi-models [24] and structured models [20]. Both specialised models are trained easily as one-step ahead predictors, but they calculate predictions for the whole prediction horizon without being used recurrently. As a result, the prediction error is not propagated which is particularly important in practice.

In spite of the fact that in practice MPC techniques are relatively easily tuned, a stable version of described algorithms based on different kinds of neural models is shortly discussed.

In order to show the effectiveness of discussed approaches to MPC a chemical reactor process is discussed. The development of appropriate models for MPC is thoroughly discussed, the control accuracy and the computational complexity of recommended MPC are also shown.

2 Model Predictive Control Problem Formulation

In MPC algorithms [29, 44, 47] at each consecutive sampling instant k a set of future control increments is calculated

$$\triangle \boldsymbol{u}(k) = \begin{bmatrix} \triangle u(k|k) \\ \vdots \\ \triangle u(k+N_u-1|k) \end{bmatrix} \tag{1}$$

It is assumed that $\triangle u(k+p|k) = 0$ for $p \geq N_u$, where N_u is the control horizon. The objective is to minimise differences between the reference trajectory $y^{\mathrm{ref}}(k+p|k)$ and predicted output values $\hat{y}(k+p|k)$ over the prediction horizon $N \geq N_u$ and to penalise excessive control increments. The minimised cost function is usually

$$J(k) = \sum_{p=1}^{N} (y^{\mathrm{ref}}(k+p|k) - \hat{y}(k+p|k))^2 + \sum_{p=0}^{N_u-1} \lambda_p (\triangle u(k+p|k))^2 \tag{2}$$

where $\lambda_p > 0$ are weighting coefficients. Only the first element of the determined sequence (1) is applied to the process, i.e. $u(k) = \triangle u(k|k) + u(k-1)$. At the next sampling instant, $k+1$, the prediction is shifted one step forward and the whole procedure is repeated. Fig. 1 illustrates the general idea of MPC.

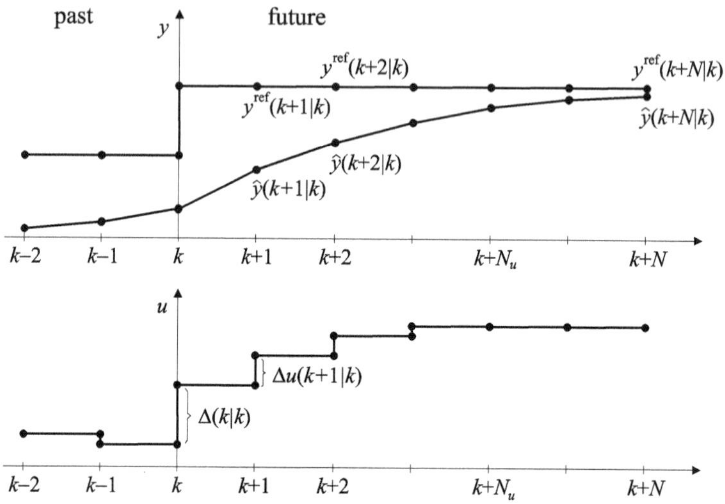

Fig. 1. The principle of predictive control

A unique feature of MPC is the fact that constraints imposed on process variables can be rigorously taken into account. More specifically, future control increments (1) are found on-line from the following optimisation problem

$$\begin{aligned}
&\min_{\triangle u(k|k)\ldots\triangle u(k+N_u-1|k)} \{J(k)\} \\
&\text{subject to} \\
&\quad u^{\min} \leq u(k+p|k) \leq u^{\max}, \quad p = 0, \ldots, N_u - 1 \\
&\quad -\triangle u^{\max} \leq \triangle u(k+p|k) \leq \triangle u^{\max}, \quad p = 0, \ldots, N_u - 1 \\
&\quad y^{\min} \leq \hat{y}(k+p|k) \leq y^{\max}, \quad p = 1, \ldots, N
\end{aligned} \tag{3}$$

where $u^{\min}$, $u^{\max}$, $\triangle u^{\max}$, $y^{\min}$, $y^{\max}$ define constraints imposed on the magnitude of the input variable, the increment of the input variable and the magnitude of the output variable, respectively.

MPC algorithms directly use an explicit dynamic model in order to predict future behaviour of the process, i.e. to calculate predicted values of the output variable, $\hat{y}(k+p|k)$, over the prediction horizon ($p = 1, \ldots, N$). Hence, the role of the model in MPC is crucial. As a result, MPC techniques are very model-based, the accuracy of the model significantly affects the quality of control.

A great advantage of MPC algorithms is the fact that they can be efficiently used for multivariable processes. Assuming that the process has n_u inputs and n_y outputs (i.e. $u(k) \in \Re^{n_u}$, $y(k) \in \Re^{n_y}$), the MPC cost function (2) becomes

$$J(k) = \sum_{p=1}^{N} \left\| y^{\text{ref}}(k+p|k) - \hat{y}(k+p|k) \right\|^2_{\boldsymbol{M}_p} + \sum_{p=0}^{N_u-1} \left\| \triangle u(k+p|k) \right\|^2_{\boldsymbol{\Lambda}_p} \tag{4}$$

where $\|x\|_A^2 = x^T A x$, $M_p \geq 0$ and $\Lambda_p > 0$ are weighting matrices of dimensionality $n_y \times n_y$ and $n_u \times n_u$, respectively. Having completed the MPC optimisation task (3), the first n_u elements of the sequence (1) are applied to the process.

3 Neural Models of the Process

Let the dynamic process under consideration be described by the following discrete-time Nonlinear Auto Regressive with eXternal input (NARX) model

$$y(k) = f(\boldsymbol{x}(k)) = f(u(k-\tau), \ldots, u(k-n_B), y(k-1), \ldots, y(k-n_A)) \tag{5}$$

where $f : \Re^{n_A+n_B-\tau+1} \longrightarrow \Re$ is a nonlinear function which describes the model, integers n_A, n_B, τ define the order of dynamics, $\tau \leq n_B$. Although, in general, the function f can be realised by various nonlinear models [39], Multi Layer Perceptron (MLP) and Radial Basis Functions (RBF) neural networks are most frequently used. Both structures are universal approximators [13] capable of approximating any smooth function to an arbitrary degree of accuracy, they have relatively a small number of parameters and a simple structure.

3.1 MLP Neural Model

When the MLP neural network with one hidden layer and a linear output [11] is used as the function f in (5), the output of the model can be expressed as

$$y(k) = f(\boldsymbol{x}(k)) = w_0^2 + \sum_{i=1}^{K} w_i^2 \varphi(z_i(k)) \tag{6}$$

where $z_i(k)$ is a sum of inputs of the ith hidden node, $\varphi : \Re \longrightarrow \Re$ is the nonlinear transfer function (e.g. hyperbolic tangent), K is the number of hidden nodes. From (5) one has

$$z_i(k) = w_{i,0}^1 + \sum_{j=1}^{I_u} w_{i,j}^1 u(k-\tau+1-j) + \sum_{j=1}^{n_A} w_{i,I_u+j}^1 y(k-j) \tag{7}$$

Weights of the network are denoted by $w_{i,j}^1$, $i = 1, \ldots, K$, $j = 0, \ldots, n_A + n_B - \tau + 1$, and w_i^2, $i = 0, \ldots, K$, for the first and the second layer, respectively, $I_u = n_B - \tau + 1$.

In the case of the multivariable processes with n_u inputs and n_y outputs the model is usually comprised of n_y independent Multiple-Input Single-Output (MISO) models. The whole nonlinear model has the general form

$$\begin{aligned} y_1(k) = f_1(&u_1(k-\tau^{1,1}), \ldots, u_1(k-n_B^{1,1}), \ldots, \\ &u_{n_u}(k-\tau^{1,n_u}), \ldots, u_{n_u}(k-n_B^{1,n_u}), y_1(k-1), \ldots, y_1(k-n_A^1)) \end{aligned} \tag{8}$$

$$\vdots$$

$$\begin{aligned} y_{n_y}(k) = f_{n_y}(&u_1(k-\tau^{n_y,1}), \ldots, u_1(k-n_B^{n_y,1}), \ldots, \\ &u_{n_u}(k-\tau^{n_y,n_u}), \ldots, u_{n_u}(k-n_B^{n_y,n_u}), y_{n_y}(k-1), \ldots, y_{n_y}(k-n_A^{n_y})) \end{aligned} \tag{9}$$

where integers n_A^m, $n_B^{m,n}$, $\tau^{m,n}$ for $m = 1, \ldots, n_y$, $n = 1, \ldots, n_u$ define the order of dynamics, $\tau^{m,n} \leq n_B^{m,n}$. Functions $f_m : \Re^{n_A^m + \sum_{n=1}^{n_u}(n_B^{m,n} - \tau^{m,n} + 1)} \longrightarrow \Re$, $m = 1, \ldots, n_y$ are realised by independent neural networks which are trained separately. Alternatively, the model can be realised by only one neural network with as many as n_y outputs, but training of such models is usually more difficult.

Unlike fundamental models, which are comprised of systems of nonlinear algebraic and differential equations, MLP and RBF neural models directly describe relations between inputs and outputs of the process. Thanks to it, neural models can be effectively used in MPC since during on-line calculation of the control policy it is not necessary to solve these equations at each sampling instant, which may be computationally complex and lead to numerical problems (e.g. stiffness, ill-conditioning).

3.2 RBF Neural Model

If the RBF neural network containing one hidden layer with Gaussian functions and a linear output is used as the function f in (5), the output of the model is

$$\begin{aligned} y(k) = f(\boldsymbol{x}(k)) &= w_0 + \sum_{i=1}^{K} w_i \exp(- \|\boldsymbol{x}(k) - \boldsymbol{c}_i\|_{\boldsymbol{Q}_i}^2) \\ &= w_0 + \sum_{i=1}^{K} w_i \exp(-z_i(k)) \end{aligned} \tag{10}$$

where K is the number of hidden nodes. Vectors $\boldsymbol{c}_i \in \Re^{n_A+n_B-\tau+1}$ and the diagonal weighting matrices $\boldsymbol{Q}_i = diag(q_{i,1}, \ldots, q_{i,n_A+n_B-\tau+1})$ describe centres and widths of nodes, respectively, $i = 1, \ldots, K$. The model (10) is sometimes named Hyper Radial Basis Function (HRBF) neural network in contrast to the ordinary RBF neural networks in which widths of nodes are constant. Let $z_i(k)$ be the sum of inputs of the i-th hidden node. Recalling the arguments of the rudimentary model (5), one has

$$z_i(k) = \sum_{j=1}^{I_u} q_{i,j}(u(k-\tau+1-j) - c_{i,j})^2 + \sum_{j=1}^{n_A} q_{i,I_u+j}(y(k-j) - c_{i,I_u+j})^2 \tag{11}$$

Although it is a well known fact that both MLP and RBF neural models are universal approximators, MLP networks are global ones whereas RBF networks are local ones. It is because in the first case all hidden nodes are used to calculate the output for a given input, in the second case only selected hidden nodes are employed, other nodes are practically inactive. A direct consequence of this fact is that for MLP networks it is difficult to establish a link between available data and parameters of hidden nodes. As a result, weights of such networks are usually initialised randomly, training should be repeated many times for different numbers of hidden nodes to find the adequate topology which gives good approximation. Conversely, training of RBF models is much more efficient because parameters of basis functions are directly found from available data. Moreover,

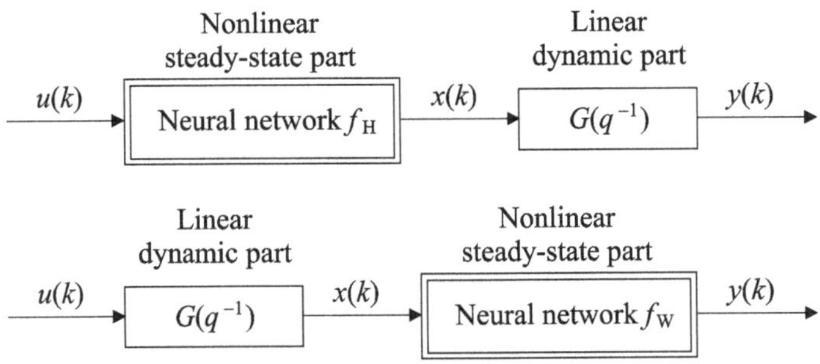

Fig. 2. The structure of neural Hammerstein (*top*) and Wiener (*bottom*) models

selection of the optimal structure of the RBF model which leads to desired approximation accuracy can be included in a training procedure, whereas selection of the structure of the MLP model usually needs training many networks.

3.3 Hammerstein and Wiener Neural Models

Neural MLP and RBF models are entirely black-box models. It means that the model structure has nothing to do with the physical nature of the process and model parameters (weights) have no physical interpretation. An interesting alternative is to use block-oriented nonlinear models which are composed of linear dynamic systems and nonlinear steady-state (static) elements. Hammerstein and Wiener models are most known and most widely implemented members of this class [15]. A model is called the Hammerstein model if the linear dynamic part follows the nonlinear steady-state one, in the Wiener model the connection order is reversed as shown in Fig. 2. Unlike black-box models, block-oriented models have a clear interpretation, the steady-state part describes the gain of the system.

In the simplest case polynomials can be used as the steady-state nonlinear part of block-oriented models. Unfortunately, some nonlinear functions need polynomials of a high order. In such cases models are complex and the model uncertainty is likely to be increased. Moreover, polynomials are likely to have oscillatory interpolation and extrapolation properties. In consequence, the application of polynomials is in practice limited [15].

A sound alternative is to use neural networks in the steady-state part of Hammerstein and Wiener models. Such an approach has a few advantages. Not only are neural networks universal approximators, but also, in contrast to polynomial approximators, neural approximations are very smooth, they do not suffer from oscillatory interpolation and extrapolation behavior. An excellent review of identification algorithms and applications of block-oriented Hammerstein and Wiener models is given in [15].

Both MLP and RBF neural networks can be used as the nonlinear steady-state part of block-oriented models. When the MLP neural network is used, the nonlinear steady-state function f_H of the Hammerstein model is described by

$$x(k) = f_{\mathrm{H}}(u(k)) = w_0^2 + \sum_{i=1}^{K} w_i^2 \varphi(w_{i,0}^1 + w_{i,1}^1 u(k)) \tag{12}$$

where $x(k)$ is an auxiliary signal. Weights are denoted by $w_{i,j}^1$, $i = 1, \ldots, K$, $j = 0, 1$ and w_i^2, $i = 0, \ldots, K$, for the first and the second layer, respectively.

The transfer function of the linear dynamic part is

$$G(q^{-1}) = \frac{\boldsymbol{B}(q^{-1})}{\boldsymbol{A}(q^{-1})} = \frac{b_\tau q^{-\tau} + \ldots + b_{n_B} q^{-n_B}}{1 + a_1 q^{-1} + \ldots + a_{n_A} q^{-n_A}} \tag{13}$$

where q^{-1} is the backward shift operator. Hence, the output of the dynamic part is

$$y(k) = \sum_{l=\tau}^{n_B} b_l x(k-l) - \sum_{l=1}^{n_A} a_l y(k-l) \tag{14}$$

Combining (12) and (14), the output of the neural Hammerstein model is

$$y(k) = \sum_{l=\tau}^{n_B} b_l \left(w_0^2 + \sum_{i=1}^{K} w_i^2 \varphi(w_{i,0}^1 + w_{i,1}^1 u(k-l)) \right) - \sum_{l=1}^{n_A} a_l y(k-l) \tag{15}$$

Analogously, for the neural Wiener model one has

$$\begin{aligned} y(k) = f_{\mathrm{W}}(x(k)) &= w_0^2 + \sum_{i=1}^{K} w_i^2 \varphi(w_{i,0}^1 + w_{i,1}^1 x(k)) \\ &= w_0^2 + \sum_{i=1}^{K} w_i^2 \varphi \left(w_{i,0}^1 + w_{i,1}^1 \left(\sum_{l=\tau}^{n_B} b_l u(k-l) - \sum_{l=1}^{n_A} a_l x(k-l) \right) \right) \end{aligned} \tag{16}$$

Similarly as rudimentary MLP and RBF neural models, block-oriented neural models have a regular structure, unlike fundamental models they do not contain differential and algebraic equations which have to be solved on-line in MPC at each sampling instant. Hence, they can be also easily used in MPC.

4 MPC with Nonlinear Optimisation (MPC-NO)

In MPC algorithms a nonlinear model is directly used to calculate predictions $\hat{y}(k+p|k)$ which are taken into account in the minimised cost function (2). The general prediction equation is

$$\hat{y}(k+p|k) = y(k+p|k) + d(k) \tag{17}$$

where quantities $y(k+p|k)$ are calculated from the model of the process. Rudimentary MLP (6), (7), RBF (10), (11) neural models or block oriented neural Hammerstein (15) and Wiener (16) models can be used. Usually, the "DMC type" disturbance model is used, in which the unmeasured disturbance $d(k)$ is assumed to be constant over the prediction horizon [47]. It is estimated from

$$d(k) = y(k) - y(k|k-1) \tag{18}$$

where $y(k)$ is measured while $y(k|k-1)$ is calculated from the model.

If for prediction in MPC a nonlinear model is used without any simplifications, predictions $\hat{y}(k+p|k)$ depend in a nonlinear way on the calculated control policy, i.e. on future control increments $\triangle \boldsymbol{u}(k)$. It means that the MPC optimisation problem (3) becomes a nonlinear task which has to be solved on-line in real time. It may be computationally demanding and time consuming. Moreover, such an approach may be not reliable in practice because the nonlinear optimisation routine is likely to terminate in a shallow local minimum. Nevertheless, simulation results of the MPC-NO algorithm based on different neural models are frequently presented in the literature, its apparent computational inefficiency and limited practical applicability are overlooked [2,3,14,16,33,37,49,50,51,52].

5 MPC with Nonlinear Prediction and Linearisation (MPC-NPL)

In this work the MPC algorithm with Nonlinear Prediction and Linearisation (MPC-NPL) [26,47,48] is recommended. At each sampling instant k a local linear approximation of the nonlinear neural model is found on-line. Thanks to linearisation, predictions of the output depend in a purely linear way on the calculated control policy. As a result, the MPC-NPL algorithm needs solving on-line a quadratic programming problem, which can be easily completed within a foreseeable time period, the necessity of nonlinear optimisation is avoided. In practice, for different technological processes the algorithm gives good closed-loop control performance, comparable to that obtained in computationally demanding MPC-NO approach with full nonlinear optimisation [20,21,22,23,24,25,26,27,28,47,48].

The linear approximation of the nonlinear neural model (5) is

$$y(k) = \sum_{l=1}^{n_B} b_l(k)(u(k-l)) - \sum_{l=1}^{n_A} a_l(k)(y(k-l)) \tag{19}$$

where $a_l(k)$ and $b_l(k)$ are coefficients of the linearised model. Using the linearised model (19) recurrently, from the general prediction equation (17) one obtains

$$\begin{aligned} \hat{y}(k+1|k) =& b_1(k)u(k|k) + b_2(k)u(k-1) + b_3(k)u(k-2) + \ldots \\ &+ b_{n_B}(k)u(k-n_B+1) \\ &- a_1(k)y(k) - a_2(k)y(k-1) - a_3(k)y(k-2) - \ldots \\ &- a_{n_A}(k)y(k-n_A+1) + d(k) \end{aligned} \tag{20}$$

$$\begin{aligned}\hat{y}(k+2|k) =& b_1(k)u(k+1|k) + b_2(k)u(k|k) + b_3(k)u(k-1) + \ldots \\ &+ b_{n_B}(k)u(k-n_B+2) \\ &- a_1(k)\hat{y}(k+1|k) - a_2(k)y(k) - a_3(k)y(k-1) - \ldots \\ &- a_{n_A}(k)y(k-n_A+2) + d(k)\end{aligned} \tag{21}$$

$$\begin{aligned}\hat{y}(k+3|k) =& b_1(k)u(k+2|k) + b_2(k)u(k+1|k) + b_3(k)u(k|k) + \ldots \\ &+ b_{n_B}(k)u(k-n_B+3) \\ &- a_1(k)\hat{y}(k+2|k) - a_2(k)\hat{y}(k+1) - a_3(k)y(k) - \ldots \\ &- a_{n_A}(k)y(k-n_A+3) + d(k)\end{aligned} \tag{22}$$

$$\vdots$$

Predictions can be expressed in a compact form as functions of future control increments

$$\hat{y}(k+1|k) = s_1(k)\triangle u(k|k) + \ldots \tag{23}$$

$$\hat{y}(k+2|k) = s_2(k)\triangle u(k|k) + s_1(k)\triangle u(k+1|k) + \ldots \tag{24}$$

$$\hat{y}(k+3|k) = s_3(k)\triangle u(k|k) + s_2(k)\triangle u(k+1|k) + s_1(k)\triangle u(k+2|k) + \ldots \tag{25}$$

$$\vdots$$

where step-response coefficients of the linearised model are determined recurrently for $j = 1, \ldots, N$ from

$$s_j(k) = \sum_{i=1}^{\min(j,n_B)} b_i(k) - \sum_{i=1}^{\min(j-1,n_A)} a_i(k)s_{j-i}(k) \tag{26}$$

Using the linearised model (19), it is possible to express the output prediction vector $\hat{\boldsymbol{y}}(k) = [\hat{y}(k+1|k) \ldots \hat{y}(k+N|k)]^T$ as the sum of two parts

$$\hat{\boldsymbol{y}}(k) = \boldsymbol{G}(k)\triangle\boldsymbol{u}(k) + \boldsymbol{y}^0(k) \tag{27}$$

The first part depends only on the future (on future control moves $\triangle\boldsymbol{u}(k)$), the second part is a free trajectory vector $\boldsymbol{y}^0(k) = \left[y^0(k+1|k) \ldots y^0(k+N|k)\right]^T$, which depends only on the past. The dynamic matrix $\boldsymbol{G}(k)$ of dimensionality $N \times N_u$ contains step-response coefficients of the local linear approximation of the nonlinear model

$$\boldsymbol{G}(k) = \begin{bmatrix} s_1(k) & 0 & \ldots & 0 \\ s_2(k) & s_1(k) & \ldots & 0 \\ \vdots & \vdots & \ddots & \vdots \\ s_N(k) & s_{N-1}(k) & \ldots & s_{N-N_u+1}(k) \end{bmatrix} \tag{28}$$

The dynamic matrix is calculated on-line from the local linearisation of the full nonlinear model taking into account the current state of the process.

Thanks to using the suboptimal prediction (27), the optimisation problem (3) becomes the following quadratic programming task

$$\begin{aligned}&\min_{\triangle \boldsymbol{u}(k)} \left\{ \left\| \boldsymbol{y}^{\mathrm{ref}}(k) - \boldsymbol{G}(k)\triangle \boldsymbol{u}(k) - \boldsymbol{y}^0(k) \right\|^2 + \left\| \triangle \boldsymbol{u}(k) \right\|^2_{\boldsymbol{\Lambda}} \right\} \\ &\text{subject to} \\ &\quad \boldsymbol{u}^{\min} \leq \boldsymbol{J}\triangle \boldsymbol{u}(k) + \boldsymbol{u}^{k-1}(k) \leq \boldsymbol{u}^{\max} \\ &\quad -\triangle \boldsymbol{u}^{\max} \leq \triangle \boldsymbol{u}(k) \leq \triangle \boldsymbol{u}^{\max} \\ &\quad \boldsymbol{y}^{\min} \leq \boldsymbol{G}(k)\triangle \boldsymbol{u}(k) + \boldsymbol{y}^0(k) \leq \boldsymbol{y}^{\max}\end{aligned} \tag{29}$$

where $\boldsymbol{y}^{\mathrm{ref}}(k) = \left[y^{\mathrm{ref}}(k+1|k)\ldots y^{\mathrm{ref}}(k+N|k)\right]^T$, $\boldsymbol{y}^{\min} = \left[y^{\min}\ldots y^{\min}\right]^T$, $\boldsymbol{y}^{\max} = \left[y^{\max}\ldots y^{\max}\right]^T$ are vectors of length N, $\boldsymbol{u}^{\min} = \left[u^{\min}\ldots u^{\min}\right]^T$, $\boldsymbol{u}^{\max} = \left[u^{\max}\ldots u^{\max}\right]^T$, $\boldsymbol{u}^{k-1}(k) = \left[u(k-1)\ldots u(k-1)\right]^T$, $\triangle \boldsymbol{u}^{\max} = \left[\triangle u^{\max}\ldots \triangle u^{\max}\right]^T$ are vectors of length N_u, $\boldsymbol{\Lambda} = diag(\lambda_0, \ldots, \lambda_{N_u-1})$, $\boldsymbol{J}$ is the all ones lower triangular matrix of dimensionality $N_u \times N_u$.

If output constraints are present, the MPC optimisation task (29) may be affected by the infeasibility problem. In such a case the original output constraints have to be softened by using slack variables [29, 47]. Using a quadratic penalty for constraint violations the MPC-NPL optimisation problem is

$$\begin{aligned}&\min_{\triangle \boldsymbol{u}(k),\ \boldsymbol{\varepsilon}^{\min},\ \boldsymbol{\varepsilon}^{\max}} \left\{ \left\| \boldsymbol{y}^{\mathrm{ref}}(k) - \boldsymbol{G}(k)\triangle \boldsymbol{u}(k) - \boldsymbol{y}^0(k) \right\|^2 + \left\| \triangle \boldsymbol{u}(k) \right\|^2_{\boldsymbol{\Lambda}} \right. \\ &\qquad\qquad \left. + \rho^{\min} \left\| \boldsymbol{\varepsilon}^{\min} \right\|^2 + \rho^{\max} \left\| \boldsymbol{\varepsilon}^{\max} \right\|^2 \right\} \\ &\text{subject to} \\ &\quad \boldsymbol{u}^{\min} \leq \boldsymbol{J}\triangle \boldsymbol{u}(k) + \boldsymbol{u}^{k-1}(k) \leq \boldsymbol{u}^{\max} \\ &\quad -\triangle \boldsymbol{u}^{\max} \leq \triangle \boldsymbol{u}(k) \leq \triangle \boldsymbol{u}^{\max} \\ &\quad \boldsymbol{y}^{\min} - \boldsymbol{\varepsilon}^{\min} \leq \boldsymbol{G}(k)\triangle \boldsymbol{u}(k) + \boldsymbol{y}^0(k) \leq \boldsymbol{y}^{\max} + \boldsymbol{\varepsilon}^{\max} \\ &\quad \boldsymbol{\varepsilon}^{\min} \geq 0, \quad \boldsymbol{\varepsilon}^{\max} \geq 0\end{aligned} \tag{30}$$

where slack variables vectors of length N are denoted by $\boldsymbol{\varepsilon}^{\min}$ and $\boldsymbol{\varepsilon}^{\max}$, $\rho^{\min}$ and $\rho^{\max} > 0$ are weights.

All things considered, at each sampling instant k of the MPC-NPL algorithm the structure of which is shown in Fig. 3, the following steps are repeated:

1. Linearisation of the neural model: obtain the matrix $\boldsymbol{G}(k)$.
2. Find the nonlinear free trajectory $\boldsymbol{y}^0(k)$ using the neural model.
3. Solve the quadratic programming task (30) to find the control policy $\triangle \boldsymbol{u}(k)$.
4. Implement the first element of the obtained policy $u(k) = \triangle u(k|k) + u(k-1)$.
5. Set $k := k + 1$, go to step 1.

The formulation of the MPC-NPL algorithm is general, different neural models can be used: MLP structures [20, 21, 22, 26, 28, 47, 48], RBF models [27], neural Hammerstein [25] and Wiener models [23]. Thanks to a simple, regular nature of these models, the implementation of the algorithm is easy. For example, taking into account the structure of the MLP neural model defined by (6) and (7),

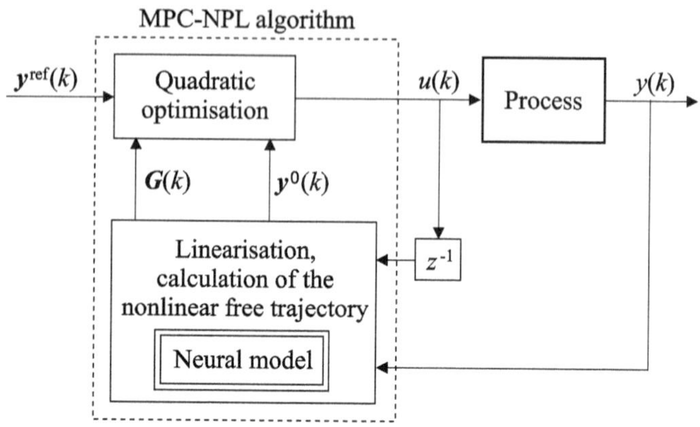

Fig. 3. The structure of the MPC-NPL algorithm

coefficients of the linearised model $a_l(k) = -\frac{\partial f(\bar{\boldsymbol{x}}(k))}{\partial y(k-l)}$ and $b_l(k) = \frac{\partial f(\bar{\boldsymbol{x}}(k))}{\partial u(k-l)}$ are calculated easily from

$$a_l(k) = -\sum_{i=1}^{K} w_i^2 \frac{\mathrm{d}\varphi(z_i(\bar{\boldsymbol{x}}(k)))}{\mathrm{d}z_i(\bar{\boldsymbol{x}}(k))} w_{i,I_u+l}^1 \tag{31}$$

where $l = 1, \ldots, n_A$, and

$$b_l(k) = \begin{cases} 0 & \text{if } l = 1, \ldots, \tau - 1 \\ \displaystyle\sum_{i=1}^{K} w_i^2 \frac{\mathrm{d}\varphi(z_i(\bar{\boldsymbol{x}}(k)))}{\mathrm{d}z_i(\bar{\boldsymbol{x}}(k))} w_{i,l-\tau+1}^1 & \text{if } l = \tau, \ldots, n_B \end{cases} \tag{32}$$

The linearisation point $\bar{\boldsymbol{x}}(k) = [\bar{u}(k-\tau) \ldots \bar{u}(k-n_B)\ \bar{y}(k-1) \ldots \bar{y}(k-n_A)]^T$ is determined by past input and output signals (measurements) corresponding to the arguments of the nonlinear model (5). If hyperbolic tangent is used as the nonlinear transfer function φ in the hidden layer of the neural model, one has $\frac{\mathrm{d}\varphi(z_i(\bar{\boldsymbol{x}}(k)))}{\mathrm{d}z_i(\bar{\boldsymbol{x}}(k))} = 1 - \tanh^2(z_i(\bar{\boldsymbol{x}}(k)))$.

For optimisation of the future control policy a linearised model is used. Although the free trajectory can be also calculated using the linearised model, it is a better idea to use the full nonlinear neural model. The nonlinear free trajectory $y^0(k+p|k)$ over the prediction horizon, i.e. for $p = 1, \ldots, N$, is calculated recursively from

$$y^0(k+p|k) = w_0^2 + \sum_{i=1}^{K} w_i^2 \varphi(z_i^0(k+p|k)) + d(k) \tag{33}$$

Quantities $z_i^0(k+p|k)$ are determined from (7) assuming no changes in the control signal from the sampling instant k onwards. One has

$$
\begin{aligned}
z_i^0(k+p|k) =& w_{i,0}^1 + \sum_{j=1}^{I_{uf}(p)} w_{i,j}^1 u(k-1) \\
&+ \sum_{j=I_{uf}(p)+1}^{I_u} w_{i,j}^1 u(k-\tau+1-j+p) \\
&+ \sum_{j=1}^{I_{yp}(p)} w_{i,I_u+j}^1 y^0(k-j+p|k) + \sum_{j=I_{yp}(p)+1}^{n_A} w_{i,I_u+j}^1 y(k-j+p)
\end{aligned}
\tag{34}
$$

where $I_{uf}(p) = \max(\min(p-\tau+1, I_u), 0)$ and $I_{yp}(p) = \min(p-1, n_A)$. Using (6) and (18), the unmeasured disturbance is estimated from

$$
d(k) = y(k) - \left(w_0^2 + \sum_{i=1}^{K} w_i^2 \varphi(z_i(k)) \right) \tag{35}
$$

6 Training of Neural Models for MPC

The neural model is usually trained off-line. During on-line control in MPC the model is not retrained. On-line model adaptation, although discussed in the literature [2, 50], has a limited applicability and in practice is not popular. For training a set of data is given (i.e. the input sequence $u(1), \ldots, u(S)$ and the output sequence $y(1), \ldots, y(S)$ where S is the number of samples) collected from measurements of the process. The objective of identification is to find the topology and parameters of the neural model in such a way that a predefined performance index which describes the accuracy of the model is minimised. A yet another set (the test set) is used in order to assess generalisation abilities of the model.

Identification consists of three phases: model structure selection, training and assessment. Typically, various models with different topology (K) and different order of dynamics (n_A, n_B, τ) should be trained and evaluated. The order of dynamics can be determined analytically [7]. The model finally chosen for MPC should have a relatively small number of parameters and be precise.

Typically, during model training the following Sum of Squared Errors (SSE) performance function is minimised

$$
\mathrm{SSE} = \sum_{k \in \texttt{data set}} (y(k|k-1) - y(k))^2 \tag{36}
$$

where $y(k|k-1)$ denotes the output of the model for the sampling instant k calculated using signals up to the sampling instant $k-1$ whereas $y(k)$ is the real value of the process output variable collected during the identification experiment.

Training is an unconstrained optimisation task in which the SSE performance index is minimised. Gradient-based algorithms are usually used, gradients of the SSE function are calculated analytically using the backpropagation scheme [11]. The optimisation direction can be found by different nonlinear optimisation methods: the steepest descent, the conjugate gradient methods (Polak-Ribiere, Fletcher-Reeves), the quasi-Newton algorithms (DFP, BFGS) or the Levenberg-Marquardt method [6]. Because of fast convergence and robustness BFGS and Levenberg-Marquardt algorithms are recommended. The optimisation problem is likely to be non-convex, the minimised objective function may have many local minima. Hence, in practice, for a given model structure the gradient-based algorithm is usually initialised randomly and the training procedure is repeated a few times (the multi-start approach). Alternatively, for initialisation global optimisation can be used (e.g. the genetic algorithm, simulated annealing).

Two configurations of dynamic models can be used: the one-step ahead prediction configuration (the series-parallel model) and the simulation configuration (the parallel model) [36]. In the first case the current value of the output signal, $y(k)$, is a function of past input and output values (i.e. real values measured at previous sampling instants). In the second case current and future output values are calculated recurrently, without using real output measurements. Naturally, in MPC the model must be used recurrently since the prediction over the prediction horizon N is considered. For the model (5) one has

$$y(k|k) = f(u(k-\tau),\ldots,u(k-n_B),y(k-1),\ldots,y(k-n_A)) \tag{37}$$

$$\begin{aligned} y(k+1|k) = f(&u(k-\tau+1),\ldots,u(k-n_B+1),\\ &y(k|k),\ldots,y(k-n_A+1|k)) \end{aligned} \tag{38}$$

$$\begin{aligned} y(k+2|k) = f(&u(k-\tau+2),\ldots,u(k-n_B+2),\\ &y(k+1|k),\ldots,y(k-n_A+2|k)) \end{aligned} \tag{39}$$

$$\vdots$$

Bearing in mind the role of the model in MPC for long-range prediction, it is obvious that recurrent training should be used [15,36,37]. In the SSE performance function the output of the model $y(k|k-1)$ should be calculated recurrently. An alternative formulation of the performance function can be also used in which all predictions over the whole horizon N are taken into account for all data samples

$$\mathrm{SSE} = \sum_{k\in\text{data set}} \sum_{p=1}^{N} (y(k+p|k-1) - y(k+p))^2 \tag{40}$$

Example: Modelling and MPC of a Polymerisation Reactor

The process under consideration is a polymerisation reaction taking place in a jacketed continuous stirred tank reactor [9] depicted in Fig. 4. The reaction is the free-radical polymerisation of methyl methacrylate with azo-bis-isobutyronitrile as initiator and toluene as solvent. The output $NAMW$ (Number Average Molecular Weight) [kg kmol^{-1}] is controlled by manipulating the inlet initiator

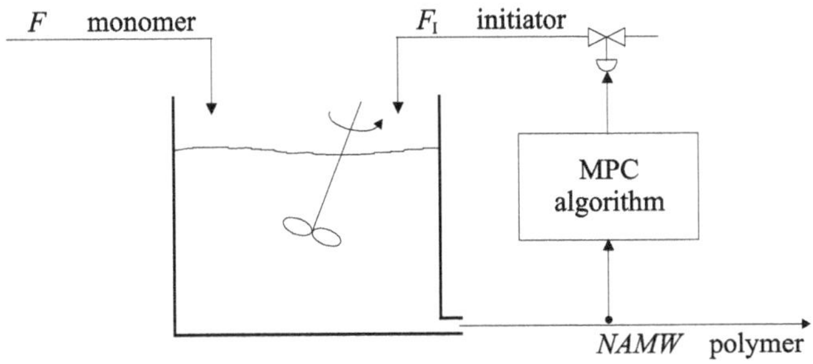

Fig. 4. The polymerisation reactor control system structure

flow rate F_{I} [$\mathrm{m}^3\ \mathrm{h}^{-1}$]. The flow rate F [$\mathrm{m}^3\mathrm{h}^{-1}$] of the monomer is a disturbance. Properties of the considered process are highly nonlinear. The reactor is frequently used as a benchmark [9, 26, 47].

The continuous-time fundamental model of the polymerisation reactor [9] is comprised of four nonlinear ordinary differential equations

$$\frac{\mathrm{d}C_{\mathrm{m}}(t)}{\mathrm{d}t} = -\left[Z_{\mathrm{P}}\exp\left(\frac{-E_{\mathrm{P}}}{RT}\right) + Z_{\mathrm{f_m}}\exp\left(\frac{-E_{\mathrm{f_m}}}{RT}\right)\right]C_{\mathrm{m}}(t)P_0(t) - \frac{F(t)C_{\mathrm{m}}(t)}{V} + \frac{F(t)C_{\mathrm{m_{in}}}}{V} \tag{41}$$

$$\frac{\mathrm{d}C_{\mathrm{I}}(t)}{\mathrm{d}t} = -Z_{\mathrm{I}}\exp\left(\frac{-E_{\mathrm{I}}}{RT}\right)C_{\mathrm{I}}(t) - \frac{F(t)C_{\mathrm{I}}}{V} + \frac{F_{\mathrm{I}}(t)C_{\mathrm{I_{in}}}}{V} \tag{42}$$

$$\frac{\mathrm{d}D_0(t)}{\mathrm{d}t} = \left[0.5Z_{\mathrm{T_c}}\exp\left(\frac{-E_{\mathrm{T_c}}}{RT}\right) + Z_{\mathrm{T_d}}\exp\left(\frac{-E_{\mathrm{T_d}}}{RT}\right)\right]P_0^2(t) + Z_{\mathrm{f_m}}\exp\left(\frac{-E_{\mathrm{f_m}}}{RT}\right)C_{\mathrm{m}}(t)P_0(t) - \frac{F(t)D_0(t)}{V} \tag{43}$$

$$\frac{\mathrm{d}D_{\mathrm{I}}(t)}{\mathrm{d}t} = M_{\mathrm{m}}\left[Z_{\mathrm{P}}\exp\left(\frac{-E_{\mathrm{P}}}{RT}\right) + Z_{\mathrm{f_m}}\exp\left(\frac{-E_{\mathrm{f_m}}}{RT}\right)\right]C_{\mathrm{m}}(t)P_0(t) - \frac{F(t)D_{\mathrm{I}}(t)}{V} \tag{44}$$

where

$$P_0(t) = \sqrt{\frac{2f^*C_{\mathrm{I}}(t)Z_{\mathrm{I}}\exp\left(\frac{-E_{\mathrm{I}}}{RT}\right)}{Z_{\mathrm{T_d}}\exp\left(\frac{-E_{\mathrm{T_d}}}{RT}\right) + Z_{\mathrm{T_c}}\exp\left(\frac{-E_{\mathrm{T_c}}}{RT}\right)}} \tag{45}$$

and one algebraic output equation

$$NAMW(t) = \frac{D_{\mathrm{I}}(t)}{D_0(t)} \tag{46}$$

Parameters of the fundamental model are given in Table 1. The initial operating conditions are: $F_{\mathrm{I}} = 0.028328\ \mathrm{m}^3\ \mathrm{h}^{-1}$, $F = 1\ \mathrm{m}^3\ \mathrm{h}^{-1}$, $NAMW = 20000$ kg

Table 1. Parameters of the fundamental model

Parameter	Value	Parameter	Value
$C_{\mathrm{I_{in}}}$	8 kmol m^{-3}	R	8.314 kJ kmol^{-1} K^{-1}
$C_{\mathrm{m_{in}}}$	6 kmol/m^{-3}	T	335 K
$E_{\mathrm{T_c}}$	$2.9442 \cdot 10^3$ kJ kmol^{-1}	$Z_{\mathrm{T_c}}$	$3.8223 \cdot 10^{10}$ m^3 kmol^{-1} h^{-1}
$E_{\mathrm{T_d}}$	$2.9442 \cdot 10^3$ kJ kmol^{-1}	$Z_{\mathrm{T_d}}$	$3.1457 \cdot 10^{11}$ m^3 kmol^{-1} h^{-1}
$E_{\mathrm{f_m}}$	$7.4478 \cdot 10^4$ kJ kmol^{-1}	$Z_{\mathrm{f_m}}$	$1.0067 \cdot 10^{15}$ m^3 kmol^{-1} h^{-1}
E_{I}	$1.2550 \cdot 10^5$ kJ kmol^{-1}	Z_{I}	$3.7920 \cdot 10^{18}$ h^{-1}
E_{P}	$1.8283 \cdot 10^4$ kJ kmol^{-1}	Z_{P}	$1.7700 \cdot 10^9$ m^3 kmol^{-1} h^{-1}
f^*	0.58	V	0.1 m^3
M_{m}	100.12 kg kmol^{-1}		

kmol^{-1}, $C_{\mathrm{m}} = 5.3745$ kmol m^{-3}, $C_{\mathrm{I}} = 2.2433 \cdot 10^{-1}$ kmol m^{-3}, $D_0 = 3.1308 \cdot 10^{-3}$ kmol m^{-3}, $D_{\mathrm{I}} = 6.2616 \cdot 10^{-1}$ kmol m^{-3}.

For identification the fundamental model (41)–(46) is used as the real process, it is simulated open-loop in order to obtain two sets of data, namely training and test data sets depicted in Fig. 5. Both sets contain 2000 samples, the sampling time is 1.8 min. The output signal contains small measurement noise. During calculations the system of differential equations comprising the fundamental model is solved using the Runge-Kutta RK45 method.

The classical MLP neural model of the process is used. A number of model candidates are trained to assess the order of dynamics and the number of hidden nodes. Finally, the second-order model is chosen containing $K = 6$ hidden nodes with the hyperbolic tangent transfer function in the hidden layer

$$y(k) = f(u(k-2), y(k-1), y(k-2)) \tag{47}$$

Because input and output process variables have different orders of magnitude, they are scaled as $u = 100(F_{\mathrm{I}} - F_{\mathrm{I0}})$, $y = 0.0001(NAMW - NAMW_0)$ where $F_{\mathrm{I0}} = 0.028328$, $NAMW_0 = 20000$ correspond to the initial operating point.

During model training the SSE performance function (36) is minimised. Because models used in MPC have to be able to make good predictions not only one

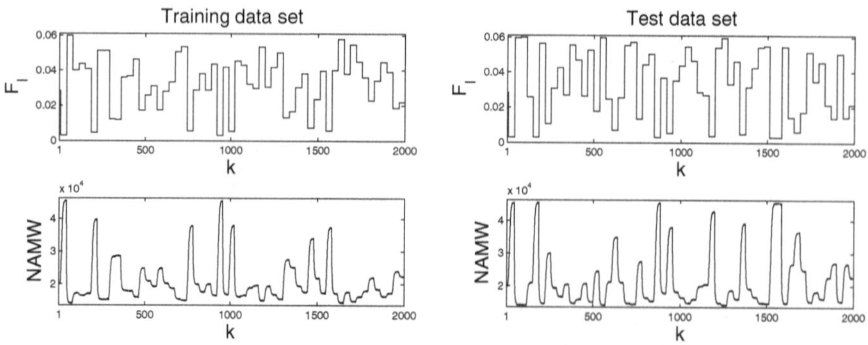

Fig. 5. Training and test data sets

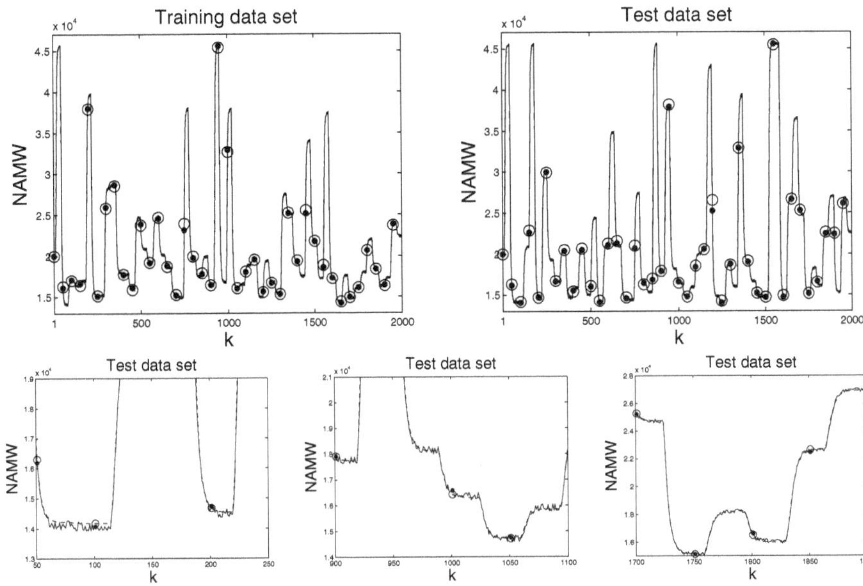

Fig. 6. The process (*solid line with dots*) vs. the neural model (*dashed line with circles*) for training and test data sets: complete data sets (*top*) and enlarged fragments of the test data set (*bottom*)

step ahead, but over the whole prediction horizon, neural networks are trained as recurrent Output Error (OE) models which make it possible to properly calculate multi-step ahead predictions [37]. The SSE performance function is minimised using the BFGS optimisation algorithm [6]. For each neural model topology training is repeated 10 times, weights of neural networks are initialised randomly. Fig. 6 shows the output of the process and the output of the chosen neural model for both training and test data sets. The accuracy of the neural model is very high. For the training data set $\mathrm{SSE} = 5.559159 \cdot 10^{-1}$, for the test data set $\mathrm{SSE} = 1.190907 \cdot 10^{0}$.

To demonstrate very high accuracy of the neural model the linear model

$$y(k) = b_2 u(k-2) - a_1 y(k-1) - a_2 y(k-2) \tag{48}$$

is found. The linear model has exactly the same arguments as the neural one. Fig. 7 shows the output of the process and the output of the linear model for both data sets. Unfortunately, because the process is really nonlinear, the accuracy of the linear model is low. For the training data set $\mathrm{SSE} = 3.402260 \cdot 10^{2}$, for the test data set $\mathrm{SSE} = 6.255822 \cdot 10^{2}$. Comparing properties of both models, one can expect that it is better to use in MPC the neural model rather than the linear one.

Compared MPC strategies are:

a) the linear MPC algorithm based on the linear model (48),

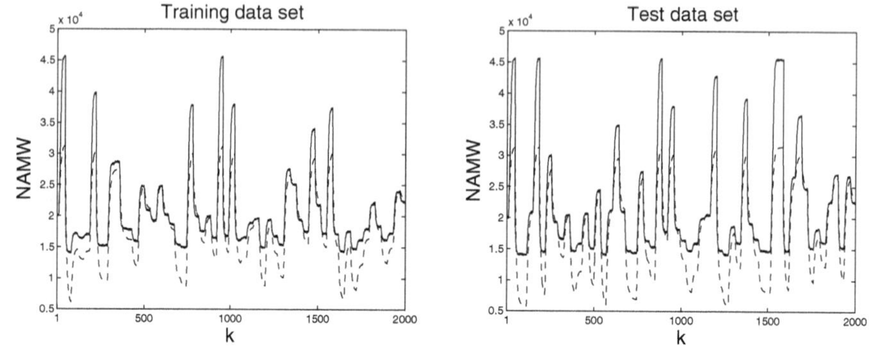

Fig. 7. The process (*solid line*) vs. the linear model (*dashed line*) for training and test data sets

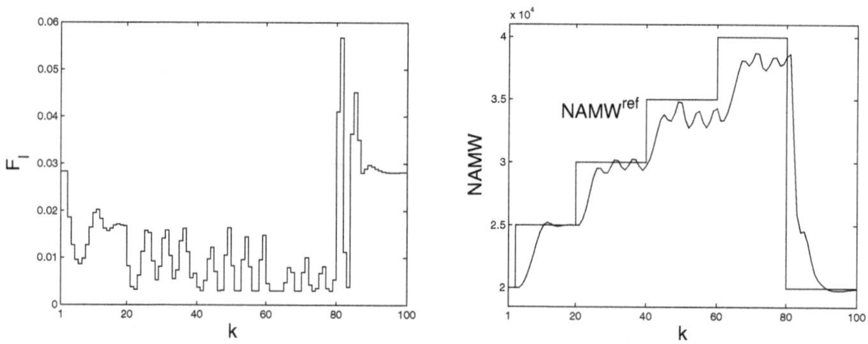

Fig. 8. Simulation results of the MPC algorithm based on the linear model

b) the nonlinear MPC-NPL algorithm based on the neural model (47),
c) the nonlinear MPC-NO algorithm based on the same neural model (47).

All algorithms are implemented in Matlab. The MPC algorithm based on the linear model and the MPC-NPL algorithm use the quadratic programming procedure whereas the MPC-NO algorithm uses the Sequential Quadratic Programming (SQP) nonlinear optimisation routine. As the initial point for MPC-NO nonlinear optimisation $N_u - 1$ control values calculated at the previous sampling instant and not applied to the process are used. The fundamental model (41)–(46) is used as the real process, it is solved using the Runge-Kutta RK45 method.

Horizons of all compared MPC algorithms are $N = 10$, $N_u = 3$, the weighting coefficients $\lambda_p = 0.2$. (As far as choosing parameters of MPC the reader is referred to the literature [29,44,45,47].) The manipulated variable is constrained: $F_{\mathrm{I}}^{\min} = 0.003$, $F_{\mathrm{I}}^{\max} = 0.06$, the sampling time is 1.8 min.

Simulation results of the MPC algorithm based on the linear model are depicted in Fig. 8. As the reference trajectory ($NAMW^{\mathrm{ref}}$) five set-point changes

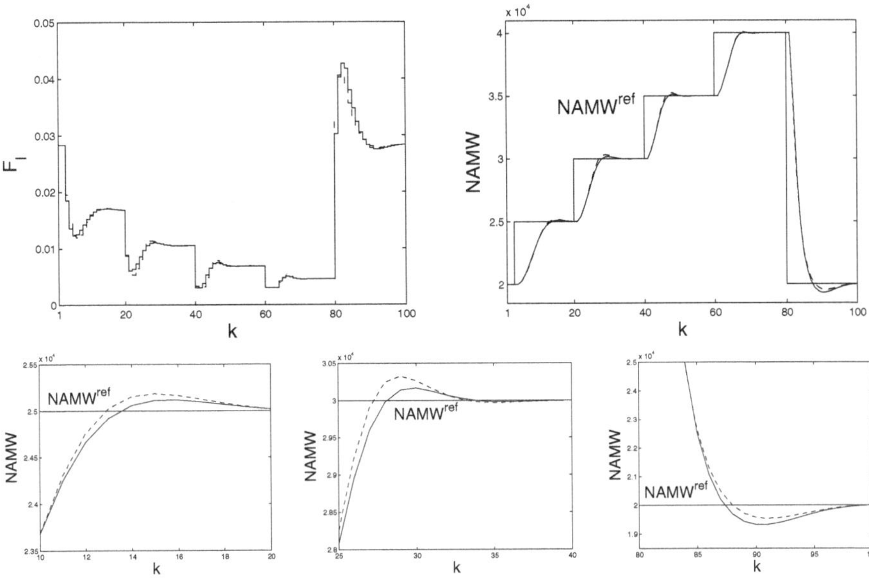

Fig. 9. Simulation results of MPC-NPL (*solid*) and MPC-NO (*dashed*) algorithms based on the same neural model: the whole simulation (*top*) and enlarged fragments (*bottom*)

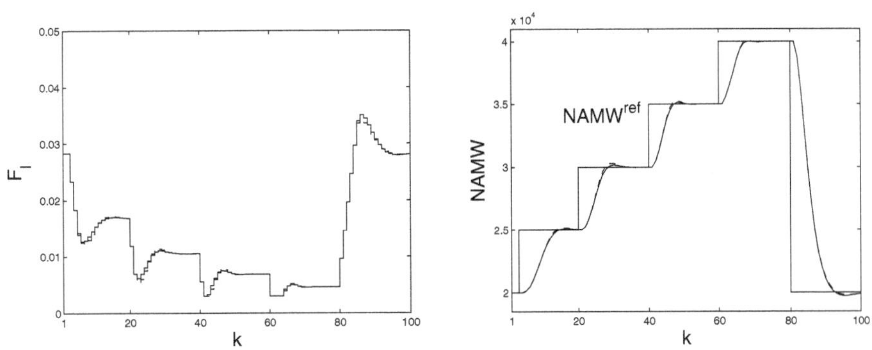

Fig. 10. Simulation results of MPC-NPL (*solid*) and MPC-NO (*dashed*) algorithms based on the same neural model with constraints imposed on increments of the manipulated variable, $\triangle F_{\mathrm{I}}^{\max} = 0.005\ \mathrm{m}^3/\mathrm{h}$

are considered. The linear algorithm works well only for the smallest set-point change, whereas for bigger ones the system becomes unstable. Simulation results of MPC-NPL and MPC-NO algorithms based on the same neural model are depicted in Fig. 9. Both nonlinear algorithms are stable. Moreover, the closed-loop performance obtained in the suboptimal MPC-NPL algorithm with quadratic programming is very close to that obtained in the computationally demanding

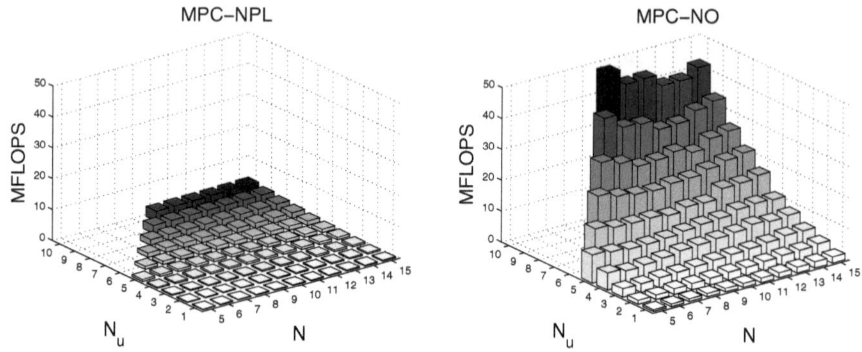

Fig. 11. The computational complexity (MFLOPS) of MPC-NPL (*left*) and MPC-NO (*right*) algorithms based on the same neural model

Table 2. The computational complexity (MFLOPS) of MPC-NPL and MPC-NO algorithms based on the same neural model

Algorithm	N	$N_u = 1$	$N_u = 2$	$N_u = 3$	$N_u = 4$	$N_u = 5$	$N_u = 10$
MPC-NPL	5	0.131	0.176	0.275	0.422	0.637	–
MPC-NO	5	0.560	1.278	2.782	5.560	9.101	–
MPC-NPL	10	0.221	0.285	0.405	0.573	0.808	3.198
MPC-NO	10	1.262	2.640	4.110	6.779	8.896	48.358
MPC-NPL	15	0.320	0.413	0.563	0.764	1.033	3.628
MPC-NO	15	1.968	3.993	5.846	8.547	11.182	42.367

MPC-NO approach, in which a nonlinear optimisation problem has to be solved on-line at each sampling instant.

Simulation results of both nonlinear algorithms with constraints imposed on increments of the manipulated variable $\triangle F_{\mathrm{I}}^{\max} = 0.005$ m^3/h are shown in Fig. 10. Such constraints are very important when changes in the reference trajectory are big, they take into account the actuator's limitations. In comparison with Fig. 9, additional constraints result in a slightly slower output profile, but technological restrictions are rigorously taken into account.

Table 2 shows the influence of control and prediction horizons on the computational complexity of MPC-NPL and MPC-NO algorithms (in terms of floating point operations MFLOPS) for $N = 5, 10, 15$, $N_u = 1, 2, 3, 4, 5, 10$. Fig. 11 depicts the computational complexity for $N = 5, \ldots, 15$, $N_u = 1, \ldots, 10$. In general, the suboptimal MPC-NPL algorithm is considerably less computationally demanding than the MPC-NO strategy. The minimal computational complexity reduction (MPC-NPL vs. MPC-NO) for all considered combinations of horizons is 4.28 times, the maximal is 15.37 times, the average reduction factor is 10.59.

7 Specialised Neural Models for MPC

In MPC the explicit model of the process is directly used to predict its future behaviour and to determine the optimal control policy. As a result, MPC algorithms are very model-based, the control performance is determined by the accuracy of predictions calculated by the model. The role of the model in MPC control cannot be ignored during model structure selection and identification. Naturally, the model has to be able to make good predictions of future behaviour of the process over the whole prediction horizon.

According to the general prediction equation (17)

$$\hat{y}(k+p|k) = y(k+p|k) + d(k) \tag{49}$$

predictions are calculated by means of the model used for the sampling instant $k+p$ at the current sampling instant k. The unmeasured disturbance estimation $d(k)$ is also calculated (18) using the model of the process. Using the prediction equation and the general nonlinear neural model (5) which represents any model discussed so far (MLP, RBF neural Hammerstein or Wiener), output predictions over the prediction horizon ($p = 1, \ldots, N$) are calculated recurrently from

$$\begin{aligned}\hat{y}(k+p|k) = f(&\underbrace{u(k-\tau+p|k), \ldots, u(k|k)}_{I_{uf}(p)}, \underbrace{u(k-1), \ldots, u(k-n_B+p)}_{I_u - I_{uf}(p)}, \\ &\underbrace{\hat{y}(k-1+p|k), \ldots, \hat{y}(k+1|k)}_{I_{yp}(p)}, \underbrace{y(k), \ldots, y(k-n_A+p)}_{n_A - I_{yp}(p)}) + d(k)\end{aligned} \tag{50}$$

Predictions $\hat{y}(k+p|k)$ depend on $I_{uf}(p) = \max(\min(p-\tau+1, I_u), 0)$ future values of the control signal (i.e. decision variables of the MPC algorithm), where $I_u = n_B - \tau + 1$, $I_u - I_{uf}(p)$ values of the control signal applied to the plant at previous sampling instants, $I_{yp}(p) = \min(p-1, n_A)$ future output predictions and $n_A - I_{yp}(p)$ plant output signal values measured at previous sampling instants. For prediction in MPC the classical NARX model (5) has to be used recurrently, because predictions depend on predictions calculated for previous sampling instants within the prediction horizon.

Quite frequently, neural models are trained using the rudimentary backpropagation algorithm which yields one-step ahead predictors. Intuitively, they are not suited to be used recurrently in MPC for long-range prediction (50) since the prediction error is propagated. It is particularly important in the case of noise, model inaccuracies and underparameterisation, i.e. the order of the model is usually significantly lower than the order of the real process or even the proper model order is unknown.

To solve the problem resulting from the inaccuracy of one-step ahead predictors in MPC two general approaches can be recommended. First of all, specialised recurrent training algorithms for neural models can be used (Section 6), but they are significantly more computationally demanding in comparison with one-step ahead predictor training. Moreover, obtained models can be sensitive to noise. An alternative is to choose the structure of the model in such a way that its role in MPC is not ignored. In this work two specialised neural structures are discussed: multi-models and structured models. In both cases, thanks to the nature of these models, the prediction error is not propagated. Both model types are easily trained as one-step ahead predictors.

7.1 Neural FIR Models

The easiest way of avoiding the necessity of using the model recurrently is to use Nonlinear Finite Impulse Response (NFIR) models

$$y(k) = f(\boldsymbol{x}(k)) = f(u(k-\tau), \ldots, u(k-n_B)) \tag{51}$$

where $f : \Re^{n_B-\tau+1} \longrightarrow \Re$. For the NFIR model, output predictions

$$\hat{y}(k+p|k) = f(\underbrace{u(k-\tau+p|k), \ldots, u(k|k)}_{I_{uf}(p)}, \underbrace{u(k-1), \ldots, u(k-n_B+p)}_{I_u - I_{uf}(p)}) + d(k) \tag{52}$$

depend only on past and future control signals. For prediction in MPC the NFIR model is not used recurrently, the prediction error is not propagated.

The NFIR model can be realised by the MLP or RBF neural network. The output of the MLP NFIR model is given by (6). In contrast to the classical MLP neural model (7), sums of inputs of hidden nodes ($i = 1, \ldots, K$) depend only on the input signal

$$z_i(k) = w_{i,0}^1 + \sum_{j=1}^{I_u} w_{i,j}^1 u(k-\tau+1-j) \tag{53}$$

Weights of the network are denoted by $w_{i,j}^1$, $i = 1, \ldots, K$, $j = 0, \ldots, n_B - \tau + 1$, and w_i^2, $i = 0, \ldots, K$, for the first and the second layer, respectively, $I_u = n_B - \tau + 1$.

Unfortunately, because of its nature, the NFIR model usually needs a high order of dynamics (determined by n_B), much higher than the classical model of a similar accuracy which depends on both past input and output signals (5).

7.2 Neural Multi-models

In the multi-model approach [24] one independent neural model is used for each sampling instant within the prediction horizon. For the sampling instant $k+1$ the following model is used

$$y(k+1) = f_1(u(k-\tau+1), \ldots, u(k-n_B), y(k), \ldots, y(k-n_A)) \tag{54}$$

For the sampling instant $k+2$ the model is

$$y(k+2) = f_2(u(k-\tau+2), \ldots, u(k-n_B), y(k), \ldots, y(k-n_A)) \tag{55}$$

In a similar way independent submodels are formulated for all sampling instants within the prediction horizon. For the sampling instant $k+N$ the model is

$$y(k+N) = f_N(u(k-\tau+N), \ldots, u(k-n_B), y(k), \ldots, y(k-n_A)) \tag{56}$$

In general, for $p = 1, \ldots, N$, all submodels can be expressed in a compact form

$$\begin{aligned} y(k+p) = f_p(\boldsymbol{x}(k+p|k)) = \\ f_p(u(k-\tau+p), \ldots, u(k-n_B), y(k), \ldots, y(k-n_A)) \end{aligned} \tag{57}$$

The multi-model is comprised of N neural networks which calculate predictions for consecutive sampling instants within the prediction horizon. Each network realises the function $f_p : \Re^{\min(n_B+p-\tau+1,n_B)+\max(p-\tau+1,0)+n_A+1} \longrightarrow \Re$.

Predictions calculated from the multi-model are

$$\hat{y}(k+p|k) = y(k+p|k) + d(k+p|k) \tag{58}$$

for $p = 1, \ldots, N$. Independent disturbance estimations are

$$d(k+p|k) = y(k) - f_p(k|k-1) \tag{59}$$

where $y(k)$ is measured while $f_p(k|k-1)$ is calculated from the multi-model used for the sampling instant k.

Using (57) and (58), output predictions for $p = 1, \ldots, N$ calculated from the multi-model are

$$\begin{aligned} \hat{y}(k+p|k) = f_p(\underbrace{u(k-\tau+p|k), \ldots, u(k|k)}_{I_{uf}(p)}, \underbrace{u(k-\max(\tau-p,1)), \ldots, u(k-n_B)}_{I_{up}(p)} \\ \underbrace{y(k), \ldots, y(k-n_A))}_{n_A+1} + d(k+p|k) \end{aligned} \tag{60}$$

where $I_{uf}(p) = \max(p-\tau+1, 0)$, $I_{up}(p) = n_B - \max(\tau-p, 1) + 1$. Analogously as in the case of the classical NARX model (5), predictions calculated by means of the multi-model (57) depend on $I_{uf}(p)$ future values of the control signal, $I_{up}(p)$ values of the control signal applied to the plant at previous sampling instants and on $n_A + 1$ values of the plant output signal measured at previous sampling instants. Unlike classical predictions (50), they do not depend on predictions calculated for previous sampling instants within the prediction horizon. As a result, the multi-model is not used recurrently, the prediction error is not propagated. Fig. 12 depicts the structure of the multi-model used for prediction in MPC.

Neural multi-model training needs finding independent N submodels. They are trained separately by means of the standard backpropagation algorithm which yields one-step ahead predictors. It is possible because for prediction one independent neural submodel is used for each sampling instant within the prediction horizon and predictions do not depend on previous predictions.

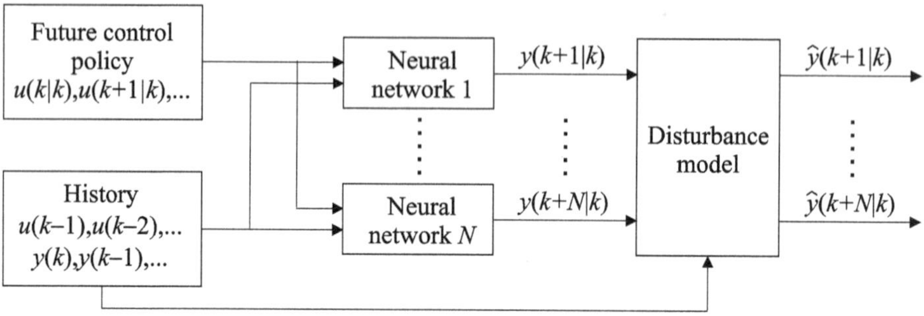

Fig. 12. The structure of the neural multi-model used for prediction in MPC

As submodels both MLP and RBF neural networks can be used. These neural networks realise functions f_p, $p = 1, \ldots, N$ in (57). Outputs of MLP submodels for the sampling instant $k + p$, $p = 1, \ldots, N$ are

$$y(k+p|k) = f_p(\boldsymbol{x}(k+p|k)) = w_0^{2,p} + \sum_{i=1}^{K^p} w_i^{2,p} \varphi(z_i^p(k+p|k)) \tag{61}$$

where $z_i^p(k+p|k)$ are sums of inputs of the ith hidden node, K^p is the number of hidden nodes. Recalling the prediction of the multi-model (60) one has

$$\begin{aligned} z_i^p(k+p|k) =& w_{i,0}^{1,p} + \sum_{j=1}^{I_{uf}(p)} w_{i,j}^{1,p} u(k-\tau+1-j+p|k) \\ &+ \sum_{j=1}^{I_{up}(p)} w_{i,I_{uf}(p)+j}^{1,p} u(k-\max(\tau-p,1)+1-j) \\ &+ \sum_{j=1}^{n_{A+1}} w_{i,I_{uf}(p)+I_{up}(p)+j}^{1,p} y(k+1-j) \end{aligned} \tag{62}$$

Weights are denoted by $w_{i,j}^{1,p}$, $i = 1, \ldots, K^p$, $j = 0, \ldots, \max(p-\tau+1, 0) - \max(\tau-p, 1) + n_A + n_B + 2$, and $w_i^{2,p}$, $i = 0, \ldots, K^p$, for the first and the second layer, respectively, p indicates the submodel, $p = 1, \ldots, N$.

The MPC-NPL algorithm based on MLP multi-models is detailed in [24].

7.3 Structured Neural Models

Rewriting the model (5) for sampling instants $k-1, \ldots, k-N+1$ one has

$$\begin{aligned} y(k-1) = f(&u(k-\tau-1), \ldots, u(k-n_B-1), \\ &y(k-2), \ldots, y(k-n_A-1)) \end{aligned} \tag{63}$$

$$\vdots$$

$$y(k-N+2) = f(u(k-\tau-N+2),\ldots,u(k-n_B-N+2), \\ y(k-N+1),\ldots,y(k-n_A-N+2)) \tag{64}$$

$$y(k-N+1) = f(u(k-\tau-N+1),\ldots,u(k-n_B-N+1), \\ y(k-N),\ldots,y(k-n_A-N+1)) \tag{65}$$

Using (65), the quantity $y(k-N+2)$ given by (64) can be expressed as

$$y(k-N+2) = f(u(k-\tau-N+2),\ldots,u(k-n_B-N+2), \\ f(u(k-\tau-N+1),\ldots,u(k-n_B-N+1), \\ y(k-N),\ldots,y(k-n_A-N+1)), \\ y(k-N),\ldots,y(k-n_A-N+2)) \tag{66}$$

which can be rewritten as the function

$$y(k-N+2) = f_{N-2}(u(k-\tau-N+2),\ldots,u(k-n_B-N+1), \\ y(k-N),\ldots,y(k-n_A-N+1)) \tag{67}$$

Model arguments rearrangement can be repeated for all quantities $y(k-N+2),\ldots,y(k)$, giving functions $f_{N-2},\ldots,f_0$. Finally, one has

$$y(k) = f(u(k-\tau),\ldots,u(k-n_B), \\ f_1(u(k-\tau-1),\ldots,u(k-n_B-N+1),\ldots, \\ y(k-N),\ldots,y(k-n_A-N+1)),\ldots, \\ f_{n_A}(u(k-\tau-n_A),\ldots,u(k-n_B-N+1), \\ y(k-N),\ldots,y(k-n_A-N+1))) \tag{68}$$

which can be rewritten as the function

$$y(k) = f_0(u(k-\tau),\ldots,u(k-n_B-N+1), \\ y(k-N),\ldots,y(k-n_A-N+1)) \tag{69}$$

The equation (69) represents the structured model, $f_0 : \Re^{n_A+n_B-\tau+N} \longrightarrow \Re$.

Using the general prediction equation (17), output predictions calculated from the structured model (69) are

$$\hat{y}(k+p|k) = f_0(\underbrace{u(k-\tau+p|k),\ldots,u(k|k)}_{I_{uf}(p)},\underbrace{u(k-1),\ldots,u(k-n_B-N+1+p)}_{I_u-I_{uf}(p)}, \\ \underbrace{y(k-N+p),\ldots,y(k-n_A-N+1+p)}_{n_A}) + d(k) \tag{70}$$

For the structured model $I_u = n_B + N - \tau$. Predictions depend on $I_{uf}(p) = \max(p-\tau+1, 0)$ future values of the control signal and $I_u - I_{uf}(p)$ values of the control signal applied to the plant at previous sampling instants. Unlike classical predictions (50), they do not depend on predictions calculated for previous sampling instants within the prediction horizon, but only on n_A values of the plant output signal measured at previous sampling instants. As a result, the structured model is not used recurrently, the prediction error is not propagated.

Both MLP and RBF neural networks can be used in the structured model to realise the function f_0 in (69). The output of the structured MLP model is

$$y(k) = f(\boldsymbol{x}(k)) = w_0^2 + \sum_{i=1}^{K} w_i^2 \varphi(z_i(k)) \tag{71}$$

where $z_i(k)$ is the sum of inputs of the ith hidden node. From (69)

$$z_i(k) = w_{i,0}^1 + \sum_{j=1}^{I_u} w_{i,j}^1 u(k-\tau+1-j) + \sum_{j=1}^{n_A} w_{i,I_u+j}^1 y(k-j-N+1) \tag{72}$$

Weights are denoted by $w_{i,j}^1$, $i = 1, \ldots, K$, $j = 0, \ldots, n_A + n_B - \tau + N$, and w_i^2, $i = 0, \ldots, K$, for the first and the second layer, respectively.

The MPC-NPL scheme based on structured MLP models is described in [20].

Example: Multi-Modelling of the Polymerisation Reactor

Two models are trained off-line: the classical NARX model (5)

$$y(k) = f(u(k-2), y(k-1)) \tag{73}$$

and the multi-model (57) for $N = 10$

$$\begin{aligned} y(k+1|k) = &f_1(u(k-1), u(k-2), \\ &y(k), y(k-1)) \end{aligned} \tag{74}$$

$$\begin{aligned} y(k+2|k) = &f_2(u(k|k), u(k-1), u(k-2), \\ &y(k), y(k-1)) \end{aligned} \tag{75}$$

$$\begin{aligned} y(k+3|k) = &f_3(u(k+1|k), u(k|k), u(k-1), u(k-2), \\ &y(k), y(k-1)) \end{aligned} \tag{76}$$

$$\begin{aligned} y(k+4|k) = &f_4(u(k+2|k), u(k+1|k), u(k|k), u(k-1), u(k-2), \\ &y(k), y(k-1)) \end{aligned} \tag{77}$$

$$\vdots$$

$$\begin{aligned} y(k+10|k) = &f_{10}(u(k+8|k), \ldots, u(k|k), u(k-1), u(k-2), \\ &y(k), y(k-1)) \end{aligned} \tag{78}$$

Both models used have the same order of dynamics determined by $\tau = n_B = 2$, $n_A = 1$. To show advantages of the multi-model both models are underparameterised, because in fact the fundamental model (41)–(46) consists of four differential equations. In order to precisely capture the nature of the process, the classical

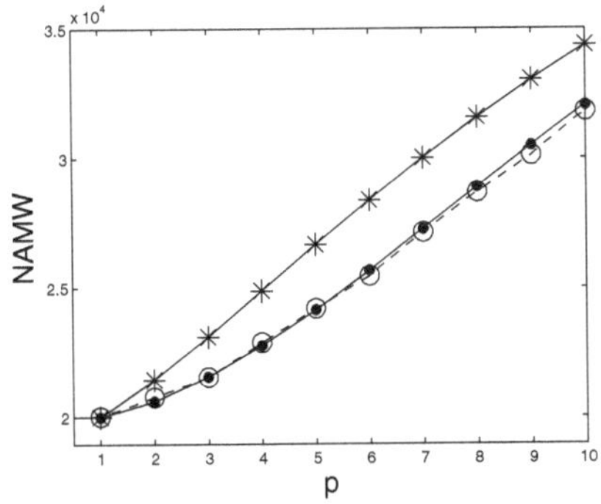

Fig. 13. Step-responses (long-range predictions) calculated recurrently by the classical NARX neural model (*solid line with asterisks*) and by the neural multi-model (*dashed line with circles*) vs. the real process (*solid line with dots*)

neural model should have at least the second order, i.e. $n_A = n_B = \tau = 2$ (in all experiments described so far the model (47), i.e. $y(k) = f(u(k-2), y(k-1), y(k-2))$ is used).

In both model types MLP neural networks are used with the hyperbolic tangent transfer function in hidden layers. All neural models are trained as one-step ahead predictors. The NARX model has $K = 6$ hidden nodes, in the case of the multi-model, six submodels have $K^p = 3$ hidden nodes (for $p = 1, 6, 7, 8, 9, 10$), four submodels have $K^p = 3$ hidden nodes (for $p = 2, 3, 4, 5$). The number of hidden nodes in submodels comprising the multi-model is adjusted in such a way that when trained and tested as one-step ahead predictors they give comparable values of the SSE performance index as the classical NARX model.

To reduce the complexity of models, the Optimal Brain Damage (OBD) pruning algorithm is used [17]. The complexity of the NARX model is reduced by only 16.0% whereas in case of the multi-model in the best case by 34.15%, in the worst case by 17.65%. An average complexity reduction factor is 29.66%.

In light of the application of models in MPC, it is interesting to compare their long range prediction accuracy. Fig. 13 shows step-responses of the process and predictions. The manipulated variable F_I changes at the sampling instant $k = 0$ from 0.028328 to 0.004602, which corresponds to changing the operating point from $NAMW = 20000$ to $NAMW = 40000$. The one-step ahead NARX neural model is used recurrently, it correctly calculates only the prediction for the first sampling instant of the prediction horizon (i.e. for $p = 1$). As a result of under-parameterisation, for next sampling instants the prediction error is propagated and consecutive predictions significantly differ from the real process. The neural multi-model is not used recurrently, the prediction error is not propagated. It correctly predicts behaviour of the process over the whole prediction horizon.

To further compare the long-range prediction accuracy and show the potential of using neural multi-models for long-range prediction in MPC, the ratio

$$R_N = \frac{1}{N}\sum_{p=1}^{N} \frac{\sum\limits_{k\in\text{data set}} (y(k+p|k) - y(k+p))^2}{\sum\limits_{k\in\text{data set}} (y_{NARX}(k+p|k) - y(k+p))^2} \tag{79}$$

is considered. It compares the average long-range prediction accuracy of the multi-model (the numerator) and the classical NARX model (the denominator). The output of the classical one-step ahead model used recurrently for long-range prediction is denoted by $y_{NARX}(k+p|k)$, the output of the multi-model is denoted by $y(k+p|k)$, $y(k+p)$ is the real data sample.

If $R_N < 1$ it is a potential for using in MPC multi-models rather than classical models recurrently. The smaller the value of R_N, the worse long-range prediction abilities of the classical model and it is more appropriate to use multi-models. For the training data set $R_{10} = 0.3241$, for the test data set $R_{10} = 0.2298$.

8 The Stabilising MPC-NPL Algorithm

In practice stability of MPC is achieved by adjusting horizons' lengths and penalty factors λ_p. MPC algorithms have gained recognition and have been successfully applied in industry for years before theoretical results concerning their stability analysis appeared. Over the years several MPC algorithms with guaranteed stability have been developed, an excellent review is given in [31]. Usually, stability is guaranteed provided that the nonlinear optimisation is used to solve the MPC optimisation task (3). An additional strong assumption under which stability is guaranteed is the necessity of finding the global solution to the nonlinear MPC optimisation problem at each algorithm iteration. On the other hand, as emphasised in this work, suboptimal MPC algorithms with on-line linearisation and quadratic programming give very good control performance and are computationally efficient. The gap between the practice and the theory is evident. Stability of suboptimal MPC are researched exceptionally infrequently.

It can be easily proved that to guarantee stability it is sufficient to bring the terminal state (i.e. at the end of the prediction horizon) to the origin [31, 32]. In other words, stability is enforced by using in the MPC optimisation task an additional equality terminal constraint

$$x(k+N|k) = 0 \tag{80}$$

The state $x(k) = [u(k-1) \ldots u(k-n_B+1)\ y(k) \ldots y(k-n_A+1)]^T$ [29] corresponds to arguments of the nonlinear model (5). Unfortunately, in nonlinear MPC such a constraint significantly increases the complexity of the nonlinear optimisation problem.

To overcome the necessity of finding the global solution to nonlinear MPC optimisation problem, the dual-mode MPC scheme [34, 46] can be used in which a terminal inequality constraint

$$x(k+N|k) \in \Omega_\alpha \tag{81}$$

guarantees stability. The terminal set Ω_α is a convex neighbourhood of the origin

$$\Omega_\alpha = \left\{ x(k) \in \Re^{n_A+n_B-1} : \|x(k)\|_{\boldsymbol{P}}^2 \leq \alpha \right\} \tag{82}$$

where the matrix $\boldsymbol{P} > 0$, $\alpha > 0$. In this approach merely feasibility, rather then optimality, is sufficient to guarantee stability. It means that it is only necessary to find a feasible solution to the nonlinear MPC optimisation problem, this solution does not need to be the global or even a local minimum to guarantee stability of the whole control algorithm. Hence, the dual-mode approach can be efficiently combined with the MPC-NPL algorithm as described in [22]. The formulation of the stabilising MPC-NPL algorithm is general, different neural structures can be used: MLP and RBF models, neural Hammerstein and Wiener models as well as neural multi-models and structured models.

In the dual-model approach the current value of the manipulated variable is calculated in two different ways, the method used depends on the current state of the process. Outside the terminal set Ω_α a nonlinear MPC algorithm is used whereas an additional, usually linear, controller is used inside this set. The objective of the additional feedback controller is to bring the state of the process to the origin. The control law is

$$u(k) = \begin{cases} u(k|k) & \text{if } x(k) \notin \Omega_\alpha \\ \boldsymbol{K}x(k) & \text{if } x(k) \in \Omega_\alpha \end{cases} \tag{83}$$

The nonlinear MPC, which is used if the current state of the plant is outside the terminal set Ω_α, takes into account all the constraints explicitly in the optimisation problem. The linear control law $u(k) = \boldsymbol{K}x(k)$, which is used if the state is inside the terminal set, despite being unconstrained, must never violate constraints. Constraints are used when the terminal set is calculated off-line.

In the stabilising dual-mode approach, the MPC-NPL optimisation problem (30) takes also into account the terminal inequality constraint (81). The main advantage of the algorithm is its suboptimality, i.e. feasibility of the nonlinear optimisation problem (30) is sufficient to guarantee stability. In other words, the determined control sequence $\boldsymbol{u}(k)$ must be always feasible, but it does not need to be the global or even a local minimum of the optimisation problem (30). Determination of the terminal set Ω_α and simulation results of the stabilising dual-model MPC-NPL algorithm are presented in [22].

9 Approximate Neural MPC

Because neural networks are universal approximators, it is an appealing idea to use a network capable of approximating the whole MPC algorithm [4,8,38]. The key idea is to calculate on-line values of the manipulated variable without any optimisation. In other words, the neural network replaces the whole MPC algorithm. The control signal is simply calculated by the neural network as a function of past input and output signals measured at previous sampling instants.

An important advantage of approximate neural MPC is its speed. On the other hand, the main problem is training. At first, a classical MPC algorithm is developed, for example the MPC-NPL one. Next, a sufficiently large number of simulation should be carried out to cover the whole operation domain (i.e. different initial conditions and reference trajectories). Finally, this data set is used to train a neural model the role of which is to replace the control algorithm.

10 Conclusions

This work emphasises two issues:

1. The proper choice of the model used in MPC is crucial as MPC algorithms are very model-based, the possible control performance is determined by the accuracy of predictions calculated by means of the model.
2. MPC algorithms used in practice should be computationally efficient.

Since fundamental models (complicated systems of nonlinear differential and algebraic equations) are usually not suitable for on-line control and their development is difficult, neural models are recommended. In this work models based on rudimentary MLP and RBF neural networks are discussed, including Hammerstein and Wiener neural structures. All discussed models can be easily used in MPC, without solving any equations which is likely to be computationally demanding and may lead to numerical problems. Neural models are trained using recorded data sets, no technological knowledge is necessary.

The role of the model in MPC cannot be ignored during model structure selection and training. The model has to be able to make good predictions of future behaviour of the process over the whole prediction horizon. Two classes of specialised neural models designed with the specific aim of using them in MPC are discussed: multi-models and structured models. Both these models are trained easily as one-step ahead predictors, but they calculate predictions for the whole prediction horizon without being used recurrently. As a result, the prediction error is not propagated which is particularly important in practice.

In this work the suboptimal MPC-NPL algorithm is recommended. The neural model is successively linearised on-line, the linear approximation is used for prediction and calculation of the future control policy. The MPC-NPL algorithm is computationally efficient, it needs solving on-line a quadratic programming task. It is also demonstrated that the algorithm gives very good control accuracy, comparable to that obtained when nonlinear optimisation is used on-line in MPC.

Acknowledgement. This work was supported by Polish national budget funds 2009-2011 for science as a research project.

References

1. Al-Duwaish, H., Karim, M.N., Chandrasekar, V.: Use of multilayer feedforward neural networks in identification and control of Wiener model. Proceedings IEE, Part D, Control Theory and Applications 143, 225–258 (1996)

2. Alexandridis, A., Sarimveis, H.: Nonlinear adaptive model predictive control based on self-correcting neural network models. AIChE Journal 51, 3495–3506 (2005)
3. Al Seyab, R.K., Cao, Y.: Nonlinear system identification for predictive control using continuous time recurrent neural networks and automatic differentiation. Journal of Process Control 18, 568–581 (2008)
4. Åkesson, B.M., Toivonen, H.T.: A neural network model predictive controller. Journal of Process Control 16, 937–946 (2006)
5. Arto, V., Hannu, P., Halme, A.: Modeling of chromato-graphic separation process with Wiener-MLP representation. Journal of Process Control 78, 443–458 (2001)
6. Bazaraa, M.S., Sherali, J., Shetty, K.: Nonlinear programming: theory and algorithms. Prentice-Hall, Englewood Cliffs (1999)
7. Bomberger, J.D., Seborg, D.E.: Determination of model order for NARX models directly from input-output data. Journal of Process Control 8, 459–468 (1998)
8. Cavagnari, L., Magni, L., Scattolini, R.: Neural network implementation of nonlinear receding-horizon control. Neural Computing and Applications 8, 86–92 (1999)
9. Doyle, F.J., Ogunnaike, B.A., Pearson, R.K.: Nonlinear model-based control using second-order Volterra models. Automatica 31, 697–714 (1995)
10. El Ghoumari, M.Y., Tantau, H.J.: Non-linear constrained MPC: real-time implementation of greenhouse air temperature control. Computers and Electronics in Agriculture 49, 345–356 (2005)
11. Haykin, S.: Neural networks – a comprehensive foundation. Prentice-Hall, Englewood Cliffs (1999)
12. Henson, M.A.: Nonlinear model predictive control: current status and future directions. Computers and Chemical Engineering 23, 187–202 (1998)
13. Hornik, K., Stinchcombe, M., White, H.: Multilayer feedforward networks are universal approximators. Neural Networks 2, 359–366 (1989)
14. Hussain, M.A.: Review of the applications of neural networks in chemical process control – simulation and online implmementation. Artificial Intelligence in Engineering 13, 55–68 (1999)
15. Janczak, A.: Identification of nonlinear systems using neural networks and polynomial models. In: A block-oriented approach. LNCIS, vol. 310. Springer, Berlin (2005)
16. Lazar, M., Pastravanu, O.: A neural predictive controller for non-linear systems. Mathematics and Computers in Simulation 60, 315–324 (2002)
17. LeCun, Y., Denker, J., Solla, S.: Optimal brain damage. In: Touretzky, D. (ed.) Advances of NIPS2, pp. 598–605. Morgan Kaufmann, San Mateo (1990)
18. Liu, G.P., Kadirkamanathan, V., Billings, S.A.: Predictive control for non-linear systems using neural networks. International Journal of Control 71, 1119–1132 (1998)
19. Luyben, W.L.: Process modelling, simulation and control for chemical engineers. McGraw Hill, New York (1990)
20. Ławryńczuk, M.: Efficient nonlinear predictive control based on structured neural models. International Journal of Applied Mathematics and Computer Science 19 (in press, 2009)
21. Ławryńczuk, M.: Modelling and nonlinear predictive control of a yeast fermentation biochemical reactor using neural networks. Chemical Engineering Journal 145, 290–307 (2008)
22. Ławryńczuk, M., Tadej, W.: A computationally efficient stable dual-mode type nonlinear predictive control algorithm. Control and Cybernetics 145, 99–132 (2008)

23. Ławryńczuk, M.: Suboptimal Nonlinear Predictive Control Based on Neural Wiener Models. In: Dochev, D., Pistore, M., Traverso, P. (eds.) AIMSA 2008. LNCS (LNAI), vol. 5253, pp. 410–414. Springer, Heidelberg (2008)
24. Ławryńczuk, M.: Suboptimal nonlinear predictive control with neural multi-models. In: Rutkowski, L., Tadeusiewicz, R., Zadeh, L.A., Zurada, J. (eds.) Computational intelligence: methods and applications, pp. 45–56. Exit, Warsaw (2008)
25. Ławryńczuk, M.: Suboptimal nonlinear predictive control with MIMO neural Hammerstein models. In: Nguyen, N.T., Borzemski, L., Grzech, A., Ali, M. (eds.) IEA/AIE 2008. LNCS (LNAI), vol. 5027, pp. 225–234. Springer, Heidelberg (2008)
26. Ławryńczuk, M.: A family of model predictive control algorithms with artificial neural networks. International Journal of Applied Mathematics and Computer Science 17, 217–232 (2007)
27. Ławryńczuk, M., Tatjewski, P.: A computationally efficient nonlinear predictive control algorithm with RBF neural models and its application. In: Kryszkiewicz, M., Peters, J.F., Rybiński, H., Skowron, A. (eds.) RSEISP 2007. LNCS (LNAI), vol. 4585, pp. 603–612. Springer, Heidelberg (2007)
28. Ławryńczuk, M., Tatjewski, P.: An efficient nonlinear predictive control algorithm with neural models and its application to a high-purity distillation process. In: Rutkowski, L., Tadeusiewicz, R., Zadeh, L.A., Żurada, J.M. (eds.) ICAISC 2006. LNCS (LNAI), vol. 4029, pp. 76–85. Springer, Heidelberg (2006)
29. Maciejowski, J.M.: Predictive control with constraints. Prentice-Hall, Englewood Cliffs (2002)
30. Marlin, T.E.: Process control. McGraw Hill, New York (1995)
31. Mayne, D.Q., Rawlings, J.B., Rao, C.V., Scokaert, P.O.M.: Constrained model predictive control: stability and optimality. Automatica 36, 789–814 (2000)
32. Mayne, D.Q., Michalska, H.: Receding horizon control of nonlinear systems. IEEE Transactions on Automatic Control 35(7), 814–824 (1990)
33. da Cruz Meleiro, L.A., José, F., Zuben, V., Filho, R.M.: Constructive learning neural network applied to identification and control of a fuel-ethanol fermentation process. Engineering Applications of Artificial Intelligence (in press, 2009)
34. Michalska, H., Mayne, D.Q.: Robust receding horizon control of constrained nonlinear systems. IEEE Transactions on Automatic Control 38, 1623–1633 (1993)
35. Morari, M., Lee, J.H.: Model predictive control: past, present and future. Computers and Chemical Engineering 23, 667–682 (1999)
36. Narendra, K.S., Parthasarathy, K.: Identification and control of dynamical systems using neural networks. IEEE Transactions on Neural Networks 1, 4–26 (1990)
37. Nørgaard, M., Ravn, O., Poulsen, N.K., Hansen, L.K.: Neural networks for modelling and control of dynamic systems. Springer, London (2000)
38. Parisini, T., Sanguineti, M., Zoppoli, R.: Nonlinear stabilization by receding-horizon neural regulators. International Journal of Control 70, 341–362 (1998)
39. Pearson, R.K.: Selecting nonlinear model structures for computer control. Journal of Process Control 13, 1–26 (2003)
40. Peng, H., Yang, Z.J., Gui, W., Wu, M., Shioya, H., Nakano, K.: Nonlinear system modeling and robust predictive control based on RBF-ARX model. Engineering Applications of Artificial Intelligence 20, 1–9 (2007)
41. Piche, S., Sayyar-Rodsari, B., Johnson, D., Gerules, M.: Nonlinear model predictive control using neural networks. IEEE Control System Magazine 20, 56–62 (2000)
42. Pottmann, M., Seborg, D.E.: A nonlinear predictive control strategy based on radial basis function models. Computers and Chemical Engineering 21, 965–980 (1997)

43. Qin, S.J., Badgwell, T.A.: A survey of industrial model predictive control technology. Control Engineering Practice 11, 733–764 (2003)
44. Rossiter, J.A.: Model-based predictive control. CRC Press, Boca Raton (2003)
45. Scattolini, R., Bittanti, S.: On the choice of the horizon in long-range predictive control – some simple criteria. Automatica 26, 915–917 (1990)
46. Scokaert, P.O.M., Mayne, D.Q., Rawlings, J.B.: Suboptimal model predictive control (feasibility implies stability). IEEE Transactions on Automatic Control 3, 648–654 (1999)
47. Tatjewski, P.: Advanced control of industrial processes, structures and algorithms. Springer, London (2007)
48. Tatjewski, P., Ławryńczuk, M.: Soft computing in model-based predictive control. International Journal of Applied Mathematics and Computer Science 16, 101–120 (2006)
49. Trajanoski, Z., Wach, P.: Neural predictive control for insulin delivery using the subcu-taneous route. IEEE Transactions on Biomedical Engineering 45, 1122–1134 (1998)
50. Yu, D.L., Yu, D.W., Gomm, J.B.: Neural model adaptation and predictive control of a chemical process rig. IEEE Transactions on Control Systems Technology 14, 828–840 (2006)
51. Yu, D.L., Gomm, J.B.: Implementation of neural network predictive control to a multivariable chemical reactor. Control Engineering Practice 11, 1315–1323 (2003)
52. Zamarreño, J.M., Vega, P., Garcia, L.D., Francisco, M.: State-space neural network for modelling prediction and control. Control Engineering Practice 8, 1063–1075 (2000)

Application of a Multi-domain Knowledge Structure: The Decisional DNA

Cesar Sanín, Leonardo Mancilla-Amaya, Edward Szczerbicki, and Paul CayfordHowell

The University of Newcastle, Faculty of Engineering and Built Environment
School of Mechanical Engineering
University Drive, Callaghan, NSW 2308
{Cesar.Sanin, Edward.Szczerbicki}@newcastle.edu.au,
{Leonardo.Mancilla,
Paul.CayfordHowell}@studentmail.newcastle.edu.au

Abstract. Knowledge engineering techniques are becoming useful and popular components of hybrid integrated systems used to solve complicated practical problems in different disciplines. Knowledge engineering techniques offer features such as: learning from experience, handling noisy and incomplete data, helping with decision making, and predicting. This chapter presents the application of a knowledge structure to different fields of study by constructing Decisional DNA. Decisional DNA, as a knowledge representation structure, offers great possibilities on gathering explicit knowledge of formal decision events as well as a tool for decision making processes. Its versatility is shown in this chapter when applied to decisional domains in finances and energy. The main advantages of using the Decisional DNA rely on: *(i)* versatility and dynamicity of the knowledge structure, *(ii)* storage of day-to-day explicit experience in a single structure, *(iii)* transportability and share ability of the knowledge, and *(iv)* predicting capabilities based on the collected experience. Thus, after showing the results, we conclude that the Decisional DNA, as a unique structure, can be applied to multi-domain systems while enhancing predicting capabilities and facilitating knowledge engineering processes inside decision making systems.

Keywords: Knowledge Engineering, Decisional DNA, Knowledge Representation, Set of Experience Knowledge Structure, Decision Making.

1 Introduction

The term knowledge engineering has been defined as a discipline that aims to offering solutions for complex problems by the means of integrating knowledge into computer systems [1]. It involves the use and application of several computer science domains such as artificial intelligence, knowledge representation, databases, and decision support systems, among others. Knowledge engineering technologies make use of the synergism of hybrid systems to produce better, powerful, more efficient and effective computer systems.

N.T. Nguyen and E. Szczerbicki (Eds.): Intel. Sys. for Know. Management, SCI 252, pp. 65–86.
springerlink.com

Among the features associated with knowledge engineering systems are human intelligence capabilities such as learning, reasoning and forecasting from current knowledge or experience. In our case, experience is the main and most appropriate source of knowledge and its use leads to useful systems with improved performance. Knowledge engineering techniques have been used in several domains and applications.

As case studies, we apply the Decisional DNA knowledge representation to the domains of energy and finance. We have chosen these two domains because of their noticeable differences in terms of collected knowledge and the great potential and extended use of intelligent techniques in such fields. Finances have been ruling the word since the industrial revolution in the 1700's, and lately, have become a cumbersome issue. Proper understanding of finances could lead to better community quality life. On the other hand, renewable energy (RE) resources have great potential and can supply the present world energy demand. RE can reduce atmospheric emissions, offer diversity in energy supply markets, and guarantee long-term sustainable energy supply [2]. In addition, the use of RE resources has been outlined as a research focus in many developed countries.

Knowledge engineering has been applied to the geothermal energy and renewable energy domains in several occasions. They comprise applications in artificial intelligence systems, knowledge base systems, expert systems, decisional support systems, heuristic systems, and the list appears without an end, just a few publications to mention include [3], [4], [5], [6], [7]. On the finance domain, the list is as similar as in the energy domain, and it goes even further in a way that there are specialized publications on the area such as the book Computational Intelligence in Economics and Finance and the Journal of Computational Intelligence in Finance; these among several publications, just a few to mention [8], [9], [10], [11], [12].

All this multiple applications perform decisions, and most of them, decision that are taken in a structured and formal way, this is what we call *formal decision events* [13]. All these formal decision events are usually disregarded once the decision is made, or even worst, if the system is queried again, the decision has to be repeated. What to do with the experience gained on taking such decisions relies on our knowledge representation structure. We propose the Decisional DNA as a unique and single structure for capturing, storing, improving and reusing decisional experience.

The present chapter explains the process of implementing a knowledge engineering technology that can collect, use and offer trustable knowledge in multiple domains, using as examples the energy and financial domains. We make use of the Set of Experience (SOE) as a knowledge structure that allows the acquisition and storage of formal decision events in a knowledge-explicit form. It comprises variables, functions, constraints and rules associated in a DNA shape allowing the construction of the Decisional DNA of an organization. Having a powerful knowledge structure such as the Set of Experience Knowledge Structure (SOEKS) in the Decisional DNA and apply it to the construction of chromosomes, such as the ones in energy and financial domains, can enrich and develop any decisional system based upon previous experience.

2 Background

A brief introduction to the reader into the elements and concepts used along the chapter is included as a way to guarantee a better understanding of the ideas and proposals exposed. However, an interested reader can gain a deeper knowledge by approaching the additional mentioned bibliography.

2.1 Knowledge

Humanity has always being accompanied by knowledge. Both have grown together to construct what we understand now as society and civilization. Hence, humankind has been trying to make knowledge part of its assets. Knowledge seems to be a valuable possession of incalculable worth and it has been considered as the only true source of a nation's economic and military strength as well as, the key source of competitive advantage of a company [14]. Thus, humankind in general and, more specifically, managers have turned to knowledge administration. They want technologies that facilitate control of all forms of knowledge because such technologies can be considered as the key for the success or failure of an organization, and subsequently, knowledge society.

One theory proposes that experienced decision-makers base most of their decisions on situation assessments [15]. In other words, decision-makers principally use experience for their decisions, i.e. when a decision event emerges, managers select actions that have worked well in previous similar situations. Then, in a brain process that is not well understood, managers extract the most significant characteristics from the current circumstances, and relate them to similar situations and actions that have worked well in the past. Therefore, this theory suggests that any mechanism that supports the process of storing previous decisions would improve the decision maker's job; and as such, it is related to a process of storing knowledge and experience.

Since our focus is more related to engineering and computer fields, we concentrate on the concept knowledge engineering (KE). KE is defined as a technique applied by knowledge engineers to construct intelligent systems. Two main movements surround KE, they are: the transfer movement and the modelling movement. The transfer movement aims for techniques to transfer human knowledge into the artificial intelligent systems. The modelling movement aims for modelling the knowledge and problem solving techniques of the domain expert into the artificial intelligent system. Our research concentrates on the modelling trend which requires the areas of knowledge representation (KR) and knowledge modelling.

Knowledge engineering came up as part of the knowledge society which arrived and brought with it all the difficulties that information faces. Characteristics such as unstructured, disintegrated, not shareable, incomplete, and uncertain information represent an enormous problem for information technologies (IT) [16], [17]. Under these circumstances, the process of transforming information into knowledge is critical and difficult, because unfortunately, knowledge depends upon information [18]. Moreover, Awad and Ghaziri [19] gives acknowledge of another difficulty when affirms that up to 95 percent of information is preserved as tacit knowledge (for tacit and explicit knowledge see Nonaka & Takeuchi [20]). Thus, one of the most complicated issues involving knowledge is its representation, because it determines how knowledge is acquired and stored; and how knowledge is transformed from tacit knowledge to explicit knowledge. Failing on doing so, knowledge is no more than a volatile and fragile implicit asset. Knowledge, then, must be obtained and represented in an understandable form for the agents that experienced it. Hence, it is evident that some kind of technology is necessary to transform information into, not just knowledge, but explicit knowledge.

Many technologies, such as Knowledge Management Systems (KMS), Data Mining (DM), Knowledge-base Systems (KbS) among others are currently working with different types of knowledge. These kinds of technologies try to collect and administer knowledge in some manner. For instance, KMS are usually related to the process of storing, retrieval, distribution, and sharing knowledge; they basically keep a huge quantity of documents organized in some way, but these documents are not considered structured or explicit knowledge, but a documents' repository. On the other hand, DM, pattern recognition, and other learning technologies acquire knowledge from what is called examples, acting more as rule-makers or inferrers. Although these technologies work with decision-making in some way, they do not keep structured knowledge of the formal decision events they participate on. For us, any technology able to capture and store formal decision events as explicit knowledge will improve the decision-making process. Such technology will help by reducing decision time, as well as avoiding repetition and duplication in the process.

Unfortunately, computers are not as clever as to form internal representations of the world, and even simpler, representations of just formal decision events. Instead of gathering knowledge for themselves, computers must rely on people to place knowledge directly into their memories. This problem suggests deciding on ways to represent information and knowledge inside computers.

Representing knowledge is, at first sight, easy to understand. It has to do with communicating in some language, written, oral or something else, a condition of the world [21]. Knowledge representation (KR) has been involved in several science fields; however, its main roots come from three specific areas: logic, ontology and computation [22]. According to Levesque [21], KR has to do with "writing down descriptions of the world in which an intelligent machine might be embedded in such a way that the machine can come to new conclusions about its world by manipulating these symbolic representations". Among the KR schemas are semantics nets, ontologies, and knowledge-bases; each one being implemented with different data structures creating a universe of KR as big as the number of applications researchers and IT companies have developed. All these knowledge representations are integrated into several kinds of technologies, from Neural Networks (NN) to Knowledge-base Systems (KBS) and the enormous diversity of combinations makes the most experienced person in the field feel humble. These technologies have been developed to make useful huge quantities of stored information by modelling knowledge in some way; however, none of them keep an explicit record of the decision events they participate on. It is necessary to define a multi-domain knowledge structure able to be adaptable and versatile as to capture all these different decision events from the day-to-day operation, to store proper characteristics of the experience acquired, to keep this experience as explicit knowledge, and to allow it for multiple technologies to be used, analyzed, and categorized.

This chapter presents three case studies that use the Set of Experience Knowledge Structure (SOEKS or SOE shortly) and the Decisional DNA as the knowledge representations. These two concepts are offered as possible solutions to be utilized for the purposes mentioned above. These two concepts fully applied certainly would improve the quality of decision-making, and could advance the notion of administering knowledge in the current decision making environment.

2.2 Set of Experience Knowledge Structure (SOEKS) and Decisional DNA

Arnold and Bowie [23] argue that "the mind's mechanism for storing and retrieving knowledge is transparent to us. When we 'memorize' an orange, we simply examine it, think about it for a while, and perhaps eat it. Somehow, during this process, all the essential qualities of the orange are stored [experience]. Later, when someone mentions the word 'orange', our senses are activated from within [query], and we see, smell, touch, and taste the orange all over again" (p. 46). Computers, on their side, must depend on human beings to enter knowledge directly into them. Thus, the problem is how to adequately, efficiently, and effectively represent information and knowledge inside a computer.

The SOEKS has been developed to keep formal decision events in an explicit way [13]. It is a model based upon existing and available knowledge, which must adjust to the decision event it is built from (i.e. it is a dynamic structure that relies on the information offered by a formal decision event); besides, it can be expressed in XML or OWL as an ontology in order to make it shareable and transportable [24], [25]. Four basic components surround decision-making events, and are stored in a combined dynamic structure that comprises the SOE (fig. 1); they are: *variables V, functions F, constraints C,* and *rules R.*

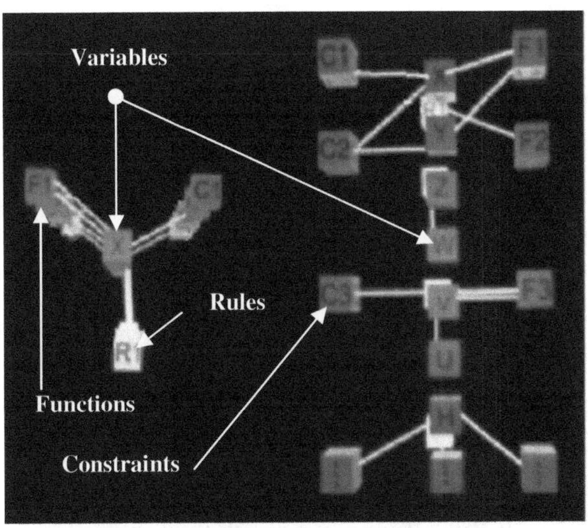

Fig. 1. Set of Experience Knowledge Structure (SOEKS)

Variables usually involve representing knowledge using an attribute-value language (i.e. by a vector of variables and values) [26]. This is a traditional approach from the origin of knowledge representation, and is the starting point for the SOEKS. Variables that intervene in the process of decision-making are the first component of the SOE. These variables are the centre root of the structure, because they are the source of the other components.

Based on the idea of Malhotra [27] who states that "to grasp the meaning of a thing, an event, or a situation is to see it in its relations to other things" (p.51), variables are related among them in the shape of functions. Functions describe associations between a dependent variable and a set of input variables; moreover, functions can be applied for reasoning optimal states, because they come out from the goals of the decision event. Therefore, the SOE uses functions, its second component, and establishes links among the variables constructing multi-objective goals, that is, multiple functions that restrict experience on decision-making by the elements of a universe of relationships.

According to Theory of Constraints (TOC), Goldratt [28] maintains that any system has at least one constraint; otherwise, its performance would be infinite. Thus, constraints are another way of relationships among the variables; in fact, they are functions as well, but they have a different purpose. A constraint, as the third component of SOEKS, is a limitation of possibilities, a restriction of the feasible solutions in a decision problem, and a factor that limits the performance of a system with respect to its goals.

Finally, rules are suitable for representing inferences, or for associating actions with conditions under which the actions should be done. Rules, the fourth component of SOEKS, are another form of expressing relationships among variables. They are conditional relationships that operate in the universe of variables. Rules are relationships between a condition and a consequence connected by the statements IF-THEN-ELSE.

Additionally, the SOEKS is organized taking into account some important features of DNA. Firstly, the combination of the four nucleotides of DNA gives uniqueness to itself, just as the combination of the four components of the SOE offer distinctiveness. Moreover, the elements of the structure are connected among themselves imitating part of a long strand of DNA, that is, a gene. Thus, a gene can be assimilated to a SOE, and in the same way as a gene produces a phenotype, a SOE produces a value of decision in terms of the elements it contains. Such value of decision can be called the efficiency or the phenotype value of the SOE [13]; in other words, the SOEKS, itself, stores an answer to a query presented. Each SOE can be categorized, and acts as a gene in DNA. A gene guides hereditary responses in living organisms. As an analogy, a SOE guides the responses of certain areas of the company.

A unique SOE cannot rule a whole system, even in a specific area or category. Therefore, more Sets of Experience should be acquired and constructed. The day-to-day operation provides many decisions, and the result of this is a collection of many different SOE.

A group of SOE of the same category comprises a decisional chromosome, as DNA does with genes. This decisional chromosome stores decisional "strategies" for a category. In this case, each module of chromosomes forms an entire inference tool, and provides a schematic view for knowledge inside an organization. Subsequently, having a diverse group of SOE chromosomes is like having the Decisional DNA of an organization, because what has been collected is a series of inference strategies related to such enterprise (Fig. 2).

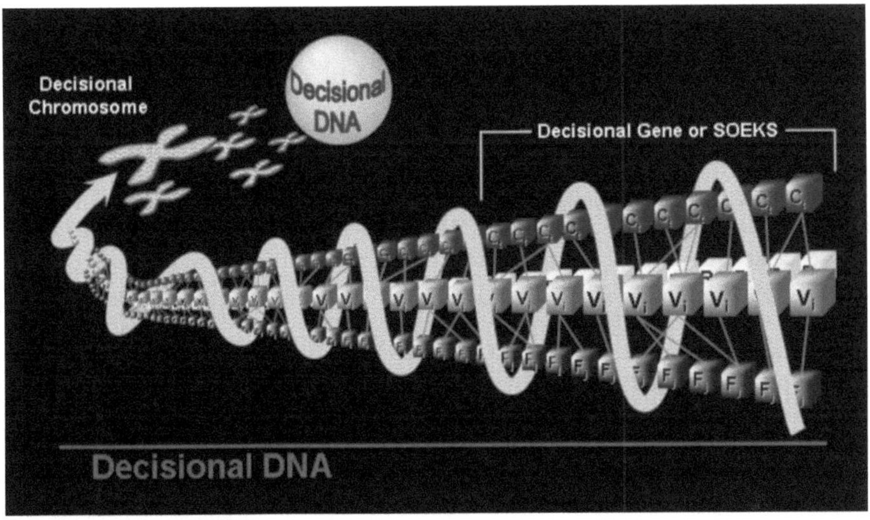

Fig. 2. SOEKS and Decisional DNA

In conclusion, the SOEKS is a compound of variables, functions, constraints and rules, which are uniquely combined to represent a formal decision event. Multiple SOE can be collected, classified, and organized according to their efficiency, grouping them into decisional chromosomes. Chromosomes are groups of SOE that can accumulate decisional strategies for a specific area of an organization. Finally, sets of chromosomes comprise what is called the Decisional DNA of the organization [13]. Moreover, the Decisional DNA can be expressed in XML as way to make it shareable and transportable. Furthermore, the Decisional DNA can be used in platforms to support decision-making, and new decisions can be made based on it. In this text a concise idea of the SOEKS and the Decisional DNA was offered, for further information Sanin and Szczerbicki [13] should be reviewed.

2.3 Ontologies and Reflexive Ontologies (RO)

Tom Gruber's [29] widespread accepted definition in the Computer Science field of Ontology sates that an ontology is the explicit specification of a conceptualization; a description of the concepts and relationships in a domain. In the context of Artificial Intelligence (AI), we can describe the ontology of a program by defining a set of representational terms. In such ontology, definitions associate names of entities in the universe of discourse with human-readable text describing what the names mean, and formal axioms that constrain the interpretation and well-formed use of these terms.

Computer programs can use ontologies for a variety of purposes including inductive reasoning, classification, and problem solving techniques, as well as communication and sharing of knowledge among different systems. In addition, emerging semantic systems use ontologies for a better interaction and understanding between different agent-based systems. Ontologies can be modelled using several languages, being the most widely used RDF and recently OWL (Ontology Web Language) [30].

In general terms, any knowledge is susceptible to be modelled as an ontology; however, no normative exists yet in order to model knowledge. This could be due to the inner nature of the object to be modelled as it is different from one schema to another. Although there are interesting approaches for a universal ontology, this outcome has not been reached yet due to the difficulty of standardization and the fact that if a universal modelling of knowledge is reachable, it will lead to a high degree of conceptualization turning into a difficulty for an inexperienced user. From the experience acquired by different working groups in ontology modelling of elements, a good starting point is to have a "well documented" schema with the typical elements in the area of knowledge that is being described.

One of the advantages of a conceptual knowledge model expressed as an ontology is the capacity of inferring semantically new derived queries. These queries relate concepts that are not explicitly specified by the user; nevertheless the concepts are relevant to the query. Modern inference engines and reasoners like Pellet and Racer deliver a highly specialized, yet efficient way to perform such queries via a JAVA compliant API. In the literature, data handling by ontology-based technology is reported by researchers in different fields [25], [31], [32].

In order to model an ontology, different tools are available such as Protégé , OilEd, Ontololingua, Swoop, among others. In our implementation, we used the protégé ontology editor and its APIs, but any other editor or API can be used as well.

Reflexivity addresses the property of an abstract structure of a knowledge base (in this case, an ontology and its instances) to "know about itself". When an abstract knowledge structure is able to maintain, in a persistent manner, every query performed on it, and store those queries as individuals of a class that extends the original ontology, it is said that such ontology is reflexive. Thus, Toro et al. [32] proposed the following definition for a Reflexive Ontology: "A Reflexive Ontology is a description of the concepts and the relations of such concepts in a specific domain, enhanced by an explicit self contained set of queries over the instances". Therefore, any RO is an abstract knowledge structure with a set of structured contents and relationships, and all the mathematical concepts of a set can be applied to it as a way of formalization and handling.

A RO is, basically, an ontology that has been extended with the concept of reflexivity and must fulfil the properties of: *Query retrieval* (storing every query performed), *integrity update* (updating structural changes in the query retrieval system), *autopoietic behaviour* (capacity of self creation), *support for logical operators* (mechanisms of set handling), and *self reasoning over the query set* (capacity of performing logical operations over the query system).

The advantage of implementing RO relies on the following main aspects: Speed on the query process, incremental nature, and self containment of the knowledge structure in a single file.

3 Constructing Decisional DNA

Decisional applications produce myriads of formal decision events, i.e. decisional experience, which in most cases is analyzed and stored; however, after a while, this decisional experience is commonly disregarded, not shared, and put behind [33], [34], [35], [36]. Little of this collected experience survives, and in some cases, over time, it becomes inaccessible due to poor knowledge engineering practices or due to technology changes in software, hardware or storage media. Knowledge and experience are lost indicating that there is a clear deficiency on experience collection and reuse. We suggest that some of the reasons are:

a) the non existence of a common knowledge-experience structure able to collect multi-domain formal decision events, and
b) the non existence of a technology able to capture, store, improve, retrieve and reuse such collected experience.

Through our project we proposed three important elements:

(i) a knowledge structure able to store and maintain experiential knowledge, that is, the SOEKS and the Decisional DNA,
(ii) a solution for collecting experience that can be applied to multiple applications from different domains, that is, a multi-domain knowledge structure, and
(iii) a way to automate decision making by using such experience, that is, retrieve collected experience by answering a query presented.

In this chapter, we introduce the reader to the three above mentioned elements throughout three different case studies and by using two different methodologies.

As a note, our tests were performed on a Windows Vista Intel Core Duo T9400 (2.53 GHz), 4 Gb RAM, 32 bit operating system, running the SUN Java virtual machine V1.6.

3.1 Case Studies

Our plan aimed for the construction and use of decisional chromosomes in the areas of geothermal energy, renewable energy and net income. Two different techniques were implemented: ontologies for geothermal energy and software engineering for renewable energy and net income. Each of these two techniques comprised a different methodology.

3.1.1 Ontology Methodology: Chromosome on Geothermal Energy

One promising renewable energy technology is geothermal energy. A geothermal system's performance that can work properly, efficiently and economically to meet the desired load requirements under local ground conditions depends upon several factors, including ambient and ground temperatures, pressures, and ground matter, among others. In any geothermal system, sizing represents an important part of the system design, e.g., the optimal pressure or the temperatures associated with those pressures which determine the size of certain components. Thus, in order to size a geothermal system, the characteristic performance of each component in the system is required.

The University of Newcastle counts with a geothermal laboratory controlled by the Geothermal Research Group (GRG) on a project sponsored by Granite Power and AusIndustry through the Renewable Energy Development Initiative (REDI) grant. The GRG develops several experiments aiming for efficient and optimal production of power cycles for generation of electricity from geothermal reservoirs.

The geothermal laboratory simulates traditional geothermal power cycles in which a working fluid is used to convert heat energy into mechanical energy. In order to collect the research knowledge and experience created for this geothermal power cycle laboratory, several instruments and sensors are positioned in the system. Due to the experimental nature of the work, the geothermal system incorporates a state of the art digital control system to continuously monitor, control, and record all the parameters of the process. The primary roles of the digital control system are monitoring the process and collecting online data from all the process sensors. Additionally, the digital control system acts as a safe-fail device that controls the process and safely shuts down the system in the event of a safety breach. The electronic system was supplied by Emerson Process Management Pty. Ltd. and consists of: a PC based control centre, a Foundation Field bus digital control system (DeltaV), 16 pressure transmitters, 24 temperature transmitters and 3 flow meters (a section of the laboratory is shown in fig. 3). The PC is connected to the field instruments via the DeltaV digital control system.

The PC based remote operator control panel operates the Emerson PlantWeb software, which provides a visual user interface. The PC is connected to the field instruments via the DeltaV and collects all the real time online information of the process sensors; this is the repository of information. This methodology involves the following steps:

a) Run experiments at the geothermal laboratory,
b) Gather online information from the 43 sensors into the DeltaV,
c) Convert information to SOEKS compliant XML (eXtensible Markup language) by parsing the information onto metadata,
d) Feed an ontology chromosome with Sets of Experience (SOE), and
e) Check Energy Decisional DNA by querying the system with samples.

At the geothermal laboratory, experiments can run continuously for about 20 hours and information from each sensor can be collected at defined intervals between 1 second and 1 hour. For testing purposes, samples from two different experiments involving 2 fluids at intervals of 1 minute were sufficient. Such experiments produced 2400 events, each one involving variables of 43 sensors at a time. This information collected by fieldbus technology arrived to the DeltaV and it is stored in comma separated values (CSV) files.

Fig. 3. Section of the geothermal laboratory with process sensors

Having these files, a parser written in Java built up the XML trees [37] which were then uploaded by using a Java API into Protégé. Once the geothermal chromosome was constructed and ready in Protégé, an additional extension for reflexive ontologies was installed [32]. The RO consists of an extension which hangs from the OWL:Thing super class and adds a new class to the base ontology with the needed schema for the reflexivity; this is the "ReflexiveOntologyQueryStorer class".

The last part of the implementation is the reflexiveness itself and it provides the ontology (programmatically) with a mechanism to perform queries and basic logic on the queries that allows the handling of the reflexiveness. For our implementation we used Protégé, its API's and the OWL-DL subspecies of the OWL specification [38]. Once this is done, the chromosome on geothermal energy is constructed and ready to be queried.

A query is used to exemplify this case study. The query is defined in the code as:

```
public static String SIMPLE_REFLEXIVE_QUERY="CLASS
variable with the PROPERTY var_name EQUALS to X1";
```

Notice that this is a value type query. Such query is written in a human-like readable form which means "retrieve all the variables of the ontology that have the variable name X1". The variable X1 represents a sensor within the set of sensors in the geothermal lab.

The execution of the code offers information about the type of query executed and the successful saving of the query executed with results within the Reflexive Ontology Structure. Following in figure 4, the results can be seen as a query

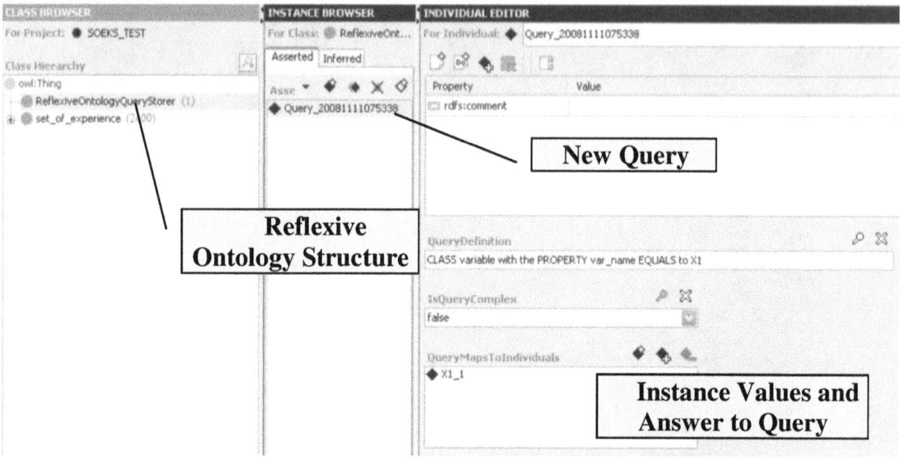

Fig. 4. Query performed on the RO geothermal chromosome

successfully executed with the new instance in the SOEKS-OWL transformed into a Reflexive Ontology:

```
---------------------------START------------------------
Testing Simple query : CLASS variable with the PROPERTY
var_name EQUALS to X1
--------------------------------------------------------
... saving successful.
File modification saved with 0 errors.
-------------------------------------END----------------
```

Such query performed over 2400 instances on the ontology took 3 seconds; however, after the first time, a new similar query reduced the query time to 0.005 seconds just by using the query retrieval functionality of the RO.

3.1.2 Software Engineering Methodology: Chromosome on Renewable Energy

Renewable energy has become a cumbersome on the energy consumption arena in the past decades. Governments and multiple organizations collect information and produce multiple reports in order to monitor and predict renewable energy production and consumption. Regularly, the US government issues a yearly report called the Annual Energy Outlook which includes energy projections. We have been taking the estimations report, Table 17 Renewable Energy Consumption by Sector and Source, from its public website [39]. This report comprises a series of estimations of renewable energy consumption from the year 2004 to 2030. In order to collect knowledge and experience, we applied linear regression to this information producing a series of functions and constraints which can be adapted to the SOEKS, and consequently, producing a renewable energy decisional chromosome.

This is the repository of experience. This methodology involves the following steps:

a) Collect energy report from the U.S. Energy Information Administration,
b) Implement the linear regressions and construct functions and constraints,
c) Build a CSV file per year with the collected information, each one representing a single formal decision event,
d) Convert these formal decision events into SOEKS programmatically by parsing the information,
e) Feed the renewable energy decisional chromosome with Sets of Experience, and
f) Check the renewable energy Decisional DNA by querying the system with samples.

After steps a), b) and c) were done, seven CSV files were constructed. Having the CSV files, a parser written in Java built up each SOEKS which was then uploaded onto a renewable energy decisional chromosome; afterwards, the renewable energy Decisional DNA was constructed and ready to be queried.

The renewable energy Decisional DNA comprised seven Sets of Experience from the years 2004, 2005, 2010, 2015, 2020, 2025 and 2030; each one including 47 variables, 29 functions and 23 constraints. For testing purposes, we chose a random SOE and used it as a query sample. Different times were taken on two execution times (T1 and T2). These were:

PT: Parsing Time	Time taken to parse the files.
DPT: Detailed Parsing Time	Time taken to parse every section (variables, functions, constraints and rules) of the files.
ZT: Zero Time	Time taken to find a similar SOE (similarity = 0).
TT: Total Time	Time taken to compare and find the similarity value between the query and all the SOE.

The results are shown below.

The parsing process of the renewable energy decisional chromosome was executed 650 times, producing an average parsing time of 11.4 ms as it can be observed from fig. 5 (average times: PT1 = 11.3723 ms, PT2 = 11.4754). This is considered a very good time taking into account that those SOE are quite complex due to the amount of variables, functions and constraints, adding up to a total of 99 key features per formal decision event.

The detailed parsing process of the renewable energy decisional chromosome produced an average parsing time per SOE of 6.8 ms (average times: DPT => Variables = 2.4120 ms, Functions = 3.8346 ms, Constraints = 0.5826 ms). It can be noticed from fig. 6 that most of the time is dedicated to the functions due to their complex knowledge construction.

The searching process of a similar SOE within the renewable energy decisional chromosome was executed 650 times, producing an average zero time usually less than 0.4 ms as it can be observed from fig. 7 (average zero time: ZT1 = 0.1953 ms, ZT2 = 0.1963). This is considered an excellent time having in mind that those SOE are quite complex due to the amount of variables, functions and constraints, and the comparisons performed among all elements in order to find a similarity value of zero.

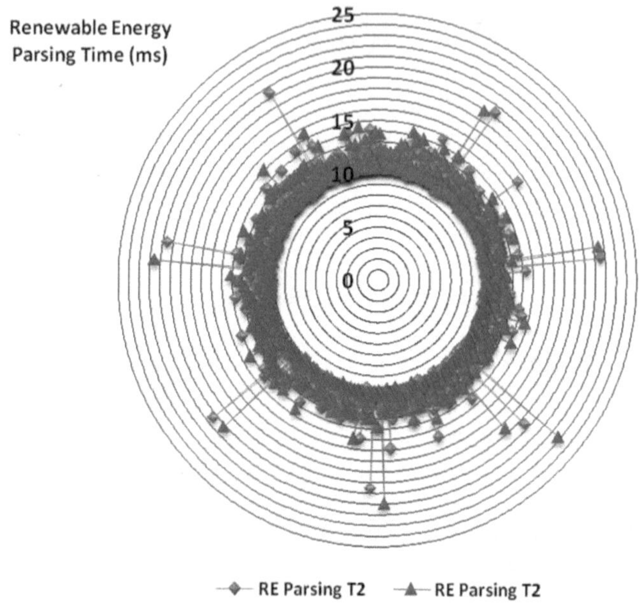

Fig. 5. Parsing time of the renewable energy decisional chromosome

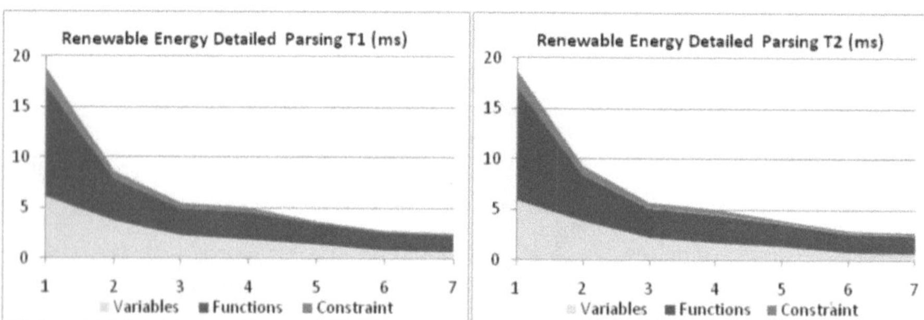

Fig. 6. Detailed parsing time of the renewable energy decisional chromosome.

Similarity comparison with the renewable energy decisional chromosome was executed 650 times, producing an total comparison time less than 0.4 ms, as it can be observed from fig. 8 (average total times: TT1 = 0.3428 ms, TT2 = 0.3516). This is an excellent time taking into account that our SOE query has to be compared with all the SOE in the renewable energy decisional chromosome.

3.1.3 Software Engineering Methodology: Chromosome on Net Income

Economics has been a major issue since human society commenced, and lately, just in the 20th century, finances began to influence human lives. Fluctuations in markets and financial analysis have made it necessary to forecast multiple variables that influence finances, and people's income is not the exception. Organizations around the world

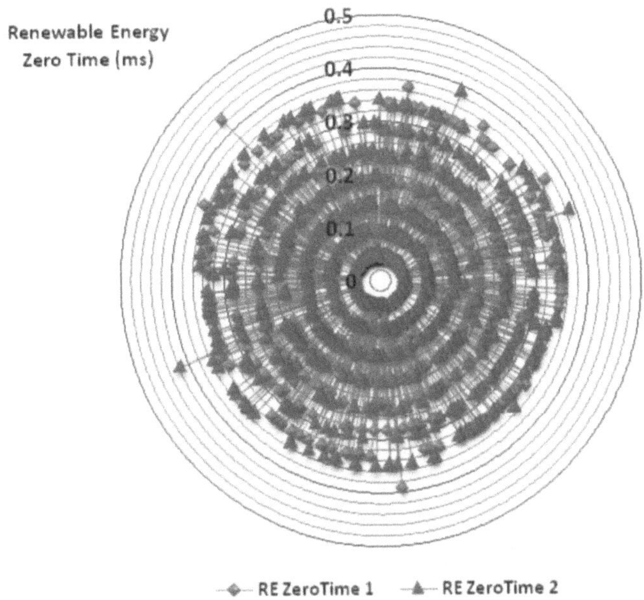

Fig. 7. Zero time of the renewable energy decisional chromosome

are interested on predicting net income in order to promote marketing strategies, analyze market behaviours and approach focus group. Therefore, several software applications and techniques are applied with such purposes.

We used a data set called "Census Income" made public at the UCI Machine Learning website [40]. This data set was created as part of a data mining and visualization research at Silicon Graphics. It comprises a set of 14 variables (key features), a class prediction variable and 14 constraints developed for data mining training purposes with 32561 formal decision events.

This methodology involves the following steps:

a) Download the net income data set,
b) Build a CSV file with the collected information,
c) Convert these formal decision events into SOEKS programmatically by parsing the information,
d) Feed the net income decisional chromosome with Sets of Experience, and
e) Check the net income Decisional DNA by querying the system with samples.

After completing steps a) and b), and having the CSV file, a parser written in Java built up each SOEKS which was then uploaded onto a net income decisional chromosome. Afterwards, the net income Decisional DNA was constructed and ready to be queried.

This net income Decisional DNA comprised 32561 Sets of Experience. For testing purposes, we chose several random SOE and used them as query samples. Different times were taken on two execution times (T1 and T2). These were:

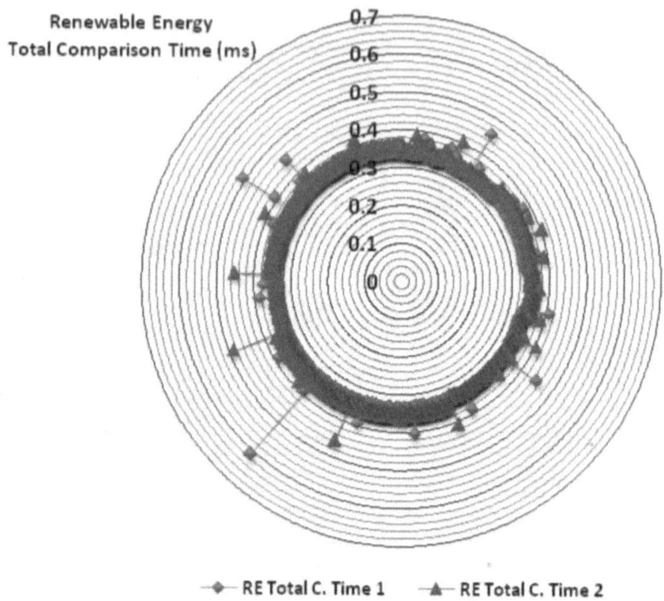

Fig. 8. Total comparison time of the renewable energy decisional chromosome

PT: Parsing Time	Time taken to parse the files.
DPT: Detailed Parsing Time	Time taken to parse every section (variables, functions, constraints and rules) of the files.
ZT: Zero Time	Time taken to find a similar SOE (similarity = 0).
TT: Total Time	Time taken to compare and find the similarity among all the SOE.

The results are shown below.

The parsing process of the net income decisional chromosome was executed 100 times, producing an average parsing time of 3.18 ms as it can be observed from fig. 9 (average value: PT1 = 3.2045 ms, PT2 = 3.1539). This is considered a very good time having in mind the amount of SOEKS loaded. Moreover, in terms of the

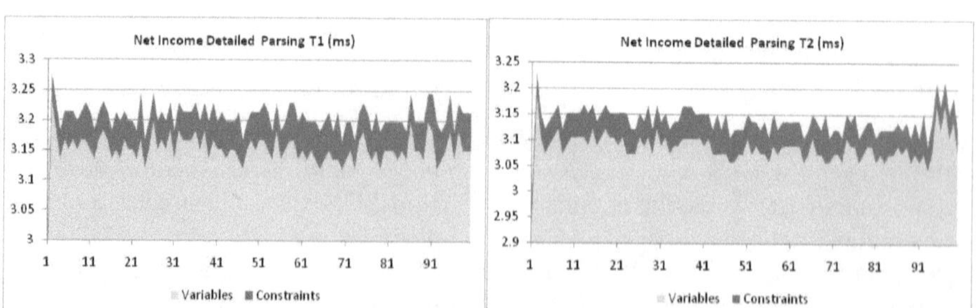

Fig. 9. Total and detailed parsing time of the net income decisional chromosome

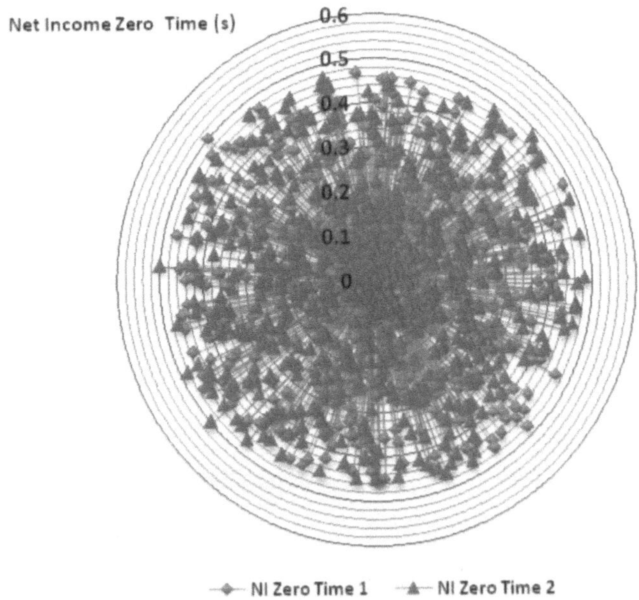

Fig. 10. Zero time of the net income decisional chromosome

detailed parsing process time, it can be observed from the graphs that variables are more time consuming than constraints (average time: Variables: DPT1 = 3.1515 ms, DPT2 = 3.0862 ms; Constraints: DPT1 = 0.0531 ms, DPT2 = 0.0515).

The searching process of a similar SOE within the net income decisional chromosome was executed 650 times, producing an average zero time around 0.23 ms (fig. 10) (average times: ZT1 = 0.2331 ms, ZT2 = 0.2393). This is considered an excellent time taking into consideration the large group of SOE and the amount of comparisons performed in order to find a similarity value of zero.

Similarity comparison with the net income decisional chromosome was executed 650 times, producing an average total time among two values, between 0.45 ms and 0.5 ms, as it can be observed from fig. 11 (average times: TT1 = 0.4722 ms, TT2 = 0.4753). Showing again an excellent time for knowledge retrieval and considering that our query answer brings out the top 5 more similar SOE.

Throughout these three case studies we argue that by using the Decisional DNA, any knowledge system can gain from the advantages it offers:

(i) versatility and dynamicity of the knowledge structure,
(ii) storage of day-to-day explicit experience in a single structure,
(iii) transportability and share ability of the knowledge, and
(iv) predicting capabilities based on the collected experience.

The Decisional DNA offers versatility as it can be used and adapted to collect and operate with multiple kinds of formal decision events, as it has been shown along this chapter by implementing three different models on the Decisional DNA. Additionally, due to the not-fixed structure of the Decisional DNA, elements of

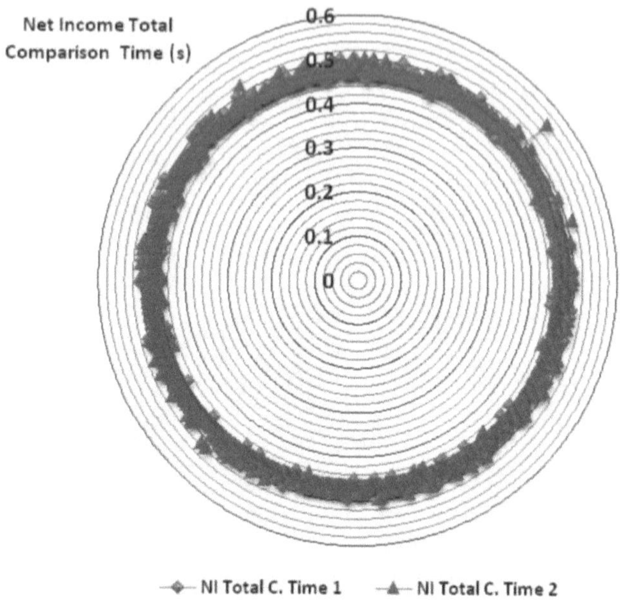

Fig. 11. Total time of the net income decisional chromosome

formal decision events (i.e. variables, functions, constraint or rules) can be incorporated or subtracted from it, adding the dynamic capability.

The Decisional DNA can collect information from the day to day operation as it was presented on the construction of the geothermal energy Decisional DNA example. Real time formal decision events can be added to the repository of experience in a single structure such as the Decisional DNA which can be improved, retrieved and reused afterwards.

Once a Decisional DNA is collected, the possibilities are increased if converted into a standard language such as XML. Thus, Decisional DNA in a XML shape offers transportability and share ability characteristics, and therefore, collected experience can be reused in different systems that conform to the Decisional DNA specifications. Finally, Decisional DNA offers predicting capabilities in terms of reusability of the experience collected in the form of formal decision events.

3.2 Time Comparison

Access times to information and response times to users are some of the most important indexes when referring to computer based services. In order to evaluate the performance of our proposed knowledge structure, response times are compared with similar systems and large scale data set retrieval. Among those, Reiner and Hahn [41] reported that large scale knowledge bases with multidimensional data sets must have an average access time of an order of magnitude between 5 ms to 12 ms.

An engineering knowledge management application developed by Chen et al. [42] was tested on indexing and retrieving entities from a database with 100,000 entities. The result of the experiment obtained by accessing an entity given 100 times each and averaging the access times is between 70ms and 80 ms, placing such system in a not very good position. On a more mathematical field, Kohlhase and Sucan [43] detailed a process by which a set of terms is stored in a special purpose data mathematical structure where common parts of the mathematical terms are shared, so as to minimize access time and storage. On testing their implementation, they used a set of over 3,400 articles with mathematical expressions to work on, representing approximately 53,000 terms. Typical query execution times took 23 ms for instance but they also clarified that for realistic examples execution time is below 50 ms. Those times can be considered acceptable for a very specialized application which focuses just on mathematical factors.

In the area of ontology systems, a similar ontology application for multi-media retrieval with 63 individuals was developed at VicomTech Institute, Spain [44]. They had a test dataset including image and video items which queried the ontology following a MPEG-7 Compliant and using the Jena API. Such ontology has been enhanced as a reflexive ontology (RO). They took times with and without the RO and gathered an average time of 49.9 ms when using RO. This time is assumed as a better response time than without the reflexivity characteristic. On the other hand, we have found that the Ontology Builder uses an object-oriented knowledge representation model based on and compatible with the Open Knowledge Base Connectivity (OKBC) model and it was implemented to exploit the best practices from different frame-based systems. Currently, no external interfaces are exposed to enable different knowledge systems to use Ontology Builder; even though, interoperability, knowledge sharing, and reuse are important goals in the future of Ontology Builder. According to Amar et al. [45], its performance, given as the response time for the user to retrieve a frame, has an average time of about 7.5 ms. This time posits this platform in a very good position in terms of knowledge retrieval.

Table 1. Time Comparison

Case Study	# SOE	Min Time (ms)	Max Time (ms)	Avg. Time (ms)	Iterations
Geothermal Energy	2,400	5.0	3000	5.0	100
Renewable Energy	7	0.0533	2.1778	0.2341	10000
Net Income	32,561	0.001	0.562	0.2205	100

Table 2. Size Comparison

Case Study	# SOE	# Variables	# Functions	# Constraints	Total Elements
Geothermal Energy	2,400	43	0	62	105
Renewable Energy	7	47	29	23	99
Net Income	32,561	15	0	14	29

Size comparison and response times obtained by using our three case studies are shown (table1 and 2). Notice the average times obtained on each of the three cases, all of them are below the performance of the Ontology Builder, that is, below 7.5 ms.

4 Conclusions and Future Work

This paper presents three case studies for the implementation of knowledge engineering. It explains, in a practical way, cases of collecting formal decision events, that is, decisional experience from: geothermal energy, renewable energy and net income; and then, uses this information for the construction of decisional chromosomes as part of Decisional DNA.

Decisional DNA and the Set of Experience were applied as a knowledge representation structure for gathering the geothermal experience. Afterwards, they were implemented in an ontology structure with reflexivity characteristics in order to use it as a tool for decision making processes that can enhance different systems with predicting capabilities and facilitate knowledge engineering processes inside decision making. Additionally, the same knowledge structures were programmatically implemented in order to gather renewable energy and net income decisional experience. Once the Decisional DNA was constructed, the experience repository acts as an enhancing tool for different systems by adding predicting capabilities and facilitate knowledge engineering processes inside decision making.

Decisional DNA and SOEKS can indeed improve the current state of the art of Knowledge Engineering Applications from diverse perspectives. In particular in this work, we proposed a knowledge representation structure for the formal decision events that would enhance knowledge retrieval and experience reuse applications. The benefits of using the Decisional DNA are evident; however, we believe that some challenges are still open. Some of such challenges are the testing of Decisional DNA within a rule-based system and its use on multimedia applications in a way that can collect formal decision events related to images and sound.

References

[1] Feigenbaum, E., McCorduck, P.: The Fifth Generation. Addison-Wesley, Reading (1983)
[2] Asif, M., Muneer, T.: Energy supply, its demand and security issues for developed and emerging economies. Renewable and Sustainable Energy Reviews 111, 388–413 (2007)
[3] Chau, K.W.: A review on the integration of artificial intelligence into coastal modelling. Journal of Environmental Management 80, 47–57 (2006)
[4] Chau, K.W.: A review on integration of artificial intelligence into water quality modelling. Marine Pollution Bulletin 52, 726–733 (2006)
[5] Kalogirou, S.: Artificial intelligence for the modeling and control of combustion processes: a review. Progress in Energy and Combustion Science 29, 515–566 (2003)
[6] Kalogirou, S.: Artificial Intelligence in energy and renewable energy systems. Nova Publisher, New York (2007)
[7] Kyung, S.P., Soung, H.K.: Artificial intelligence approaches to determination of CNC machining parameters in manufacturing: a review. Engineering Applications of Artificial Intelligence 12, 121–134 (1998)

[8] Pavlidis, N.G., Tasoulis, D.K., Plagianakos, V.P., Vrahatis, M.N.: Computational intelligence methods for financial time series modeling. International Journal of Bifurcation and Chaos 16(7), 2053–2062 (2006)
[9] Liping, L., Shenoy, C., Shenoy, P.P.: Knowledge representation and integration for portfolio evaluation using linear belief functions. IEEE Transactions on Systems, Man and Cybernetics, Part A: Systems and Humans 36(4), 774–785 (2006)
[10] Kirkos, E., Spathis, C., Manolopoulos, Y.: Data Mining techniques for the detection of fraudulent financial statements. Expert Systems with Applications: An International Journal 32(4), 995–1003 (2007)
[11] Lee, C.H.L., Liu, A., Chen, W.S.: Pattern discovery of fuzzy time series for financial prediction. IEEE Transactions on Knowledge and Data Engineering 18(5), 613–625 (2006)
[12] Kim, K.J.: Artificial neural networks with evolutionary instance selection for financial forecasting. Expert Systems with Applications 30(3), 519–526 (2006)
[13] Sanin, C., Szczerbicki, E.: Experience-based Knowledge Representation SOEKS. Cybernetics and Systems 40(2), 99–122 (2009)
[14] Drucker, P.: The Post-Capitalist Executive: Managing in a Time of Great Change. Penguin, New York (1995)
[15] Noble, D.: Distributed situation assessment. In: Arabnia, P.H.R. (ed.) FUSION 1998, pp. 478–485. University of Georgia (1998)
[16] Deveau, D.: No brain, no gain: Knowledge management. Computing Canada 28, 14–15 (2002)
[17] Ferruci, D., Lally, A.: Building an example application with the unstructured information management architecture. IBM Systems Journal 43(2), 455–475 (2004)
[18] Sanin, C., Szczerbicki, E.: Knowledge supply chain system: A conceptual model. In: Szuwarzynski, A. (ed.) Knowledge management: Selected issues, Gdansk, Poland, pp. 79–97. University Press (2004)
[19] Awad, E., Ghaziri, H.: Knowledge management. Prentice Hall, Englewood Cliffs (2004)
[20] Nonaka, I., Takeuchi, H.: The knowledge-creating company: How Japanese companies create the dynamics of innovation. Oxford University Press, New York (1995)
[21] Levesque, H.: Knowledge representation and reasoning. Annual Review of Computer Science 1, 255–287 (1986)
[22] Sowa, J.F.: Preface to knowledge representation, `http://www.jfsowa.com/krbook/krpref.htm`
[23] Arnold, W., Bowie, J.: Artificial Intelligence: A Personal Commonsense Journey. Prentice Hall, New Jersey (1985)
[24] Sanin, C., Szczerbicki, E.: Extending set of experience knowledge structure into a transportable language extensible markup language. International Journal of Cybernetics and Systems 37(2-3), 97–117 (2006)
[25] Sanin, C., Toro, C., Szczerbicki, E.: An OWL ontology of set of experience knowledge structure. Journal of Universal Computer Science 13(2), 209–223 (2007)
[26] Lloyd, J.W.: Logic for learning: Learning comprehensible theories from structure data. Springer, Berlin (2003)
[27] Malhotra, Y.: From information management to knowledge management: Beyond the 'hi-tech hidebound' systems. In: Srikantaiah, K., Koening, M.E.D. (eds.) Knowledge management for the information professional, Information Today Inc., New Jersey, pp. 37–61 (2000)
[28] Goldratt, E.M., Cox, J.: The Goal. Grover, Aldershot (1986)

[29] Gruber, T.R.: Toward Principles for the Design of Ontologies Used for Knowledge Sharing. International Journal of Human-Computer Studies 43(5-6), 907–928 (1995)
[30] Antoniou, G., Harmelen, F.V.: Web ontology language: OWL. In: Handbook on Ontologies in Information Systems, pp. 67–92. Springer, Heidelberg (2003)
[31] Sevilmis, N., Stork, A., Smithers, T., et al.: Knowledge Sharing by Information Retrieval in the Semantic Web. In: Gómez-Pérez, A., Euzenat, J. (eds.) ESWC 2005. LNCS, vol. 3532, pp. 471–485. Springer, Heidelberg (2005)
[32] Toro, C., Sanín, C., Szczerbicki, E., Posada, J.: Reflexive Ontologies: Enhancing Ontologies with self-contained queries. Cybernetics and Systems: An International Journal 39, 1–19 (2008)
[33] Blakeslee, S.: Lost on Earth: Wealth of Data Found in Space. New York Times, C1, March 20 (1990)
[34] Corti, L., Backhouse, G.: Acquiring qualitative data for secondary analysis. Forum: Qualitative Social Research 6, 2 (2005)
[35] Humphrey, C.: Preserving research data: A time for action. In: Preservation of electronic records: new knowledge and decision-making: postprints of a conference - symposium 2003, pp. 83–89. Canadian Conservation Institute, Ottawa (2004)
[36] Johnson, P.: Who you gonna call? Technicalities 10(4), 6–8 (1990)
[37] Sanin, C., Szczerbicki, E.: Extending Set of Experience Knowledge Structure into a Transportable Language XML (eXtensible Markup Language). Cybernetics and Systems 37(2), 97–117 (2006)
[38] Zhang, Z.: Ontology query languages for the semantic Web. Master's thesis. University of Georgia, Athens (2005)
[39] Energy Information Administration. Official Energy Statistics from the US Government, http://www.eia.doe.gov/oiaf/servicerpt/stimulus/aeostim.html
[40] UCI Machine Learning Repository. Adult Data Set, http://archive.ics.uci.edu/ml/datasets/Adult
[41] Reiner, B., Hahn, K.: Optimized Management of Large-Scale Data Sets Stored on Tertiary Storage Systems. IEEE Distributed Systems Online 5(5), 1–8 (2004)
[42] Chen, Y.J., Chen, Y.M., Chu, H.C., Kao, H.Y.: On technology for functional requirement-based reference design retrieval in engineering knowledge management. Decision Support Systems 44, 798–816 (2008)
[43] Kohlhase, M., Sucan, I.: A Search Engine for Mathematical Formulae. In: Calmet, J., Ida, T., Wang, D. (eds.) AISC 2006. LNCS (LNAI), vol. 4120, pp. 241–253. Springer, Heidelberg (2006)
[44] Cobos, Y., Toro, C., Sarasua, C., Vaquero, J., Linaza, M.T., Posada, J.: An Architecture for Fast Semantic Retrieval in the Film Heritage Domain. In: 6th International Workshop on Content-Based Multimedia Indexing (CBMI), London, UK, pp. 272–279 (2008)
[45] Amar, K.D., Wei, W., McGuinness, D.L.: Industrial Strength Ontology Management. In: Cruz, I., et al. (eds.) The Emerging Semantic Web, vol. 75, pp. 101–118. IOS Press, Amsterdam (2002)

Intelligent-Based Mobility Models for Data Management in Wireless Sensor Networks

Samer Hanoun and Saeid Nahavandi (SMIEEE)

Centre for Intelligent Systems Research
Deakin University
Australia
{Samer.Hanoun,Saeid.Nahavandi}@deakin.edu.au

Abstract. Wireless sensor networks (WSNs) are proposed as powerful means for fine grained monitoring in different classes of applications at very low cost and for extended periods of time. Among various solutions, supporting WSNs with intelligent mobile platforms for handling the data management, proved its benefits towards extending the network lifetime and enhancing its performance. The mobility model applied highly affects the data latency in the network as well as the sensors' energy consumption levels. Intelligent-based models taking into consideration the network runtime conditions are adopted to overcome such problems. In this chapter, existing proposals that use intelligent mobility for managing the data in WSNs are surveyed. Different classifications are presented through the chapter to give a complete view on the solutions lying in this domain. Furthermore, these models are compared considering various metrics and design goals.

1 Introduction

Wireless sensor networks (WSN) are composed of large numbers of small sensing self-powered nodes which are densely deployed either inside the phenomenon or very close to it [1], [2]. The capabilities of these inexpensive, low-power communication devices ensure serving in a wide range of application domains. Environmental monitoring, healthcare applications, surveillance systems, military applications, agriculture, and commercial applications are only few examples [3]. Optimising energy consumption for extending the network lifetime is one of the most important requirements for many WSN applications. Since energy resources are scarce and battery replacement is not an option for networks with thousands of physically embedded sensor nodes, energy optimisation for individual nodes is required as well as the entire network. Exploiting the mobility of some of the network components has been recently observed to be among the most promising ways of improving the performance of WSNs in terms of crucial metrics such as its lifetime and data latency. Mobile elements act as mechanical data carriers, taking the advantage of mobility capacity [4] and physically approaching the sensors for collecting their data. This approach trades data delivery latency for the reduction of energy consumption of sensors; however, it shows remarkable improvement in the network lifetime.

N.T. Nguyen and E. Szczerbicki (Eds.): Intel. Sys. for Know. Management, SCI 252, pp. 87–106.
springerlink.com

This chapter surveys existing work in the area of mobility in wireless sensor networks. Different classifications are presented through the chapter to give the reader a complete view on the solutions lying in this domain. Section 2 presents the types of mobility regarding which of the entities in the network are mobile. In Section 3, the purposeful solutions acting on enhancing the network performance in terms of reducing the energy consumption and extending the lifetime are surveyed. The formulation of the problem of designing the mobile collector route is detailed in Section 4 and a proposed heuristic based solution in Section 5. The experimental methodology and results are discussed in Section 6. Finally, the chapter concludes in Section 7.

2 Types of Mobility

A sensor network is composed of a set of sensor nodes and a central base station which acts as the gateway to interact with the network. Based on these two types of entities active in the network, various combinations are possible depending on which of these entities are mobile. Figure 1 presents these combinations.

1. *Static sensor nodes and static base station*: in this case there is no motion neither among the sensors nor the base station. An example is a group of sensors spread for temperature sensing. For these types of sensor networks, several routing protocols have shown [5], [6], [7] the ability to create a path between the sensors and the base station for relaying the sensory information and maintain this path under dynamic conditions while meeting the application requirements of low energy, low latency, high accuracy, and fault tolerance. Also, several studies have shown that localised algorithms [8], [9] can be used in an effective way as sensors communicate with nodes in their locality and an elected node relays a summary of the local observations to the base station, perhaps through one on more levels of hierarchy. Such algorithms extend the lifetime of the sensor network because they trade-off local computation for communication [8].
2. *Static sensor nodes and mobile base station*: this is the most common scenario in highly partitioned wireless ad hoc networks. Sensors are sparsely deployed; therefore, communication links may not exist as sensors are apart with distances greater than their radio range. Communication can be achieved when the sensor nodes are in the vicinity of the mobile base station. The Message Ferrying scheme [10] employs the mobility of the base station to act as a relay taking responsibility of carrying messages between disconnected nodes. Ferries move around the deployed area according to known routes, collect messages from the sending nodes and deliver messages to their destinations or other ferries. Also in [11], a mobile router is employed to act as an intermediate mobile base station to collect sensors data for later delivery to the user. The same ideas are applied in [12] where vehicles are used to transport data between villages and cities using a store and forward mechanism.

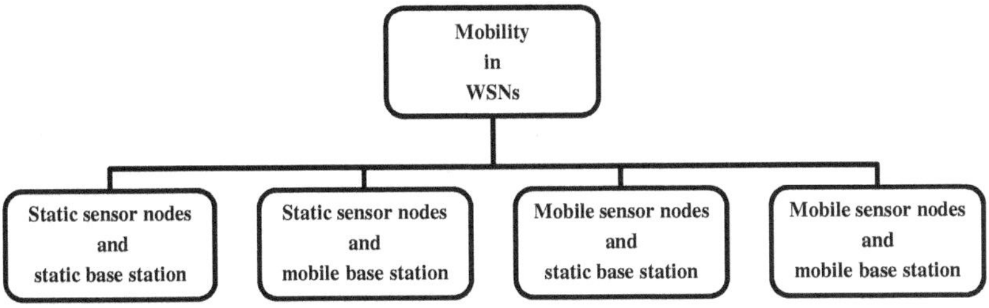

Fig. 1. Mobility in WSNs: a classification

3. *Mobile sensor nodes and static base station*: here, sensor nodes are attached to moving entities and for the information to reach the base station, the mobile sensor node should come in its vicinity. An example of such a system is the Shared Wireless Infostation Model (SWIM) [13]. Data collected by sensors on whales is replicated when two sensors are in the vicinity of each other and ultimately relayed to a small number of static on-shore base-stations when the whales come to the surface. The Infostation architecture provides strong radio signal quality to small disjoint geographical areas and, as a result, offers very high rates to users in these areas. However, due to the lack of continuous coverage, this high data rate comes at the expense of providing intermittent connectivity only.
4. *Mobile sensor nodes and mobile base station*: this is similar to the above, but the base station collecting the sensory information is mobile. In the ZebraNet project [14], data sensed by sensors attached to zebras is collected by humans as they drive by in a vehicle or by a low flying aircraft acting as the base station. When either a vehicle or the aircraft is in the vicinity of the sensor, it uploads its data directly. The project aims to study and learn the mobility patterns of the zebras.

Accordingly, the application itself may influence the behaviour of the sensors and the base station. An example would be the problem of monitoring a tornado. One option would be to fly airplanes to sense the tornado. In such a case, the sensors are mobile while the base station is either static and fixed on the ground, or mobile following the tornado. Another would be to have a sensor grid statically placed on the ground that reports data as the tornado passes through. Thus, both the sensors and the base station are static. Yet another would be to release lightweight sensors into the tornado, therefore the sensors are mobile and the base station may be static or mobile.

3 Purposeful Mobility

The second type of classification scheme presented in the previous section considers sensor networks with static sensor nodes and mobile base station[1]. The

[1] The terms sink, collector, base station and element will be used interchangeably in the chapter.

mobility of the sink is employed in several ways to enhance the network performance and extend its lifetime. Further classification can be made based on the purpose of moving the sink. One approach considers moving the sink for the purpose of balancing the energy consumption among the sensors in the network, while another employs sink mobility for the purpose of data collection and enhancing the network connectivity.

3.1 Mobility for Distributing Energy Consumption

One major problem exists with static sensor networks when using multihop routing for relaying the sensors data to the sink. As the sink is static, the nodes around the sink become hotspots and die faster than other sensors as they have to relay huge amounts of data for other sensors. As a result, these nodes become bottleneck nodes and their lifetime upper bounds the lifetime of the network. Relocating the sink from time to time [15], [16], [17], [18], [19], [20], [21], [22] can distribute over time the role of bottleneck nodes and thus even out the load and balance the energy consumption among all sensors in the network.

In [15], the authors investigate the potential of sink (gateway) repositioning for enhancing the network performance in terms of energy, delay and throughput. They address issues related to when should the sink be relocated, where it would be moved to and how to handle its motion without negative effect on data traffic. Energy metrics are considered to motivate and justify the advantage of sink relocation. The approach checks the traffic density of the nodes that are one-hop away from the gateway and their proximity to the gateway. Once the total transmission power for such nodes is guaranteed to be reduced more than a certain threshold and the overhead of moving the sink is tolerable, the sink starts to move to the new position. The results show that such repositioning of the sink increases the average lifetime of the nodes by decreasing the average energy consumed per packet. Moreover, the network throughput is positively impacted.

The authors in [16] present heuristics for controlling the sink movements. The Greedy Maximum Residual Energy (GMRE) heuristic moves the sink from its current location to a new site as if drawn towards the area where nodes have the highest residual energy. This work demonstrates that controlling sink mobility is effective in prolonging lifetime of the network up to six times than when the sink does not move. Similarly, [23], [17], [24] present a strategy for moving the sink proactively towards the nodes that have the most residual energy to balance energy consumption among all sensor nodes in the network. The authors describe the sink as an energy mower that regards the residual energy of the sensor nodes as an uneven grassy lawn and tries to make it smooth by cutting the tallest grass. It is shown that the proposed moving strategy can extend network lifetime notably and provide better adaptability to irregular-shaped networks.

Joint mobility and routing for elongating the lifetime of the network is presented in [20]. The sink moves along the periphery of the network and routing towards the sink is based on a combination of round routes and short paths. Different sink mobility strategies are investigated in this work and routing is fine-tuned to leverage on the trajectory of the base station; particularly, the nodes located at the periphery of the network. Linear programming formulation for the joint problems of determining the movement of the sink and the sojourn time at different points in the network is presented in [19], [22]. The objective acts on maximising the overall network lifetime rather than minimizing the energy consumption at the nodes. The authors verify that the sink movement results in a fair balancing of the energy depletion among the network nodes especially those around the sink.

3.2 Mobility for Data Collection

Sink mobility can be utilised in a different way rather than moving to balance the energy consumption among the sensors in the network. As the hotspot problem is related to relaying high amounts of data by the nodes around the sink, the sink eliminates this problem by approaching the sensors and collecting their data buffers through single hop communication rather than relaying the buffers using multihop routing. In such cases, data delivery latency is traded for the reduction of energy consumption of sensors [4], [25], however, there is high dependency on the mobility regime employed by the sink. The mobility regime can be classified based on how the sink moves. Figure 2 shows the types of data collection mobility regimes.

1. *Random mobility regime*: in this case, the base station is mounted on entities moving randomly by nature in the network deployment area. For example, in [26], [27] random moving humans and animals act as "data mules", collect data opportunistically from sensor nodes when entering their communication ranges. The concept of a Mobile Ubiquitous LAN Extension (MULE) stems from the idea of deploying mobile agents for carrying information [28], especially among nodes in possibly disconnected networks. MULEs are resource-rich nodes that roam freely throughout the network. When, as a result of its motion, a MULE gets to be in the radio proximity of a sensor, it receives all the packets generated by that sensor so far (if any). Communication in the MULEs architecture is single-hop: a node stores the sensed data until a MULE passes by. Once the MULE arrives, the node transmits all the stored packets to the MULE. Given the limited storage capacity of the sensor nodes, chances are that if a MULE does not pass by, some of the packets have to be discarded. When the MULE passes by the central sink it transmits to it all the packets it has collected. This end-to-end routing is performed using a Seek and Focus strategy [29].

   Assuming all nodes are mobile, random mobility has shown to improve data capacity [4], and similar results were also shown to hold for mobility constrained to a single dimension [30]. However, in all cases, the worst-case latency of data delivery cannot be bounded. This unbounded latency may lead to excessive data caching at MULE, resulting in buffer overflows and data in transit may have to be dropped before being delivered to the destination, making it harder to provide transport layer reliability.

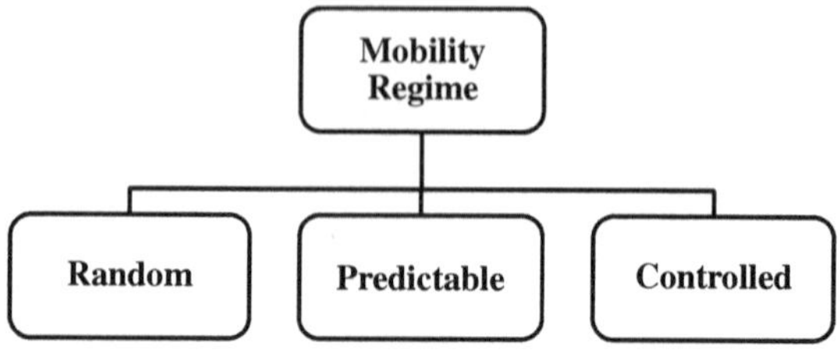

Fig. 2. Data collection mobility regimes

2. *Predictable mobility regime*: here, there is no control over the motion of the mobile entity; however, its motion can be predicted. This approach is investigated in [31], where a network access point is mounted on a public transportation bus moving with a periodic schedule. The sensor nodes learn the times at which they have connectivity with the bus, and wake up accordingly to transfer their data. It is assumed that the mobile bus comes within direct radio range of all the static nodes, thus in practice, this requires that the trajectory of the bus is efficiently designed for data collection purposes. A queuing model is introduced to accurately model the data collection process. Using this queuing system model, the success rate of data collection and power consumption are analysed. It is observed that exploiting predictable mobility can help save energy in WSNs.

   Another example of this scheme is presented in [32]. A mobile sink traverses the network, where the static sensor nodes (called moles), that are one hop from the path of the mobile sink, learn its movement pattern over time and statistically characterise it as a probability distribution function. Whenever a sensor node needs to send data, it sends towards the mole, which should be in the vicinity of the mobile sink. The scheme uses reinforcement learning to locate the mobile sink efficiently at any point of time.

   Also in [33], a kind of mobility is presented that models situations where the mobile sink can not execute complex movements. Movement on the terrain is possible only in certain areas or the mobile sink does not have enough energy to traverse the whole network area. The sink moves in a linear trajectory consisting of a horizontal or vertical line segment passing through the centre of the network and the sensors predict the sink locations over time and use multihop routing for delivering data to the sink.

3. *Controlled mobility regime*: this is the case where the motion of the sink is controlled and programmed to achieve certain tasks in the network. Topology adaptivity, increasing network capacity, fault isolation, and prolonging the network lifetime while containing packet end-to-end delay [34], [35] are among the factors that favour the controlled mobility regime.

   In this regime, the motion strategy of the mobile sink is formulated as a scheduling problem. The schedule arranges visits to sensor nodes such that to avoid overflow in the sensor's buffer. In the case of generating an offline schedule, all possible rearrangements of the visits are explored to find an existing schedule that satisfies no buffer overflow. However, it is not the case when working online, as some sensor nodes may have buffer overflow, but the objective is to minimise the overall buffers overflow in the network.

Three different schedules, Round-Robin, Rate-Based, and Min Movement [36] are used by the mobile element to collect data from cluster head nodes in the network. In the Round Robin schedule, the collector visits each cluster head to collect data in a round robin manner, while in the Data Rate based schedule, the collector visits the cluster heads preferentially with a frequency proportional to the aggregate data rate from all the nodes in the cluster. In the third schedule, the Min Movement schedule, the collector visits the cluster heads in the proportion of the aggregate data rate, but also with the goal of optimising the distance traversed. The first two schedules outperform in minimising the data delivery latency and the third minimises the energy expended by the sensors.

In [37], an offline heuristic called Partitioning Based Scheduling (PBS) is presented to compute the trajectory of the mobile element based on the knowledge of sensors' data aggregation rates and locations. Sensor nodes are partitioned into several groups, called Bins, such that nodes in the same bin have similar deadlines and are geographically close to each other. The solution to the Travelling Salesman Problem is used to produce a schedule for each bin, then the schedules for individual groups are concatenated to form the entire schedule. The speed of the mobile element is studied to guarantee no buffer overflow occurs along the computed path.

Also in [38], the mobile element schedule is formed based on minimum spanning tree construction. A mathematical formulation of the route design problem is presented to show that the general problem is computationally intractable. The sensors are assigned with different weights according to their types and importance of data, where the main objective is to minimise the weighted mobile element inter-arrival time among all sensors.

In [39], the authors showed that the mobile element schedule in wireless sensor networks is NP-complete and proposed three heuristic algorithms. The first one is the Earliest Deadline First (EDF) algorithm, where the node with the closest deadline is visited first. To improve EDF, the second algorithm, EDF with k-lookahead, is proposed. Instead of visiting a node whose deadline is the earliest, this algorithm considers the k! permutations of the k nodes with smallest deadlines, and chooses the next node which leads to the earliest finish time. The third algorithm is the Minimum Weight Sum First (MWSF) algorithm, which accounts for the weights of deadlines as well as distances between nodes in determining the visiting schedule. The MWSF algorithm performs the best among the three proposed algorithms, however, the new deadline for the node's future visit is updated, once it is visited and depends mainly on knowing the node's buffer size and sensing rate to compute its next overflow deadline.

3.3 Mobility for Enhancing Network Functionality

It is important to note that the mobility of network nodes can help to improve the performance of wireless ad hoc and sensor networks in other aspects rather than network lifetime. There are some more works that do not fall into the classifications mentioned in the previous sections.

In their very recent work [21], a mobile node is proposed as an approach to prolong network lifetime. The basic idea is that a few powerful mobile nodes can be deployed to replace different (heavily loaded) static nodes. In their approach, a mobile node inherits the responsibilities of a static node with which it is co-located, such that the static node can shut down for energy saving. In [40], node mobility is considered by moving the intermediate nodes along a route, so that the distances between nodes are minimised and lower energy is used to transmit over a shorter range. Also, node mobility facilitates sensor deployment [41]; movement-assisted protocols can re-shape a WSN to improve the network coverage. Additionally, moving sensor nodes reduces sensing uncertainty [42]; for achieving the same level of sensing fidelity, it can be much cheaper to deploy a few mobility capable nodes instead of a large number of static nodes. Finally, using mobile beacons can facilitate the positioning of network nodes in their deployment areas [43].

4 Mobile Collector Route Design Problem

This section clearly models our problem and outlines all assumptions. A formal definition of the problem is presented to give an insight into the problem objectives. The problem is modelled as follows:

- The sensor network is modelled as a fully connected graph $G(V, E)$ of n nodes, where every edge (s_i, s_j) has an associated cost $c(i, j)$, and all the costs form a matrix $C = [c(i, j)]_{i,j=1}^{n}$.
- Sensor nodes remain stationary and are deployed uniformly at random in the sensing field where the phenomenon of interest is to be monitored. The sensing field dimension is $A = L * L$ (m^2).
- Each sensor node has a deployment location *(X, Y)*, buffer size of K bytes, radio communication radius R in metres and generates a sample every T_s in seconds.
- The time required for the buffer to be full is $\geq K * T_s$, where the next time the buffer becomes full is unknown and differs from time to time. This is modelled by generating a random number between 0.0 and 1.0 to specify whether the current sample is qualified to be added to the buffer or not. If the random number is greater than a pre-specified threshold φ then the sample is added to the buffer, otherwise, it is discarded. This ensures that the time the sensor's buffer becomes full is variable which reflects different network operation modes (i.e. periodic sensing, event driven, query based). This also simulates the presence of data aggregation algorithms operating locally on the sensor for reducing the redundancy in the sensor measurements which

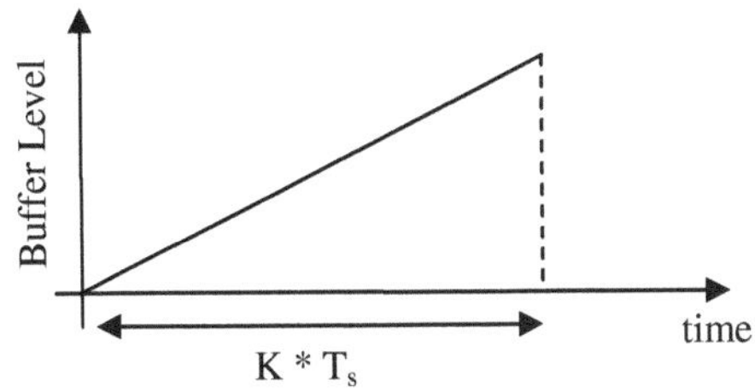

(a) Fixed with time

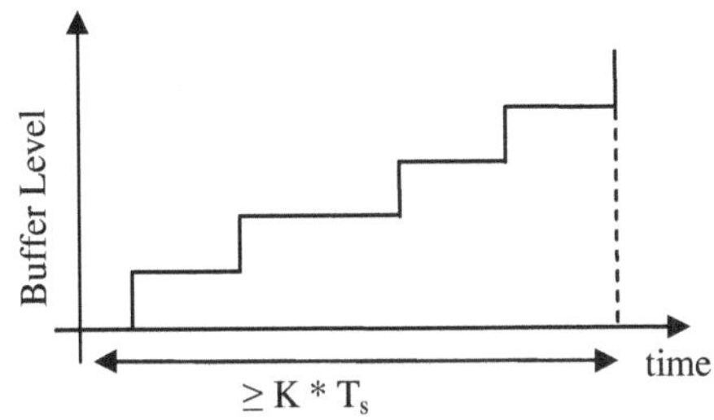

(b) Variable with time

Fig. 3. The sensor buffer level accumulation rate

alters the buffer overflow time. Figure 3 shows the buffer level accumulation with time for different models.

- Once the sensor's buffer is full, the sensor disseminates a collection request Q with fields $\{ID, X, Y\}$ and goes to an idle state and sleeps waiting for the arrival of the collector. While idle, the sensor does not sense any new samples but it relays collection requests sent by others.
- The mobile collector starts at a central position in the sensing field, has reasonably high amount of energy that can last beyond the network lifetime, and its memory size can accommodate the data generated by the sensors during the network operational time. The mobile collector moves with a speed $v \leq S_{max}$ and it can change its speed during the data collection operation.

The following assumptions are made:

- At time t = 0 all the buffers of the sensor nodes start filling up.
- The actual data transfer time from the sensor node to the mobile element is negligible.

The **Mobile Collector Route Design (MCRD)** problem is the problem of finding a sequence of visits to nodes requesting the collection of their data buffers to minimise the collection time required. Once a node is visited, its buffer is transferred to the mobile collector internal memory and its normal sensing operation is resumed. This is mathematically formulated as follows:

N: is the set of sensor nodes in the network, where $N = \{s_1, s_2, \ldots, s_n\}$,
S: is the set of sensor sites, i.e. deployment locations of the sensor nodes, where $S = \{(x_1, y_1), \ldots, (x_n, y_n)\}$,
d_{ij}: Euclidean distance (meters) between sensor sites, where $i, j \in S$,
R: is the set of requests to be serviced, where $R = \{r_1, r_2, \ldots, r_m\}$,
A_k: is the time taken for servicing request k, where $k \in R$, i.e. time taken for collecting request k generated by sensor i, where $k \in R, i \in N$,
v: is the moving speed of the mobile collector in m/sec,

Mobile Collector Route Design (MCRD):

$$\min T = \sum_{\forall k} A_k \tag{1}$$

where T is the overall route period, and each A_k is computed based on the distance cost between requests k and *(k+1)* and the speed selected by the mobile collector for servicing this request.

The problem of finding the optimum order of arranging the requests to form the least cost route is *NP*-complete. The way is to simply enumerate each possible route, and pick the one with minimum total cost. Having m requests to service, gives $(m-1)!$ possible routes. This leads to an *O(m!)* algorithm to produce the optimum route, which is not efficient to use.

5 Dynamic Insertion Route Construction (DI-R)

The problem of designing the mobile collector route is NP-complete. This section presents a heuristic based algorithm for designing the route to be followed by the mobile collector for collecting the data buffers of the sensor nodes requesting the collection service. An heuristic algorithm provides feasible solution to an optimisation problem, which may or may not be optimal. Good heuristics give solutions that are close to the optimal solution, and usually are efficient in terms of theoretical or practical running time. A desirable property of an heuristic is that the worst-case quality of the solution provided can be guaranteed. For example, some heuristics guarantee that the solution they identify is at worst η-optimal; that is, if the optimal cost is C*, the heuristic finds a solution no greater than ηC* for minimisation problems. In our problem, the route constructed by ordering the requests based on a timely manner according to their arrival is considered as an upper bound on the solution provided by the heuristic.

An algorithm can be designed which gives weights to the distance cost the mobile collector travels and the time which the sensor remains sleeping until the collection is done. Let $r_1, r_2, \ldots, r_m$ be the requests to be collected by the mobile collector on its route R, and, C_{jk} is the extra cost of placing request k after request j on R, then it is required to:

Algorithm 1. Dynamic Insertion Route Construction Algorithm (DI-R)

Input:

Request (r) with fields $\{ID, X, Y\}$

Define:

R: current route consisting of k requests;

m_p: mobile collector position;

S: sensor sleeping time;

r_f , r_l : first and last requests in R;

$C_b^a = \sqrt{(X_a - X_b)^2 + (Y_a - Y_b)^2}$

Initialize:

Cost[1..(k+1)] = 0

Body:

if (R = ∅) **then**

  $R = r$

else

  $S_1 = C_{m_p}^{r_f}$

  Cost[1] $= \alpha * (C_r^{m_p} + C_{r_f}^r - C_{r_f}^{m_p}) + (1 - \alpha) * S_1$

  Repeat for $i = 2$ *to* k

    $S_i = S_{i-1} + C_{k_i}^{k_{i-1}}$

    Cost[i] $= \alpha * (C_r^{k_i} + C_{k_{i-1}}^r - C_{k_i}^{k_{i-1}}) + (1 - \alpha) * S_i$

  end for

  $S_{k+1} = S_k + C_{r_l}^r$

  Cost[k+1] $= \alpha * C_{r_l}^r + (1 - \alpha) * S_{k+1}$

  Set j = index of minimum in Cost[1..(k+1)]

  Insert r in R at position j

end if

End Body

$$\min \sum_{j=1}^{m} C_{jk} * X_{jk} \tag{2}$$

where $X_{jk} = 1$, if the placement of request k follows request j; 0 otherwise, and

$$\sum_{j=1}^{m} X_{jk} = 1, \quad \text{for any k} \tag{3}$$

The insertion heuristic I1 equations [44] are applied to each collection request for dynamically inserting it in its minimum insertion position on the collection route.

Table 1. Effect of α on the Insertion Point

$\alpha \in [0,1]$	Result
1	Insertion point is determined only by the distance cost.
0	Insertion point is determined only by the sensor sleeping time.
> 0.5	Distance cost contributes higher in the insertion point.
≤ 0.5	Sensor sleeping time contributes higher in the insertion point.

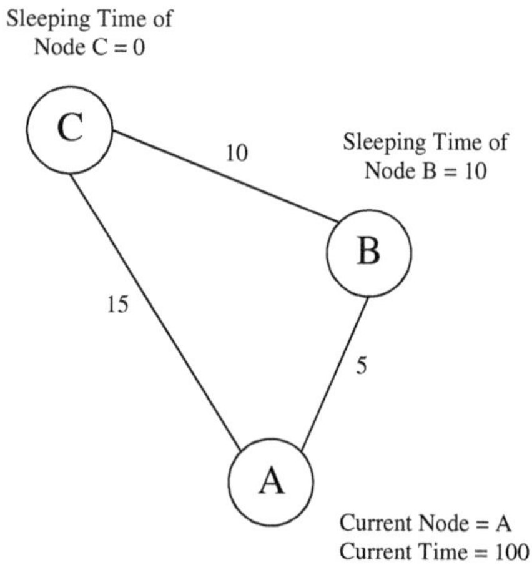

Fig. 4. An example to illustrate the effect of α on the mobile collector route

Algorithm 1 presents the procedure of finding the best insertion point on the mobile collector route for the current received collection request. The sensor sleeping time S is based on the point of insertion along the collection route and is not computed absolute, but relative to the current time. Table 1 shows the effect of different α values. To illustrate the contents of the table, consider the scenario presented in Figure 4. Suppose the mobile collector just serviced node A and the next node on its route is node B. Before initiating the service to node B, a collection request is received from node C. Nodes B and C have distance costs 5 and 15 respectively, and the sleeping time of node B is 10 time units. The insertion position of node C is either before node B or after node B. When $\alpha \leq 0.5$, node C is inserted before node B and when $\alpha > 0.5$, node C will be inserted after node B. When $\alpha = 0$, the mobile collector will always favour newly arriving requests, which can result in high delays in servicing older requests, causing those nodes to wait longer times until they are serviced.

The DI-R construction algorithm requires an $O(m)$ to compute the cost of insertion along the current route and an $O(\log n)$ to find the minimum insertion point, resulting in an $O(m)$ running time complexity. It should be noted that the DI-R algorithm works totally online and adapts the current route according to each received request.

6 Experimental Methodology and Results

The previous section described a heuristic based algorithm for constructing the route of the mobile collector. This section presents the evaluation of the algorithms through simulation.

6.1 Methodology

As the parameter space is huge, some parameters are fixed as follows:

- Sensing Field: a square area of dimension 100 x 100 m^2 is considered for sensors deployment. A number of sensor nodes varying from 50 to 100 are distributed uniformly at random, preserving homogenous node density among all areas in the sensing field.
- Sensor Parameters: each sensor has a radio communication radius of 25m and a buffer of size 1 Kbyte.
- Mobile Collector Parameters: initially the mobile collector is localised at the centre of the sensing field. The mobile collector moves with a fixed speed of 1m/s, and has a communication radius of 50m. The mobile collector achieves communication with the sensor node when it is in the sensor's radio communication range.
- Simulation Time: each simulation run lasts for 50000 time units, and all results are averaged over 20 different independent networks.
- Sensors Sampling Frequency: the sampling frequency determines the time required for the sensor's buffer to become full and therefore controls the distribution of the collection requests over the simulation time. The first option assumes that events occur regarding some point of interest located at the centre of the sensing field. In such a case, a higher number of requests come from a specific part of the network, as the sensors closer to the centre sample more frequently. The requests distribution in this case follow Gaussian distribution with small variance. A concentric topology for the sensors' sampling rates, shown in Figure 5, is used to model this case. A sequence of n concentric circles divides the sensing field into several ring shaped regions; Region 1 to Region n. The radius of each concentric circle is denoted by $R_1, R_2, R_3, \ldots, R_n$, where $R_1 = 10m$. The value of each radius is calculated as:

$$R_i = i * R_1, \quad i = \{1, \ldots, n\} \tag{4}$$

where

$$n = \frac{L}{2 * R_1} + 1 \tag{5}$$

The sensors in the innermost region are assigned sensing rates in the range [1, *baserate*] and the sensing rates of sensors in regions radically outwards are calculated as:

$$Sensingrate_i = \quad [\, baserate * (i-1) + 1 \,,\; baserate * i \,], \quad i = \{1, \ldots, n\} \tag{6}$$

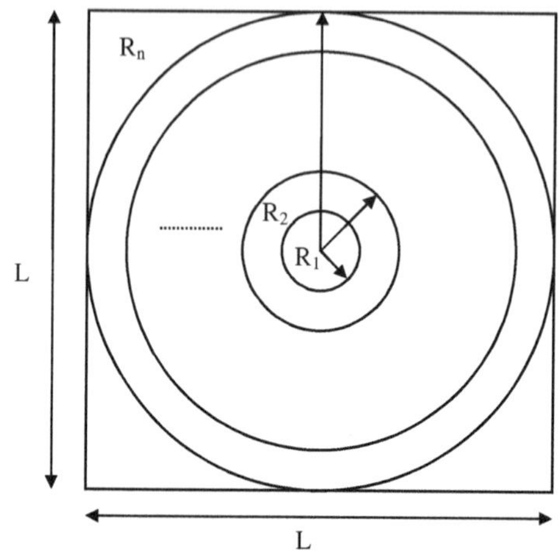

Fig. 5. Concentric Topology: the first type of topology considered in the simulation

Table 2. Experiments: Set I

Label	*Sensing Rate Topology*	*Number of Sensors*	*Baserate*
A1	Concentric	50	2 sec
A2	Concentric	75	3 sec
A3	Concentric	100	4 sec
B1	Random	50	2 sec
B2	Random	75	3 sec
B3	Random	100	4 sec

When the *baserate* is chosen to be 2 seconds, the sensors in region R_2 will have a sensing rate in the interval [3,4] which means that some sense a sample every 3 seconds and others sense a sample every 4 seconds.

The second option assumes random occurrence of events independently from one another in the sensing field. In this case, the requests distribution follow Poisson distribution. The sensing rate of each sensor node is chosen randomly in the interval $[1, baserate * n]$.

Putting everything together, the experiments shown in Table 2 are run. The experiments are labelled for referencing purposes. For each of these, results are presented based on an average of 20 runs. The parameters that changed from run to run are the location of the sensor nodes, and correspondingly, the distance cost and the sensing rates.

For the purpose of evaluation, the following performance metrics are considered:

- **Data Collection Time:** this is defined as the period from the time the sensor sends the collection request to the time of arrival of the mobile collector to collect the sensor's buffer. This is averaged across all the nodes in the network.

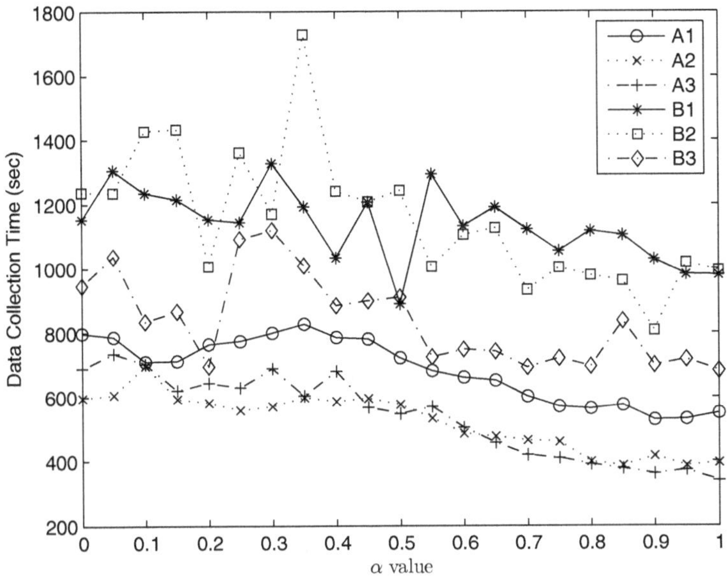

Fig. 6. Data Collection Time of DI-R for experiments A1-A3 and B1-B3

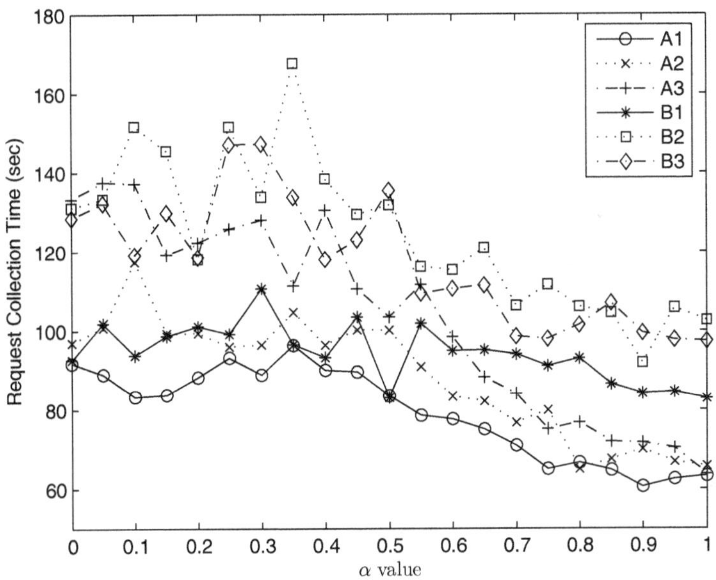

Fig. 7. Request Collection Time of DI-R for experiments A1-A3 and B1-B3

- **Request Collection Time:** this is computed as the overall data collection time to the number of requests collected from all nodes in the network.
- **Sleeping Sensors Percentage:** this measure is calculated as the number of sleeping sensors awaiting the arrival of the mobile collector to the total number of sensors in the network over the simulation time of the experiment.
- **Distance Ratio:** this is defined as the ratio of the distance travelled by the mobile collector to the number of requests collected.

6.2 Results

Figure 6 and 7 show the result of running the DI-R construction algorithm on the set of experiments described in Table 2. α ranged from 0 to 1 with steps of 0.5. The data collection time and the request collection time decrease as α goes near to 1. This is because the request is inserted on the mobile collector route in the position resulting in the minimum extra travelled distance, as explained in Table 1. Therefore, the forth and back travelling between nodes is minimised which results in less waiting time for the sensors, thus less sleeping sensors as in Figure 9. The performance on experiments A1-A3 is better than B1-B3. This is obvious because most of the events come from a specific part of the network which helps the mobile collector to optimise its collection route more than when the events are scattered as in experiments B1-B3. This also appears on the distance ratio of the mobile collector shown in Figure 8. As events are scattered randomly over the sensing field, the mobile collector is required to travel further for the collection which appears on the distance ratio for experiments B1-B3. It is hard

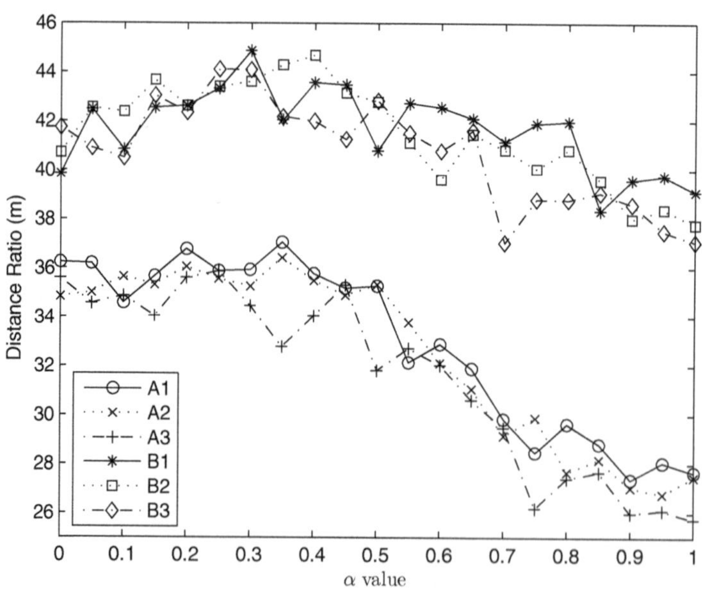

Fig. 8. Mobile Collector Distance Ratio of DI-R for experiments A1-A3 and B1-B3

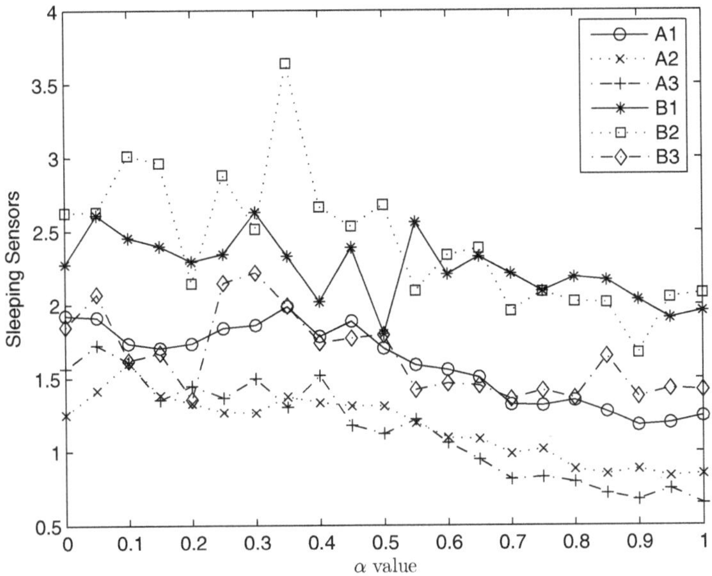

Fig. 9. Percentage of Sleeping Sensors of DI-R for experiments A1-A3 and B1-B3

to conclude about the dependence of α value on the results, however, α around 0.9 leads to minimum results for the performance metrics described.

7 Conclusion

This chapter presents different classifications for the mobility models used for data management in wireless sensor networks. The types of mobility are highly depend on which of the network entities are mobile. Different mobility regimes are adopted for static sensor nodes and mobile based station; however, the controlled regime is the most effective one in optimising the network performance. A heuristic based algorithm is presented to show this advantage. There are many directions in which this work may be pursued further. Statistical measures are required to measure the buffer filling rate and so the sensor can send its collection request before its buffer is full, giving an extra advantage to the mobile collector. Also, applying multiple mobile collectors can enhance the performance of the algorithm with efficient control schemes for coordination.

References

1. Akyildiz, I.F., Su, W., Sankarasubramaniam, Y., Cayirci, E.: Wireless sensor networks: A survey. Computer Networks 38, 393–422 (2002)
2. Culler, D., Estrin, D., Srivastava, M.: Introduction: Overview of sensor networks. IEEE Computer 37, 41–49 (2004)

3. Estrin, D., Govindan, R., Heidemann, J., Kumar, S.: Next century challenges: Scalable coordination in sensor networks. In: Proceedings of the 5th Annual ACM/IEEE International Conference on Mobile Computing and Networking (MobiCom), pp. 263–270 (1999)
4. Grossglauser, M., Tse, D.N.C.: Mobility increases the capacity of ad hoc wireless networks. IEEE/ACM Transactions on Networking 10, 477–486 (2002)
5. Al-Karaki, J.N., Kamal, A.E.: Routing techniques in wireless sensor networks: A survey. IEEE Wireless Communications 11, 6–28 (2004)
6. Datta, N.N., Gopinath, K.: A survey of routing algorithms for wireless sensor networks. Journal of Indian Institute of Science 86, 569–598 (2006)
7. Jiang, Q., Manivannan, D.: Routing protocols for sensor networks. In: Proceedings of the 1st IEEE Consumer Communications and Networking Conference (CCNC), January 2004, pp. 93–98 (2004)
8. Heinzelman, W.R., Chandrakasan, A., Balakrishnan, H.: An application-specific protocol architecture for wireless microsensor networks. IEEE Transactions on Wireless Communications 1, 660–670 (2002)
9. Intanagonwiwat, C., Govindan, R., Estrin, D., Heidemann, J., Silva, F.: Directed diffusion for wireless sensor networking. IEEE/ACM Transactions on Networking 11, 2–16 (2003)
10. Zhao, W., Ammar, M.H.: Message ferrying: Proactive routing in highly-partitioned wireless ad hoc networks. In: 9th IEEE Workshop on Future Trends of Distributed Computing Systems (FTDCS) (May 2003)
11. Kansal, A., Somasundara, A.A., Jea, D.D., Srivastava, M.B., Estrin, D.: Intelligent fluid infrastructure for embedded networks. In: Proceedings of the 2nd International Conference on Mobile Systems, Applications, and Services, MobiSys (2004)
12. Pentland, A., Fletcher, R., Hasson, A.: Daknet: Rethinking connectivity in developing nations. Computer 37, 78–83 (2004)
13. Small, T., Haas, Z.J.: The shared wireless infostation model: a new ad hoc networking paradigm (or where there is a whale, there is a way). In: Proceedings of the 4th ACM International Symposium on Mobile Ad hoc Networking and Computing, MobiHoc (2003)
14. Juang, P., Oki, H., Wang, Y., Martonosi, M., Peh, L.S., Rubenstein, D.: Energy-efficient computing for wildlife tracking: Design tradeoffs and early experiences with zebranet. ACM SIGOPS Operating Systems Review 36, 96–107 (2002)
15. Akkaya, K., Younis, M., Bangad, M.: Sink repositioning for enhanced performance in wireless sensor networks. Computer Networks 49, 512–534 (2005)
16. Basagni, S., Carosi, A., Melachrinoudis, E., Petrioli, C., Wang, Z.M.: Controlled sink mobility for prolonging wireless sensor networks lifetime. ACM/Elsevier Wireless Networks (to appear) (2007)
17. Bi, Y., Sun, L., Ma, J., Li, N., Khan, I.A., Chen, C.: Hums: An autonomous moving strategy for mobile sinks in data-gathering sensor networks. EURASIP Journal on Wireless Communications and Networking 2007 (2007)
18. Chen, J.M.C., Yu, K.: Designing energy-efficient wireless sensor networks with mobile sinks. In: Workshop on World-Sensor-Web (WSW) at ACM SenSys (2006)
19. Gandham, S.R., Dawande, M., Prakash, R., Venkatesan, S.: Energy efficient schemes for wireless sensor networks with multiple mobile base stations. In: Proceedings of the IEEE Global Telecommunications Conference, GLOBECOM (2003)
20. Luo, J., Hubaux, J.P.: Joint mobility and routing for lifetime elongation in wireless sensor networks. In: Proceedings of the 24th Annual Joint Conference of the IEEE Computer and Communications Societies (INFOCOM) (March 2005)

21. Wang, W., Srinivasan, V., Chua, K.: Using mobile relays to prolong the lifetime of wireless sensor networks. In: Proceedings of the 11th Annual International Conference on Mobile Computing and Networking, MobiCom (2005)
22. Wang, Z.M., Basagni, S., Melachrinoudis, E., Petrioli, C.: Exploiting sink mobility for maximizing sensor networks lifetime. In: Proceedings of the 38th Annual Hawaii International Conference on System Sciences, HICSS (2005)
23. Bi, Y., Niu, J., Sun, L., Huangfu, W., Sun, Y.: Moving schemes for mobile sinks in wireless sensor networks. In: Proceedings of the IEEE International Performance, Computing, and Communications Conference, IPCCC (2007)
24. Sun, L., Bi, Y., Ma, J.: A moving strategy for mobile sinks in wireless sensor networks. In: 2nd IEEE Workshop on Wireless Mesh Networks, WiMesh (2006)
25. Jun, H., Zhao, W., Ammar, M.H., Zegura, E.W., Lee, C.: Trading latency for energy in densely deployed wireless ad hoc networks using message ferrying. Ad Hoc Networks 5, 444–461 (2007)
26. Jain, S., Shah, R.C., Brunette, W., Borriello, G., Roy, S.: Exploiting mobility for energy efficient data collection in wireless sensor networks. Mobile Networks and Applications 11, 327–339 (2006)
27. Shah, R.C., Roy, S., Jain, S., Brunette, W.: Data mules: Modeling and analysis of a three-tier architecture for sparse sensor networks. Ad Hoc Networks 1, 215–233 (2003)
28. Vahdat, A., Becker, D.: Epidemic routing for partially connected ad hoc networks. Technical Report CS-200006, Duke University, Durham, NC (April 2000)
29. Spyropoulos, T., Psounis, K., Raghavendra, C.S.: Single-copy routing in intermittently connected mobile networks. In: Proceedings of the 1st Annual IEEE Communications Society Conference on Sensor and Ad Hoc Communications and Networks, IEEE SECON (2004)
30. Diggavi, S., Grossglauser, M., Tse, D.: Even one-dimensional mobility increases ad hoc wireless capacity. In: Proceedings of the IEEE International Symposium on Information Theory, ISIT (2002)
31. Chakrabarti, A., Sabharwal, A., Aazhang, B.: Using predictable observer mobility for power efficient design of sensor networks. In: 2nd International Workshop on Information Processing in Sensor Networks, IPSN (2003)
32. Baruah, P., Urgaonkar, R., Krishnamachari, B.: Learning enforced time domain routing to mobile sinks in wireless sensor fields. In: Proceedings of the 29th Annual IEEE International Conference on Local Computer Networks (2004)
33. Chatzigiannakis, I., Kinalis, A., Nikoletseas, S.: Sink mobility protocols for data collection in wireless sensor networks. In: 4th ACM International Workshop on Mobility Management and Wireless Access, MobiWac (2006)
34. Kansal, A., Rahimi, M., Estrin, D., Kaiser, W., Pottie, G.J., Srivastava, M.B.: Controlled mobility for sustainable wireless sensor networks. In: Proceedings of the 1st Annual IEEE Communications Society Conference on Sensor and Ad Hoc Communications and Networks, IEEE SECON (2004)
35. Basagni, S., Carosi, A., Petrioli, C.: Controlled vs. uncontrolled mobility in wireless sensor networks: Some performance insights. In: Proceedings of the IEEE 66th Vehicular Technology Conference (2007)
36. Tirta, Y., Lau, B., Malhotra, N., Bagchi, S., Li, Z., Lu, Y.: Controlled mobility for efficient data gathering in sensor networks with passively mobile nodes. Sensor Network Operations 3.2, 92–113 (2006)

37. Gu, Y., Bozdag, D., Ekici, E., Ozguner, F., Lee, C.: Partitioning based mobile element scheduling in wireless sensor networks. In: Proceedings of the 2nd Annual IEEE Communications Society Conference on Sensor and Ad Hoc Communications and Networks, IEEE SECON (2005)
38. Ngai, E.C.H., Lin, J., Lyu, M.R.: Delay-minimized route design for wireless sensor-actuator networks. In: Proceedings of the IEEE Wireless Communications and Networking Conference, WCNC (2007)
39. Somasundara, A.A., Ramamoorthy, A., Srivastava, M.B.: Mobile element scheduling with dynamic deadlines. IEEE Transactions on Mobile Computing 6, 395–410 (2007)
40. Goldenberg, D., Lin, J., Morse, A.S., Rosen, B., Yang, Y.R.: Towards mobility as a network control primitive. In: Proceedings of the 5th ACM International Symposium on Mobile Ad hoc Networking and Computing, MobiHoc (2004)
41. Wang, G., Cao, G., Porta, T.L.: Movement-assisted sensor deployment. In: Proceedings of the 23rd IEEE Conference on Computer Communications (2004)
42. Batalin, M.A., Rahimi, M., Yu, Y., Liu, D., Kansal, A., Sukhatme, G.S., Kaiser, W.J., Hansen, M., Pottie, G.J., Srivastava, M., Estrin., D.: Call and response: Experiments in sampling the environment. In: Proceedings of the 2nd ACM SenSys (2004)
43. Priyantha, N.B., Balakrishnan, H., Demaine, E., Teller, S.: Mobile-assisted localization in wireless sensor networks. In: Proceedings of the 24th Annual Joint Conference of the IEEE Computer and Communications Societies, INFOCOM (2005)
44. Solomon, M.: Algorithms for the vehicle routing and scheduling problem with time window constraints. Operations Research 35 (1987)

Abductive Inference Based Approach to DSS Designing for Project Portfolio Planning

Grzegorz Bocewicz and Zbigniew A. Banaszak

Dept. of Computer Science and Management, Koszalin University of Technology,
Śniadeckich 2, 74-453 Koszalin, Poland
bocewicz@ie.tu.koszalin.pl

Abstract. The reference model employing constraint programming (*CP*) paradigm describes both an enterprise and a set of project-like production orders. Moreover, encompassing consumer orders requirements and available production capabilities, the model provides the formal framework allowing one to develop a class of decision support systems aimed at interactive production process planning subject to multi-project environment constraints. In that context our contribution provides a knowledge-based and *CP*-driven approach to renewable and nonrenewable resources allocation assuming precise character of decision variables. The proposed concept of a context dedicated constraint satisfaction problem can be seen as a new formal framework for developing a task oriented Decision Support Tool for Project Portfolio Prototyping (DST4P3). The tool provides a prompt and interactive service to a set of routine queries formulated either in straight or reverse way.

Keywords: Decision Support System, Constraint Programming, abductive inference, projects portfolio scheduling.

1 Introduction

An optimal assignment of available resources to production steps in a multi-product job shop is often economically indispensable. The goal is to generate a plan/schedule of production orders for a given period of time while minimize the cost that is equivalent to maximization of profit. In that context executives want to know how much a particular production order will cost, what resources are needed, what resources allocation can guarantee due time production order completion, and so on. So, a manager needs might be formulated in a form of standard, routine questions, such as: Does the production order can be completed before an arbitrary given deadline? What is the production completion time following assumed robots operation time? Is it possible to undertake a new production order under given (constrained in time) resources availability while guaranteeing disturbance-free execution of the already executed orders? What values and of what variables guarantee the production order will completed following assumed set of performance indexes?

The problems standing behind of the quoted questions belong to the class of so called project scheduling ones. In turn, project scheduling can be defined as the process of allocating scarce resources to activities over a period of time to perform a set of

N.T. Nguyen and E. Szczerbicki (Eds.): Intel. Sys. for Know. Management, SCI 252, pp. 107–129.
springerlink.com

activities in a way taking into account a given set of performance measures. Such problems belong to NP-complete ones. Therefore, the new methods and techniques addressing the impact of real-life constraints on the decision making is of great importance, especially in case of interactive and task oriented Decision Support Systems (DSSs) designing [3], [8].

Several techniques have been proposed in the past fifty years, including MILP, Branch-and-Bound [7] or more recently Artificial Intelligence. The last sort of techniques concentrates mostly on fuzzy set theory and constraint programming frameworks. Constraint Programming/Constraint Logic Programming (*CP*/*CLP*) languages [6], [7], [15], [17] seems to be well suited for modeling of real-life and day-to-day decision-making processes in an enterprise [1]. In turn, applications of fuzzy set theory in production management [5] show that most of the research on project scheduling has been focused on fuzzy *PERT* and fuzzy *CPM* [10], [11].

In this context, the contribution provides the framework allowing one to take into account distinct data describing modeled objects and then to treat them in terms of the constraint satisfaction problem (*CSP*) [2].The approach proposed concerns of Logic-Algebraic Method *(LAM)* based and *CP*-driven methodology aimed at interactive decision making based on distinct and imprecise data. The paper can be seen as continuation of our former works concerning projects portfolio prototyping [1], [2].

The following two classes of standard routine queries are usually considered and formulated in:

a straight way (i.e. corresponding to the question: What results from premises?

- What the portfolio makespan follows from the given project constraints specified by activity duration times, resources amount and their allocation to projects' activities?
- Does a given resources allocation guarantee the production orders makespan do not exceed the given deadline?
- Does the projects portfolio can be completed before an arbitrary given deadline?
- and so on.

a reverse way (i.e. corresponding to the question: What implies conclusion?)

- What activity duration times and resources amount guarantee the given production orders portfolio makespan do not exceed the deadline?
- Does there exist resources allocation such that production orders makespan do not exceed the deadline?
- Does there exist a set of activities' operation times guaranteeing a given projects portfolio completion time will not exceed the assumed deadline?

Above mentioned categories encompass the different reasoning perspectives, i.e. deductive and abductive ones. The corresponding queries can be stated in the same model that can be treated as composition of variables and constraints, i.e. assumed sets of variables and constraints limiting their values. In that context both an enterprise and the portfolio of production orders can be specified in terms of distinct and/or imprecise variables, discrete and/or continuous variables, renewable and/or non-renewable resources, limited and/or unlimited resources, and so on.

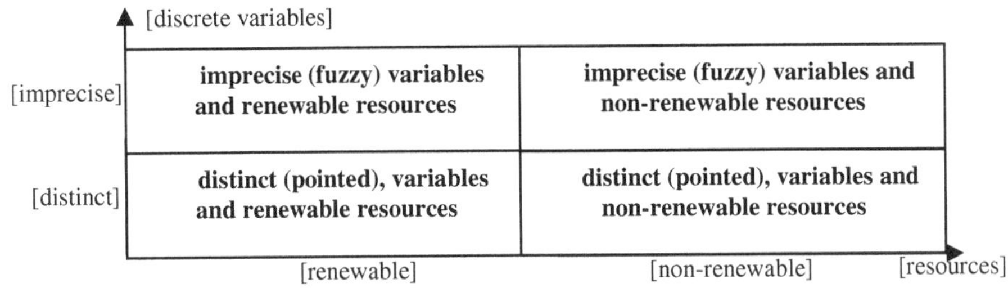

Fig. 1. Elementary decision problems

Therefore, an approach proposed assumes a kind of reference model encompassing open structure enabling one to take into account different sorts of variables and constraints as well as to formulate straight and reverse kind of project planning problems. So, the elementary as well as hybrid models can be considered, see the Fig. 1. Of course, the most general case concerns of the hybrid model specified by discrete distinct and/or imprecise (e.g., fuzzy) variables and renewable and/or non-renewable resources.

The assumed model enabling descriptive way of a problem statement encompasses constraint satisfaction problem structure and then allows implementing the problem considered in the constraint programming environment. In this paper the application of approach proposed is limited to the problems specified by distinct and both renewable and non-renewable resources.

Since the constraint programming treated as programming paradigm enables to specify both variables and relations between them in the form of constraints, then problems specified in that way can be easily implemented in the one of popular constraint logic languages such as: **CHIP V5**, **ECLiPSe**, and **SICStus** [19], or imperative constraint programming languages (assuming that a statement computation results in a program state change) such as: **Choco** [18], **ILOG** [13], and **python-constraint** [20], or public domain concurrent constraint programming language as **OzMozart** [15].

In order to illustrate the approach proposed let us focus on a reference model of decision problem encompassing equilibrium between possible expectations regarding potential orders completion (e.g. following a set of routine queries) and available production capabilities. The considered problem of projects portfolio scheduling concerns of resources conflict resolution, i.e. conflicts arising in case different activities simultaneously request their access to renewable and non renewable resources of limited quantity.

In this paper a new *CP* based approach to *DSS* designing aimed at multi-product manufacturing subject to renewable and non-renewable resources allocation assuming precise character of decision variables constraints. The paper is organized as follows. In Section 2, the problem statement of considered project portfolio scheduling is introduced. The consistency examination is the main aim of the context dedicated constraint satisfaction problem aimed at consistency-checking procedure designing. Then, some details of the modelling framework assumed, in particular the reference model employed are described in Section 3. In Section 4, illustrative examples of the possible application of the approach proposed are presented. We conclude with some results and lesson learned in Section 5.

2 Constraint Programming Techniques

2.1 Constraint Satisfaction Problem

Constraint satisfaction problem (*CSP*) is determined by the set of decision variables $X = \{x_1, x_2,...,x_n\}$, the family of variables domains $D = \{D_i \mid D_i = (d_{i,1}, d_{i,2},..., d_{i,j},...,d_{i,m}), i = 1,...,n\}$, and the set of constraints $C = \{c_i \mid i = 1,2,..,lc\}$ encompassing relations linking variables. Each constraint c_i can be seen as the relation defined on the relevant subset of variables $X_i \subset X = \{x_1, x_2,...,x_n\}$. Consequently the *CSP* is denoted as follows: $CS = ((X,D),C)$.

Consider the vector $V = (v_1, v_2,...,v_n) \in D_1 \times D_2 \times ... \times D_n$. The vector V such that the all constraints hold is treated as the admissible solution of *CS*.

Let us suppose, the constraint c_i defined on the subset $X_i = \{x_l, x_k,...,x_m\}$ follows the logic value equal to "true" (noted as $w(c_i) = 1$) in this case there exists $V_i \in D_l \times D_k \times ....\times D_m$ such that c_i holds. In that context the set of admissible solutions is defined as follows:

$$\surd = \{V = (v_1, v_2, ..., v_n) \mid v_i \in D_i, i = 1, ..., n, w(c_1) = w(c_2) = ... = w(c_{lc}) = 1\}.$$

Therefore, *CSP* can be seen also as the triple (data, constrains, query), i.e. the set of variables and family of variables domains, the set of constraints, and the question: Does there exist nonempty set $\surd$? Of course, in general case, instead of admissible solution an optimal one can be searched for, as well.

Depending on variables and constraints character different classes of *CSP* are distinguished. The problems considered can be specified either in finite or infinite domains, by discrete, e.g. Boolean, or continuous variables, as well as the subject to linear oand/or nonlinear constraints [14].

Solution strategies are based on two subsequently used mechanisms, i.e. constraints propagation and variables distribution. Variables distribution can be executed either through systematic (e.g. breath-first-search) or stochastic search of the whole or constrained state space of potential solutions obtained from constraints propagation. The searching strategies are implemented in constraint logic programming or constraint programming languages such as CHIP, OzMozart, ILOG, and so on.

Example of *CSP* formulation.

Given $CS = ((X, D), C)$

where: $X = \{x_1, x_2, x_3, x_4, x_5, x_6, x_7\}$

$D = \{D_1, D_2, D_3, D_4, D_5, D_6, D_7\}$;

$D_1 = D_2 = D_3 = D_4 = D_5 = D_6 = D_7 = \{1,2,...,10\}$,

$C = \{c_1, c_2, c_3, c_4, c_5\}$;

c_1: $x_1 = 2x_2 - 1$; c_2: $x_2 + x_3 = 10$; c_3: $x_2 + 2x_3 = 2x_4$;
c_4: $2x_6 + x_5 = 8$; c_5: $x_6 + 7 = x_7$;

Admissible solutions obtained from OzMozart implementation are of the following form:

$\surd$= {(3, 2, 8, 9, 6, 1, 8),
(7, 4, 6, 8, 6, 1, 8),
(3, 2, 8, 9, 4, 2, 9),
(7, 4, 6, 8, 4, 2, 9),
(3, 2, 8, 9, 2, 3, 10),
(7, 4, 6, 8, 4, 3, 10)}.

Let us note, however the solutions obtained can be misleading ones. For instance, let us consider two solutions (7, 4, 6, 8, 6, 1, 8) and (7, 4, 6, 8, 4, 3, 10). In the case the questions considered are:
What are the values of variables x_1, x_2, x_3, and x_4 guaranteeing the variable $x_7 \leq 9$?,
and
What are the values of variables x_1, x_2, x_3, and x_4 guaranteeing the variable $x_7 > 9$?
one may observe just the same common solution $x_1 = 7$, $x_2 = 4$, $x_3 = 6$, $x_4 = 8$. In fact, such the response means just: "I do not know".

Such confusing result can be easily explained as following from the fact that there is no any link between groups of constraints c_1: $x_1 = 2x_2 - 1$, c_2: $x_2 + x_3 = 10$, c_3: $x_2 + 2x_3 = 2x_4$, and c_4: $2x_6 + x_5 = 8$; c_5: $x_6 + 7 = x_7$. In other words that means the variable x_7 does not depend on variables x_1, x_2, x_3, x_4. The arising question is how to check in advance whether *CSP* formulation takes into account further detailed requests, e.g. as above mentioned? ■

The example provided proves the necessity of the context specification of *CSP* models considered. That means *CSP* specification should be dedicated to the questions guaranteeing that the prompt, i.e. YES or NO, response there exists. In other words, the problem specification we are looking for has to encompass the context (i.e. ability to respond to the standard questions) of routine queries as to avoid the response I DO NOT KNOW (in fact meaning YES and NO, simultaneously).

2.2 Concept of the Context Dedicated Constraint Satisfaction Problem (CCS_q)

Let us assume that any considered, routine question q ($q \in Q$) have the following standard structure:

q: *Does there exist any nonempty set of values Vu of variables U, such that the set of constraints Cy holds?*

Note that to any routine question q corresponds to the relevant *CSP*, i.e. CS_q determined as follows: $CSq = ((U \cup Y, D_{UY}), Cy)$
where: U – the set of the input variables,
Y – the set of the output variables, $U \cap Y = \varnothing$,
D_{UY} – the domain of decision variables (i.e. $U \cup Y$),
$Cy = \{cy_1, cy_2, \ldots, cy_{ly}\}$ – the set of constraints determining the output variables Y domains,

That means, the question q can be seen as a kind of *CSP*, i.e. CS_q , determined by sets of variables *U*, *Y* and constraints *Cy* specifying the question.

2.2.1 Consistency Examination

Due to the Logic-Algebraic Method (*LAM*) [9], [8] examination of the *CSP* consistency has to follow the four-stage procedure. The procedure proposed consists of the following steps:

1) For the given *CS* and *CSq*, formulate the relevant *CCS* and CCS^*, where $CCS = ((X, D), C \cup Cy)$ and $CCS^* = ((X, D), C \cup (Cy^*))$ consist of:

$$Cy = \{ cy_1 \wedge cy_2 \wedge \ldots \wedge cy_{ly}\},\ \ Cy^* = \{cy^*_1 \wedge cy^*_2 \wedge \ldots \wedge cy^*_{lq}\}$$
$$(cy^*_1 \wedge cy^*_2 \wedge \ldots \wedge cy^*_{lq}) \Leftrightarrow [\neg(cy_1 \wedge cy_2 \wedge \ldots \wedge cy_{ly})].$$

2) Consider any available *CLP/CP* language implementation of the *CCS* and CCS^*, for instance OzMozart,
3) Determine the sets Vu^1 and Vu^2,
4) Find out the intersection $Vu^1 \cap Vu^2$.

   If $Vu^1 \cap Vu^2 = \varnothing$, then *CSS* is consistent.
   If $Vu^1 \cap Vu^2 \neq \varnothing$, then *CSS* is inconsistent.

That means *CSS* is consistent or alternatively speaking *CS* is consistent with the question q if:

$$Vu^1 \cap Vu^2 = \varnothing,$$

where: Vu^1 – the set of values *U* treated as solution to *CCS*,
Vu^2 – the set of values *U* treated as solution to CCS^*.

In order to illustrate the introduced consistency-checking procedure let us consider the following example.

Consider *CS* given in an "Example of *CSP* formulation" and the question q_1: *Does there exist nonempty set of values Vu of variables* $U = \{x_1, x_2, x_3, x_4\}$, *for which the constraint* $x_7 \leq 9$ *holds*?

As it was already mentioned the solution does not exists because the relationship linking the variables *U* and variable x_7 does not exist, too.

In turn, let us consider the consistency between *CS* and the question q_1. The relevant problems *CCS* and CCS^* are of the following form:

$$CCS = ((X, D), C \cup Cy)$$
$$CCS^* = ((X, D), C \cup (Cy^*))$$

where: *X*, *D*, *C* – as assumed in an "Example of *CSP* formulation",
$Cy = \{cy_1\}$, cy_1: $x_7 \leq 9$,
$Cy^* = \{cy^*_1\}$, cy^*_1: $x_7 > 9$,

The table shown in Fig. 2 presents values of variables *X* following constraints *C* (see Table A) as well as two tables presenting values of variables *U* being the solutions of *CCS* (see the Table B) as well as CCS^* (see the Table C). The solutions obtained through OzMozart implementation of the problems considered are as follows: $Vu^1 = \{(3, 2, 8, 9), (7, 4, 6, 8)\}$ (i.e the values for which the constraint $x_7 \leq 9$

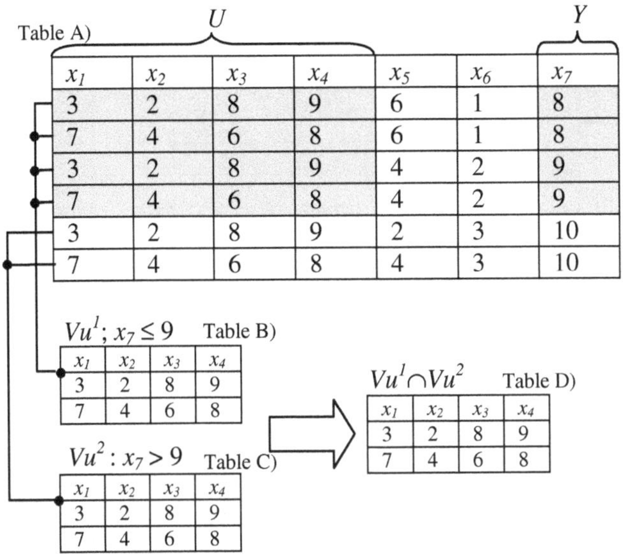

Fig. 2. Admissible solutions for *CCS* and *CCS** in the question q_1 context.

holds), and $Vu^2 = \{(3, 2, 8, 9), (7, 4, 6, 8\}$ (i.e the values for which the constraint $x_7 \leq 9$ does not hold). Note that $Vu^1 \cap Vu^2 = \{(3, 2, 8, 9), (7, 4, 6, 8)\}$ (see the Table D) is not an empty set, and consists of solutions where both constraints $x_7 \leq 9$ and $x_7 > 9$ hold simultaneously. Note that *CS* is not consistent with the question q_1, i.e. $Vu^1 \cap Vu^2$ can be seen as the set consisting solutions resulting in ambiguous responses (in other words the intersection can be seen as the set containing solutions where relation among U and x_7 does not exist). Consequently, lack of consistency implies lack of unambiguous response to the question consider.

Consider the next question stated as follows:

q_2*: Does there exist nonempty set of values* Vx_1 *of the variable* x_1*, such that the constraint* $x_4 = 9$ *holds*?

In the case considered the following value of the variable $x_1 = 3$ implies that $x_4 = 9$. Note that between variables x_1 and x_4 there exists the following variables-based linkage of constraints: c_1: $x_1 = 2x_2 - 1$; c_2: $x_2 + x_3 = 10$; c_3: $x_2 + 2x_3 = 2x_4$. That means in the case of the opposite question: *Does there exist nonempty set of values* Vx_1 *of the variable* x_1*, such that the constraint* $x_4 \neq 9$ *holds*? the relevant solution is $x_1 \neq 3$.

Let us consider the consistency between *CS* and the question q_2. The relevant problems *CCS* and *CCS** are of the following form:

$$CCS = ((X, D), C \cup Cy),$$
$$CCS^* = ((X, D), C \cup (Cy^*)),$$

where: X, D, C – as assumed in an "Example of *CSP* formulation",

$$Cy = \{cy_1\}, \qquad cy_1: x_4 = 9,$$
$$Cy^* = \{cy^*_1\}, \qquad cy^*_1: x_4 \neq 9.$$

Table A) U ... Y

x_1	x_2	x_3	x_4	x_5	x_6	x_7
3	2	8	9	6	1	8
7	4	6	8	6	1	8
3	2	8	9	4	2	9
7	4	6	8	4	2	9
3	2	8	9	2	3	10
7	4	6	8	4	3	10

Vu^1; $x_4 = 9$

Table B)

x_1
3

Vu^2; $x_4 \neq 9$

Table C)

x_1
7

$\Rightarrow$ $Vu^1 \cap Vu^2 = \varnothing$

Fig. 3. Admissible solutions for *CCS* and CCS* in the question q_2 context.

The table shown in Fig. 3 presents values of variables X following constraints C (see Table A) as well as two tables presenting values of variables U being the solutions of *CCS* (see the Table B) as well as CCS^* (see the Table C). The solutions obtained through OzMozart implementation of problems considered are as follows: $Vu^1 = \{3\}$, $Vu^2 = \{7\}$. Note that $Vu^1 \cap Vu^2 = \varnothing$. That means there is a lack of solution leading to ambiguous responses, i.e. solutions where constraint $x_4 = 9$ can hold and cannot be satisfied simultaneously. Note that consistency *CS* and the question q_2 implies the variables x_1 and x_4 are linked, i.e. dependent each other. So, the solution to the question q_2 there exists.

2.2.2 Context Dedicated Constraint Satisfaction Problem

From the last example it follows that a response to the question q_i stated in terms of *CS* there exists only if the result of consistency-checking procedure in the case of the relevant *CCS* is positive. That means the *CCS* consistency enables to consider the concept of so called context dedicated *CSP*, denoted as CCS_Q i.e. the *CSP* guaranteeing solutions to the all questions from the set Q.

Consider $CS = ((X, D), C)$,

where: the set of decision variables X consists of the input $U \subset X$ and the output $Y \subset X$, variables such that: $U \cap Y = \varnothing$,

D – the family of domains of variables X,

C – the set of constraints limiting values of variables X.

Consider the set of questions $Q = \{q_i \mid i = 1,2,...,r\}$ such that for any $q \in Q$ there exists CSq and $CCS = ((X, D), C \cup Cy)$, where Cy – the set of constraints corresponding to the question q and limiting values of the output variables Y.

Consequently, *CSP* is CCS_Q only if it follows assumptions imposed on *CCS* structure and for any $q \in Q$ the result of consistency-checking procedure is positive.

It should be noted, that the meaning of "context dedicated" is determined by the set of considered routine queries (solutions to which are guaranteed).

Solution to any particular question $q \in Q$ is the vector of values of variables $Vu = (vu_1, vu_2, \ldots, vu_{lu}) \in Du_1 \times Du_2 \times \ldots \times Du_{lu}$ (where $vu_i \in Du_1$ is the value of variable $u_i \in U$) such that all the constraints from the set $C \cup Cy$ are satisfied.

In case of questions $q \notin Q$ for which CCS_p do not guarantee solutions there arise a new problem: What is the set of additional constraints Cu guaranteeing their consistency? In other words, these question concerns of constraints "linking" the input and output variables which follow from the question context.

The main advantage of the concept proposed lies in providing the solid, i.e. *CSP* based, framework allowing one to formulated models that can be directly implemented in decision support tools. Illustration of such capability in terms of projects portfolio planning, examined with the help of the Decision Support Tool for Project Portfolio Prototyping (DST4P3) [4] aimed at Small and Medium Sized Enterprises (SME) provides the Section below.

3 Reference Model of an "Enterprise - Project Portfolio"

Let us consider the reference model of a decision problem concerning of multi-resource task allocation in a multi-product job shop assuming imprecise character of decision variables. The model specifies both the job shop capability and production orders requirement in a unified way, i.e., through the description of determining them sets of variables and sets of constraints restricting domains of discrete variables. Some conditions concerning the routine questions are included in the set of constraints. That means in case such conditions hold the response to associated questions is positive.

Therefore, the reference model considered specifies both SMEs and projects portfolio in terms of describing them variables and constraints.

Decision variables *X*:

- **SME – a job-shop perspective.**
  Given amount lz of renewable discrete resources ro_i specified by (e.g. workforce, machine tools, automated guided vehicles, etc.): $Ro = (ro_1, ro_2, \ldots, ro_z)$. Given amounts $zo_{i,k}$ of available renewable resources $zo_i = (zo_{i,1}, zo_{i,2}, \ldots, zo_{i,h})$, where $zo_{i,k}$ – limited amount of the i-th renewable resource available at the k-th moment of H: $H = \{0, 1, \ldots, h_{max}\}$, specified by $Zo = (zo_1, zo_2, \ldots, zo_{lz})$.
  Given amount ln of non-renewable resources (i.e. money) rn_i specified by: $Rn = (rn_1, rn_2, \ldots, rn_{ln})$. Given amounts zn_i of available non-renewable resources rn_i specified by: $Zn = (zn_1, zn_2, \ldots, zn_{ln})$, where zn_i denotes amount of the resource rn_i being available at the beginning of time horizon H.
- **Projects portfolio.**
  Given a set of projects $P = \{P_1, P_2, \ldots, P_{lp}\}$, where P_i is specified by the set composed of lo_i activities, i.e., $P_i = \{O_{i,1}, \ldots, O_{i,lo_i}\}$, where:

$$O_{i,j} = (s_{i,j}, t_{i,j}, Tp_{i,j}, Tz_{i,j}, Dp_{i,j}, Tr_{i,j}, Ts_{i,j}, Cr_{i,j}, Cw_{i,j}), \tag{1}$$

$s_{i,j}$ – means the starting time of the activity $O_{i,j}$, i.e., the time counted from the beginning of the time horizon H,

$t_{i,j}$ – the duration of the activity $O_{i,j}$,

$Tp_{i,j} = (tp_{i,j,1}, tp_{i,j,2},...,tp_{i,j,lz})$ – the sequence of moments the activity $O_{i,j}$ requires new amounts of renewable resources: $tp_{i,j,k}$ – the time counted since the moment $s_{i,j}$ of the $dp_{i,j,k}$ amount of the k-th resource allocation to the activity $O_{i,j.}$ That means a resource is allotted to an activity during its execution period: $0 \le tp_{i,j,k} < t_{i,j}$; $k = 1,2,\ldots,lz$.

$Tz_{i,j} = (tz_{i,j,1}, tz_{i,j,2},...,tz_{i,j,lz})$ – the sequence of moments the activity $O_{i,j}$ releases the subsequent resources, $tz_{i,j,k}$ – the time counted since the moment $s_{i,j}$ the $dp_{i,j,k}$ amount of the k-th renewable resource was released by the activity $O_{i,j}$. That is assumed a resource is released by activity during its execution: $0 < tz_{i,j,k} \le t_{i,j}$ and $tp_{i,j,k} < tz_{i,j,k}$; $k = 1, 2, \ldots, lz$.

$Dp_{i,j} = (dp_{i,j,1}, dp_{i,j,2},...,dp_{i,j,lz})$ – the sequence of the k-th resource amounts $dp_{i,j,k}$ are allocated to the activity $O_{i,j}$, i.e., $dp_{i,j,k}$ – the amount of the k-th resource allocated to the activity $O_{i,j}$. That assumes: $0 \le dp_{i,j,k} \le zo_k$; $k = 1, 2, \ldots, lz$.

$Cr_{i,j} = (cr_{i,j,1}, cr_{i,j,2},...,cr_{i,j,ln})$ – the sequence of non-renewable resources amount required by activity $O_{i,j}$, : $cr_{i,j,k}$ – the amount of the k-th resource required by the activity $O_{i,j}$, $cr_{i,j,1} \le 0$; $k = 1, 2,\ldots,ln$, $cr_{i,j,k} = 0$ means the activity does not consume the k-th resource.

$Cw_{i,j} = (cw_{i,j,1}, cw_{i,j,2},...,cw_{i,j,ln})$ – the sequence of amounts of non-renewable resources released by activity $O_{i,j}$, $cw_{i,j,k}$ – the amount of the k-th resource involved by activity $O_{i,j}$, $cw_{i,j,1} \ge 0$; $k = 1,2,\ldots,ln$, $cr_{i,j,k} = 0$ means the activity does not inflow the k-th resource.

$Tr_{i,j} = (tr_{i,j,1}, tr_{i,j,2},...,tr_{i,j,ln})$ – the sequence of moments the determined amounts of subsequent non renewable resources are required by activity $O_{i,j}$: $tr_{i,j,k}$ – the time counted since the moment $s_{i,j}$ the $dp_{i,j,k}$ amount of the k-th non renewable resource was released by the activity $O_{i,j}$. That is assumed a resource is collected by activity during its execution: $0 \le tr_{i,j,k} < t_{i,j}$; $k = 1, 2, \ldots, ln$,

$Ts_{i,j} = (ts_{i,j,1}, ts_{i,j,2},...,ts_{i,j,ln})$ - the sequence of moments the determined amounts of subsequent non renewable resources are generated (released) by activity $O_{i,j}$: $ts_{i,j,k}$ - the time counted since the moment $s_{i,j}$ the $cw_{i,j,k}$. amount of the k-th non renewable resource was generated by the activity $O_{i,j}$. That is assumed the resource is generated during activity execution, however not earlier than beginning of its collection, i.e.: $0 \le ts_{i,j,k} < t_{i,j}$; $k = 1, 2, \ldots, ln$, as well as $tr_{i,j,k} \le ts_{j,k}$; $k = 1, 2,\ldots,ln$,

NPV – the Net Present Value used to measure project's efficiency and calculated due to the following formulae:

$$NPV = \sum_{t=0}^{n} \frac{CF_t}{(1+k)^t} \quad , \tag{2}$$

where: CF_t – the money netto flow expected in the year t,
k – the discount rate (alternative capital investment cost),
n – the period of a project exploitation [years].

Consequently, each activity $O_{i,j}$ is specified by the following sequences of:

- starting times of activities in the activity network P_i:
  $S_i = (s_{i,1}, s_{i,2}, \ldots, s_{i,lo_i}),\ 0 \leq s_{i,j} < h,\ i = 1, 2, \ldots, lp;\ j = 1, 2, \ldots, lo_i,$
- duration of activities in the activity network P_i:
  $T_i = (t_{i,1}, t_{i,2}, \ldots, t_{i,lo_i}),$

Elements of sequences: $Tp_{i,j}$, $Tz_{i,j}$, $Dp_{i,j}$, $Cr_{i,j}$, $Cw_{i,j}$, $Tr_{i,j}$,$Ts_{i,j}$ specify the activity network P_i:

- starting times the j-th resource is allocated to the k-th activity in the activity network P_i:
  $TP_{i,j} = (tp_{i,1,j}, \ldots, tp_{i,k,j}, \ldots, tp_{i,lo_i,k}),$
- starting times the j-th resource is released by the k-th activity in the P_i:
  $TZ_{i,j} = (tz_{i,1,j}, \ldots, tz_{i,k,j}, \ldots, tz_{i,lo_i,j}),$
- the sequence of moments the j-th non-renewable resource is collected by activities of the projects P_i:
  $TR_{i,j} = (tr_{i,1,j}, \ldots, tr_{i,k,j}, \ldots, tr_{i,lo_i,j}),$
- the sequence of moments the j-th non-renewable resource is released by activities of the project P_i:
  $TS_{i,j} = (ts_{i,1,j}, \ldots, ts_{i,k,j}, \ldots, ts_{i,lo_i,j}),$
- amounts of the j-th resources allotted to the k-th activity in the route P_i:
  $DP_{i,j} = (dp_{i,1,j}, \ldots, dp_{i,k,j}, \ldots, dp_{i,lo_i,j}).$
- sequences of amounts of the j-th non-renewable resource consumed by activities of the route P_i:
  $CR_{i,j} = (cr_{i,1,j}, \ldots, cr_{i,k,j}, \ldots, cr_{i,lo_i,j}),$
- sequences of amounts of the j-th non-renewable resource involved by activities of the route P_i:
  $CW_{i,j} = (cw_{i,1,j}, \ldots, cw_{i,k,j}, \ldots, cw_{i,lo_i,j}).$

Constraints *C*:
Given projects portfolio and available amounts of renewable and non-renewable resources as well as the above mentioned sequences: T_i, $TP_{i,j}$, $TZ_{i,j}$, $DP_{i,j}$.

Given the time horizon $H=\{0,1,\ldots,h_{max}\}$, the projects portfolio should be completed. That is assumed the activities cannot be suspended during their execution, and moreover:

- each activity can request any kind and quantity (not exceeding the resource's limited amount) of any resource
- each resource can be uniquely used by an activity,
- the quantity of renewable resource used by an activity cannot be changed or allotted to other activity,
- an activity can start its execution only if required amounts of renewable and non-renewable resources are available at the moments given by $Tp_{i,j}$, $Ts_{i,j}$.

The project P_i is represented by activity-on-node networks, where nodes represent activities and arcs determine an order of activities execution. Consequently, the following activities order constraints are considered [4]:

- the $O_{i,k}$ activity follows the $O_{i,j}$ - th one:

$$co^{1}_{i,k}\colon s_{i,j} + t_{i,j} \leq s_{i,k}, \tag{3}$$

- the $O_{i,k}$ activity follows other activities:

$$co^{2}_{i,k}\colon (s_{i,j}+t_{i,j} \leq s_{i,k}) \wedge (s_{i,j+1}+t_{i,j+1} \leq s_{i,k}) \wedge (s_{i,j+2}+t_{i,j+2} \leq s_{i,k}) \wedge \ldots \wedge (s_{i,j+n}+t_{i,j+n} \leq s_{i,k}), \tag{4}$$

- the $O_{i,k}$ activity is followed by other activities:

$$co^{3}_{i,k}\colon (s_{i,k}+ t_{i,k} \leq s_{i,j}) \wedge (s_{i,k}+ t_{i,k} \leq s_{i,j+1}) \wedge (s_{i,k}+t_{i,k} \leq s_{i,j+2}) \wedge \ldots \wedge (s_{i,k}+t_{i,k} \leq s_{i,j+n}). \tag{5}$$

Each activity $O_{i,k}$ should be finished before *h-th* unit of time (h is a completion time of projects portfolio P), i.e. should follow the constraint:

$$co^{4}_{i,k}\colon s_{i,k} + t_{i,k} \leq h. \tag{6}$$

Constraints $co^{1}_{i,k}$, $co^{2}_{i,k}$, $co^{3}_{i,k}$, $co^{4}_{i,k}$, encompassing relations existing among activities $O_{i,k}$ produce the set of precedence constraints $Co=\{co^{1}_{i,k}, co^{2}_{i,k}, co^{3}_{i,k}, co^{4}_{i,k}\}$.

Note that limited resources amount may cause resource conflicts occurrence, i.e. requiring selection of relevant dispatching rules deciding about the order of resources allocation. In order to avoid such conflict the relevant constraints have to be taken into account. The set of conflict avoidance constraints $Cr=\{cr^{1}_{m,n,k}, cr^{2}_{m,n,k}\}$ introduced in [4] consist of:

$$cr^{1}_{m,n,k}: \sum_{i=1}^{lp}\sum_{j=1}^{lo_i}\left[dp_{i,j,k}\cdot \bar{1}(s_{m,n}+tp_{m,n,k}, s_{i,j}+tp_{i,j,k}, s_{i,j}+tz_{i,j,k})\right] \leq zo_{k,x_{m,n}+tp_{m,n}-1} \tag{7}$$

$\forall\, (m,n) \in \{(a,b) \mid a = 1,2,\ldots,lp,\ b = 1,2,\ldots,lo_a\}$,

where: lp – the number of projects, lo_a – the number of activities in the project P_a,

$\bar{1}(u,a,b)$ – an unary function determining the time of the resource occupation

$$\bar{1}(u,a,b) = 1(u-a) - 1(u-b)$$

$1(u)$ - the unit step function.

and

$$cr^{2}_{m,n,k}: \sum_{i=1}^{lp}\sum_{j=1}^{lo_i}\left[dp_{i,j,k}\cdot \bar{1}(vg_{k,d}, s_{i,j}+tp_{i,j,k}, s_{i,j}+tz_{i,j,k})\right] \leq zo_{k,vp_{k,d}-1} \tag{8}$$

$\forall d \in \{1,2,\ldots,q\}$

where: $vg_{k,i}$ - the moments $vg_{k,i} \in H$ at which the available number of the k-th resource units is changed, q – number of the characteristic points.

Note that $Cr=\{cr^{1}_{m,n,k}, cr^{2}_{m,n,k}\}$ regards of renewable resources. The similar ones should be considered in the case of nonrenewable resources, for instance concerning the money. The set of conflict avoidance constraints $Cn=\{cn^{1}_{m,n,k}\}$ introduced in [4] consist of:

$$cn^1_{m,n,k} : zn_k - \sum_{i=1}^{lp}\sum_{j=1}^{lo_i}\left[cr_{i,j,k} \cdot 1(s_{m,n} - s_{i,j} - tr_{i,j,k})\right] + \sum_{i=1}^{lp}\sum_{j=1}^{lo_i}\left[cw_{i,j,k} \cdot 1(s_{m,n} - s_{i,j} - ts_{i,j,k})\right] \geq 0 \quad (9)$$

$\forall(m,n) \in \{(a,b) \mid a = 1,2,\ldots,lp;\ b = 1,2,\ldots,lo_a\}$
where: lp –the number of projects, lo_a – the number of activities in the project P_a.

Finally, the considered set of constraints has the following form: $C = Co \cup Cr \cup Cn$.

Therefore, the reference model considered can be seen as the constraint problem $CS=((X,D),C)$. Consequently, depending on the questions stated the relevant context dedicated constraint satisfaction problem can be considered. The standard questions can be formulated either in the straight or reverse way. So, the new problems can be aimed both at determination of [1]:

- the criteria values implied by the assumed variables and constraints, for instance:

  Do the given activities' times guarantee completion of the project portfolio within assumed time horizon H?

- the variables guaranteeing expected values of the assumed goal functions, for example:

  What are the beginning times T_i of activities guaranteeing the project portfolio completion time does not exceed a given time horizon H?

The above questions belong to the class of so called problems formulated in a reverse way, i.e. problems our considerations are focused on. Some examples illustrating the above mentioned two sorts of context dedicated *CS* are discussed in the section below.

4 Illustrative Example

In order to illustrate such possibility let us concentrate on the decision problems formulated for resources allocation conflict resolution. Let us assume the activities compete with the access to the discrete resources of the both kind's renewable and non-renewable ones. As a programming environment enabling the reference model implementation the OzMozart language is used [15].

4.1 Routine Query: What Are the Moments of Activities Beginning?

Given the following projects portfolio, i.e. the set of projects $P = \{P_1, P_2, P_3, P_4\}$. Activities $O_{i,j}$ of projects are specified by corresponding sets: $P_1 = \{O_{1,1},\ldots,O_{1,10}\}$, $P_2 = \{O_{2,1},\ldots,O_{2,12}\}$, $P_3 = \{O_{3,1},\ldots,O_{3,11}\}$, $P_4 = \{O_{4,1},\ldots,O_{4,13}\}$. The relevant activity networks [1], [4] are shown on the following figures: Fig. 4, Fig. 5, Fig. 6 and Fig. 7.

Given the time horizon $H = \{0,1,\ldots,40\}$ ($h_{max} = 40$). Operation times for particular projects P_1, P_2, P_3, P_4 are determined by the following sequences:

$T_1 = (1, 2, 3, 4, 4, 8, 3, 2, 1, 6)$, $T_2 = (3, 1, 6, 3, 2, 5, 1, 5, 2, 4, 2, 1)$,
$T_3 = (3, 7, 2,7, 2, 1,8, 3, 3, 4, 8)$, $T_4 = (3, 3, 2, 8, 3, 1, 4, 1, 8, 4, 3, 3, 8)$.

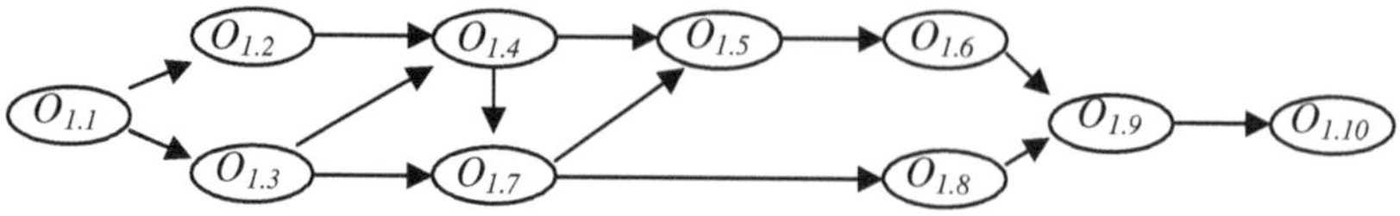

Fig. 4. Activity network of the project P_1

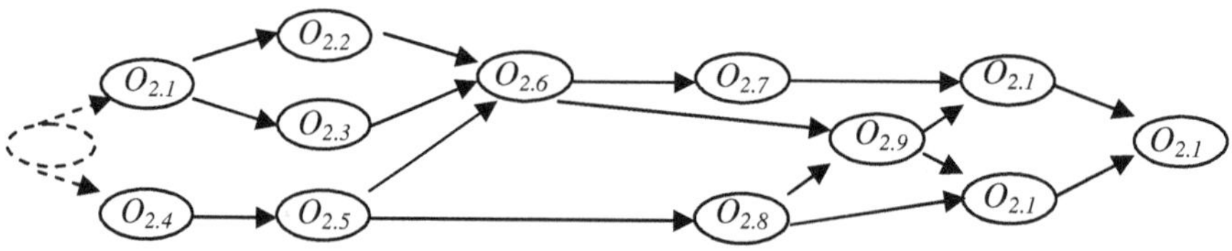

Fig. 5. Activity network of the project P_2

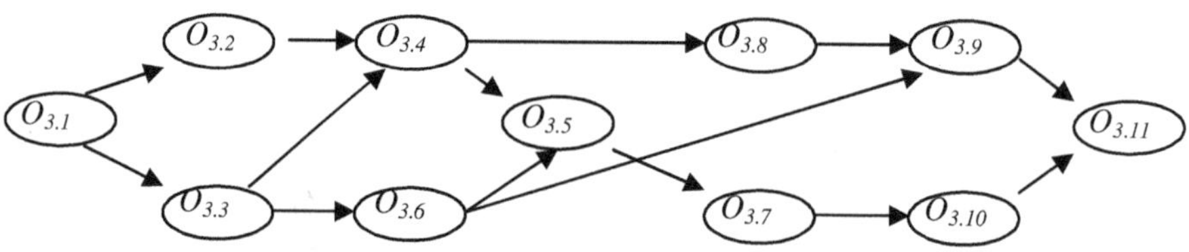

Fig. 6. Activity network of the project P_3

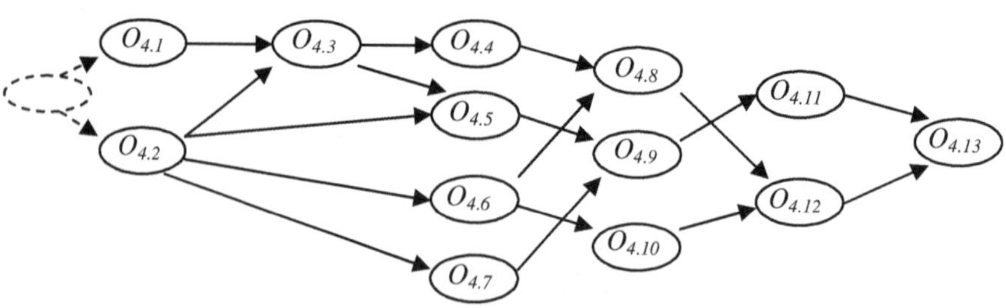

Fig. 7. Activity network of the project P_4

Given are three kinds of renewable resources ro_1, ro_2, ro_3. Resources amounts are limited by following number of units: 11, 14, 12, respectively. Resource amounts are constant in whole time horizon H. That is assumed an amount of resources allocated to the activity at the moment of its beginning can be released only by this activity and only at the moment of its completion. The amounts of particular resources required by projects' P_1, P_2, P_3, P_4 activities are given in the following tables: Table 1, Table 2, Table 3, and Table 4.

Table 1. Amounts of resources required by activities of the project P_1 (the sequences $DP_{1,1}$, $DP_{1,2}$, $DP_{1,3}$)

	$O_{1,1}$	$O_{1,2}$	$O_{1,3}$	$O_{1,4}$	$O_{1,5}$	$O_{1,6}$	$O_{1,7}$	$O_{1,8}$	$O_{1,9}$	$O_{1,10}$
$DP_{1,1}$	3	1	1	1	1	1	2	1	2	1
$DP_{1,2}$	2	1	2	1	1	2	3	3	1	1
$DP_{1,3}$	2	2	3	1	1	1	1	1	2	1

Table 2. Amounts of resources required by activities of the project P_2 (the sequences $DP_{2,1}$, $DP_{2,2}$, $DP_{2,3}$)

	$O_{2,1}$	$O_{2,2}$	$O_{2,3}$	$O_{2,4}$	$O_{2,5}$	$O_{2,6}$	$O_{2,7}$	$O_{2,8}$	$O_{2,9}$	$O_{2,10}$	$O_{2,11}$	$O_{2,12}$
$DP_{2,1}$	4	3	2	2	1	1	1	3	1	2	2	2
$DP_{2,2}$	1	2	3	1	2	1	2	1	1	2	1	1
$DP_{2,3}$	2	1	1	1	3	1	2	2	2	1	1	1

Table 3. Amounts of resources required by activities of the project P_3 (the sequences $DP_{3,1}$, $DP_{3,2}$, $DP_{3,3}$)

	$O_{3,1}$	$O_{3,2}$	$O_{3,3}$	$O_{3,4}$	$O_{3,5}$	$O_{3,6}$	$O_{3,7}$	$O_{3,8}$	$O_{3,9}$	$O_{3,10}$	$O_{3,11}$
$DP_{3,1}$	2	4	1	2	2	2	1	2	2	1	3
$DP_{3,2}$	2	1	3	2	2	2	1	1	1	2	2
$DP_{3,3}$	2	4	1	2	2	2	1	2	2	1	3

Table 4. Amounts of resources required by activities of the project P_4 (the sequences $DP_{4,1}$, $DP_{4,2}$, $DP_{4,3}$)

	$O_{4,1}$	$O_{4,2}$	$O_{4,3}$	$O_{4,4}$	$O_{4,5}$	$O_{4,6}$	$O_{4,7}$	$O_{4,8}$	$O_{4,9}$	$O_{4,10}$	$O_{4,11}$	$O_{4,12}$	$O_{4,13}$
$DP_{4,1}$	1	2	3	4	3	2	2	1	1	1	3	1	4
$DP_{4,2}$	1	1	1	2	1	2	1	3	2	2	2	1	2
$DP_{4,3}$	1	2	2	1	1	2	4	1	2	2	2	1	2

Table 5. Amount of used up (*CR*) and generated (*CW*) non-renewable resources required by activities of the project P_1 (the sequences $CR_{1,1}$, $CR_{1,2}$, $CW_{1,1}$, $CW_{1,2}$)

	$O_{1,1}$	$O_{1,2}$	$O_{1,3}$	$O_{1,4}$	$O_{1,5}$	$O_{1,6}$	$O_{1,7}$	$O_{1,8}$	$O_{1,9}$	$O_{1,10}$
$CR_{1,1}$	1	1	2	1	2	1	3	1	1	1
$CR_{1,2}$	1	2	1	1	1	0	1	0	1	1
$CW_{1,1}$	3	2	0	2	4	4	2	0	2	4
$CW_{1,2}$	1	2	3	2	2	2	0	2	1	2

It is assumed some activities besides of renewable resources require also non-renewable ones. Given are two kinds of non-renewable resources rn_1, rn_2. Initial amount of the resource rn_1 is equal to 10 units, and of the resource rn_2 is equal to 7 units. Activities may use up and generate some number of resources rn_1, rn_2 units. It is assumed each activity uses up some resource units at the beginning and generates some resource units at the activity's end. The amounts of used up and generated resource rn_l units determine sequences: $CR_{i,j}$, $CW_{i,j}$ respectively in the following tables: Table 5, Table 6, Table 7, and Table 8.

Table 6. Amount of used up (*CR*) and generated (*CW*) non-renewable resources required by activities of the project P_2 (the sequences $CR_{2,1}$, $CR_{2,2}$, $CW_{2,1}$, $CW_{2,2}$)

	$O_{2,1}$	$O_{2,2}$	$O_{2,3}$	$O_{2,4}$	$O_{2,5}$	$O_{2,6}$	$O_{2,7}$	$O_{2,8}$	$O_{2,9}$	$O_{2,10}$	$O_{2,11}$	$O_{2,12}$
$CR_{2,1}$	1	0	1	2	1	1	1	3	1	0	1	1
$CR_{2,2}$	3	2	1	2	0	2	3	2	2	2	1	2
$CW_{2,1}$	3	2	0	2	1	2	0	2	0	2	0	1
$CW_{2,2}$	3	2	1	2	0	2	3	2	2	2	1	2

Table 7. Amount of used up (*CR*) and generated (*CW*) non-renewable resources required by activities of the project P_3 (the sequences $CR_{3,1}$, $CR_{3,2}$, $CW_{3,1}$, $CW_{3,2}$)

	$O_{3,1}$	$O_{3,2}$	$O_{3,3}$	$O_{3,4}$	$O_{3,5}$	$O_{3,6}$	$O_{3,7}$	$O_{3,8}$	$O_{3,9}$	$O_{3,10}$	$O_{3,11}$
$CR_{3,1}$	1	1	2	1	1	1	0	1	3	1	1
$CR_{3,2}$	0	1	1	0	2	1	1	1	3	1	0
$CW_{3,1}$	2	3	2	0	2	1	2	2	2	3	2
$CW_{3,2}$	3	2	1	2	0	2	3	2	2	2	1

Table 8. Amount of used up (*CR*) and generated (*CW*) non-renewable resources required by activities of the project P_4 (the sequences $CR_{4,1}$, $CR_{4,2}$, $CW_{4,1}$, $CW_{4,2}$)

	$O_{4,1}$	$O_{4,2}$	$O_{4,3}$	$O_{4,4}$	$O_{4,5}$	$O_{4,6}$	$O_{4,7}$	$O_{4,8}$	$O_{4,9}$	$O_{4,10}$	$O_{4,11}$	$O_{4,12}$	$O_{4,13}$
$CR_{4,1}$	1	1	2	1	1	1	0	1	3	1	1	1	1
$CR_{4,2}$	0	1	1	0	2	1	1	1	3	1	0	1	1
$CW_{4,1}$	2	3	2	0	2	1	2	2	2	3	2	3	2
$CW_{4,2}$	3	2	1	2	0	2	3	2	2	2	1	2	2

In the context of the above assumed data, i.e. following requirements of the standard *CS* the following question is considered.

q_1: *Does there exist a schedule following constraints assumed on availability of renewable and non-renewable resources and NPV > 0 such that production orders completion time not exceeds the deadline H?*

Note that the schedule we are looking for is determined by moments $s_{i,j}$ the activities start their execution [1], [2].

Solution to the problem results in determination of moments the activities start their execution $s_{i,j}$. So, the solution we are searching for has the form of the following sequences: $S_1 = (s_{1,1},\dots,s_{1,10})$, $S_2 = (s_{2,1},\dots,s_{2,12})$, $S_3 = (s_{3,1},\dots,s_{3,11})$, $S_4 = (s_{4,1},\dots,s_{4,13})$.

Of course, the elements of sequences S_1, S_2, S_3, S_4 have to follow activities order constraints from Fig. 4 – Fig. 7 as well as constraints assumed on renewable (see tables, Table 1 - Table 4) and non-renewable (see tables Table 5 - Table 8) resources allocation (guaranteeing deadlock avoidance). Constraints have a form similar to the formulas (2), (3), (4), (5), (6), (7), (8).

The question considered implies the relevant context dedicated CS.

Given $CS = ((X, D), C)$:

where:

- the set of decision variables X containing,
  the input variables: $U = \{s_{1,1}, s_{1,2}, \ldots, s_{1,lo1}, s_{2,1}, \ldots s_{lp,lolp}\}$
  the output variables: $Y = \{h, NPV\}$
- the family of domains D, where the domains of variables T_i, $Tp_{i,j}$, $Tz_{i,j}$, $Dp_{i,j}$, $Cr_{i,j}$, $Cw_{i,j}$, $Tr_{i,j}$, $Ts_{i,j}$ are determined by Table. 1 – Table. 8, and variables corresponding to the beginning times of activities $s_{i,j} \in [0,50]$
- the set of constrains C.

The considered problem CS and the question q_1 can be specified in terms of CCS:

$$CCS = ((X, D), C \cup Cy),$$

where: Cy - the set of output constraints corresponding to q_1: $Cy = \{cy_1, cy_2\}$,

$$cy_1: h \le 40; \quad cy_2: NPV > 0.$$

Result of CCS examination by the consistency-check procedure is positive, so there exists the solution Vu following the all constraints $C \cup Cy$.

The results obtained from OzMozart implemented procedure consists of the non-empty set of solutions Vu. Admissible values of considered variables $s_{i,j}$ have the following values:

$S_1 = (0, 1, 1, 4, 11, 15, 8, 11, 23, 24)$, $S_2 = (0, 3, 7, 10, 13, 15, 20, 17, 23, 25, 25, 29)$,
$S_3 = (0, 3, 3, 10, 17, 5, 19, 17, 20, 27, 31)$, $S_4 = (0, 0, 3, 5, 5, 3, 3, 13, 8, 6, 14, 16, 19)$.

The NPV index value calculated for projects: P_1, P_2, P_3, P_4 follows the requirement $NPV > 0$, i.e. $NPV_{P1} = 0.3649$, $NPV_{P2} = 2.4775$, $NPV_{P3} = 1.3248$, $NPV_{P4} = 0.8134$.

The graphical representation of the projects portfolio schedule is show in the Fig. 8. The schedule obtained follows all constrains imposed by an enterprise capability and projects execution requirements. The system considered allows one to obtain the Gantt's-like chart illustrating the rates of resources usage both renewable and non-renewable ones. An example of graphical representation of the resource zo_1 usage rate containing assumed resource's limit in whole time horizon is shown on Fig. 9. It can be observed the assumed resource's limit was not exceeded, the same regards of resources zo_2, zo_3.

In turn, the Fig. 10 illustrates changes regarding the rate of resource usage concerning of the non-renewable resource zn_2. That is easy to note that the assumed minimal level of resource usage equal to 0 was never exceeding in whole time horizon. The same remark concerns the resource zn_1.

Therefore, the example presented illustrates the capability of an interactive multi-criteria project planning (e.g. taking into account a particular project deadline, projects portfolio deadline, resources limits, and so on) approach to projects prototyping issues. The problem of the size just considered took less than 5 minutes (i.e. finding of the first solution Vu) (the AMD Athlon(tm)XP 2500 + 1.85 GHz, RAM 1,00 GB platform has been used).

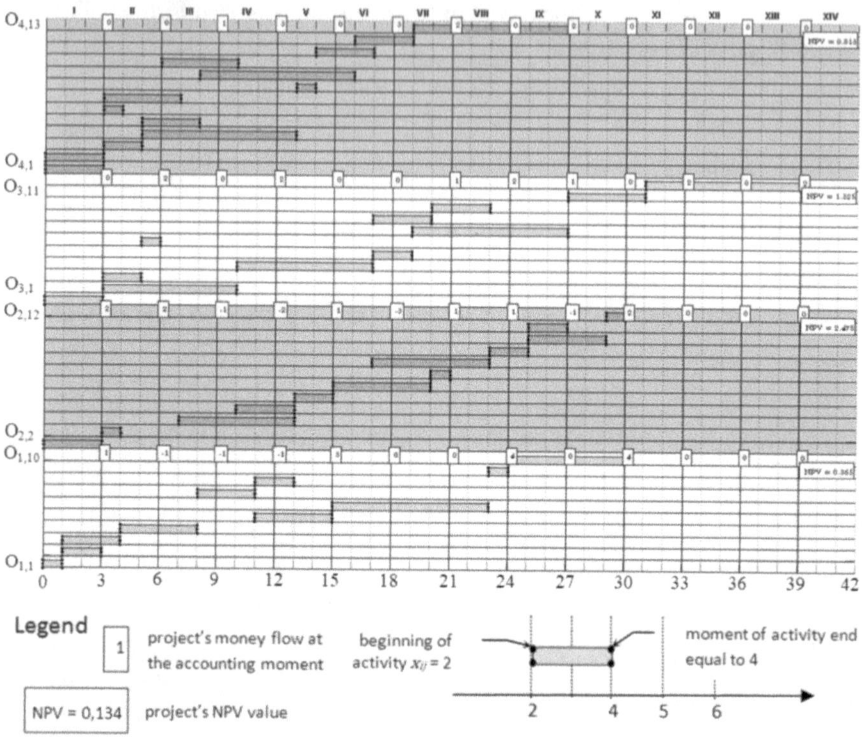

Fig. 8. Projects portfolio schedule

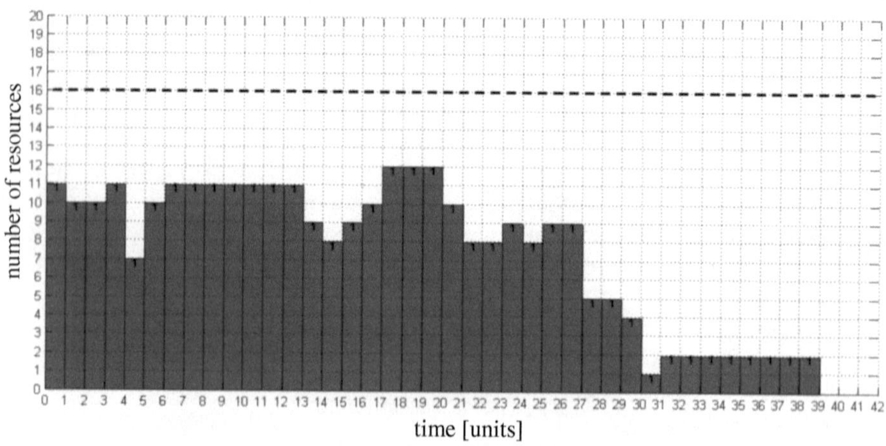

Fig. 9. Gantt's-like chart of the renewable resource zo_1 usage

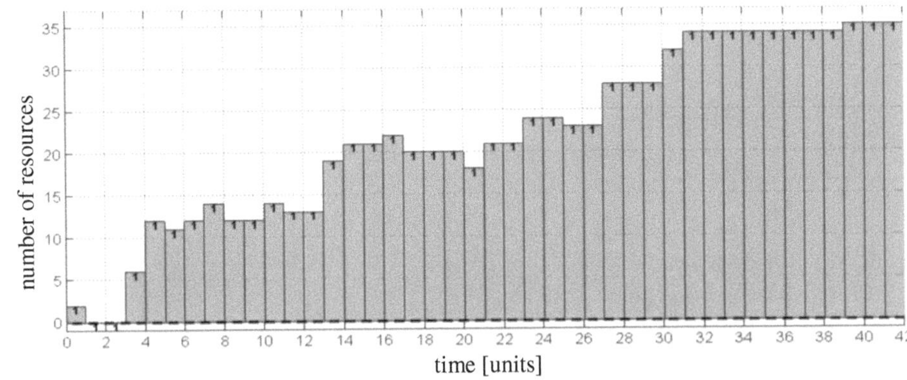

Fig. 10. Gantt's-like chart of the non-renewable resource zn_2 usage

4.2 Routine Query: What Are the Times of Activities Duration?

Given the following projects portfolio, i.e. the set of projects $P = \{P_1, P_2, P_3, P_4\}$ specified by the same activity networks (see Fig. 4, Fig. 5, Fig. 6 and Fig. 7) and resources allocations (see the Table 2 – Table 9) as in the Example 1. However, the new time horizon $H = \{0,1,\ldots,36\}$ is considered.

Given the projects port folio containing the following projects P_1, P_2, P_3, P_4. The makespan admissible cannot exceed 36 units of time. Activities operations times are not known, however constraints determining their execution constraints are given. For instance, the following constraint: $t_{3,7} + t_{3,9} = 11$ that means the activities operation times are tightly linked, i.e. increasing of activity time associated to one operation (for example to $O_{3,7}$) results in decreasing of the another one (in this case $O_{3,9}$). The set of constraints considered are contained In the Table 9.

Table 9. Constraints linking activities $O_{i,j}$ execution times.

	Constraint		Constraint
ct_1	$t_{3,7} + t_{3,9} = 11$	ct_6	$t_{3,3} + t_{3,4} = 9$
ct_2	$t_{4,12} + t_{4,13} = 11$	ct_7	$t_{2,3} + t_{2,4} = 9$
ct_3	$t_{4,3} + t_{4,4} = 11$	ct_8	$t_{2,3} + t_{2,4} = 9$
ct_4	$t_{1,5} + t_{1,6} = 12$	ct_9	$2t_{2,5} + t_{3,3} = 8$
ct_5	$t_{1,9} + t_{1,10} = 7$	ct_{10}	$2t_{4,1} + t_{2,8} = 12$

Therefore, taking into account above mentioned assumptions as well as the other ones used in the point 4.1, the problem considered now reduces to the question:

q_2: *What values and of what variables T_1, T_2, T_3, T_4 guarantee the makespan of the projects portfolio does not exceed a given deadline subject to limits imposed on available amounts of renewable and non-renewable resources as well as NPV>0?*

In order to response to this question the values of the following sequences are sought: $T_1 = (t_{1,1},....,t_{1,10})$, $T_2 = (t_{2,1},....,t_{2,12})$, $T_3 = (t_{3,1},...,t_{3,11})$, $T_4 = (t_{4,1},...,t_{4,13})$ and $S_1 = (s_{1,1},....,s_{1,10})$, $S_2 = (s_{2,1},....,s_{2,12})$, $S_3 = (s_{3,1},...,s_{3,11})$, $S_4 = (s_{4,1},...,s_{4,13})$.

Taking into account the data assumed consider the following formulation of the relevant *CCS*.

Given $CS = ((X, D), C)$,

where:

- the set of decision variables *X*, containing:
  the input variables:
  $U = \{ t_{1,1}, t_{1,2},...,t_{1,lo1}, t_{2,1},...t_{lp,lolp}, s_{1,1}, s_{1,2},...,s_{1,lo1}, s_{2,1},...,s_{lp,lolp}\}$
  the output variables: $Y = \{h, NPV\}$
- the family of domains *D*, where the domains of variables T_i, $Tp_{i,j}$, $Tz_{i,j}$, $Dp_{i,j}$, $Cr_{i,j}$, $Cw_{i,j}$, $Tr_{i,j}$, $Ts_{i,j}$ are determined by Table. 1 – Table. 8, and variables corresponding to the beginning times of activities $s_{i,j} \in [0,50]$, and variables corresponding to activities duration times $t_{i,j} \in [1,15]$,
- the set of constrains $C \cup \{ct_1, ct_2,...,ct_{10}\}$ (following the Table 9),

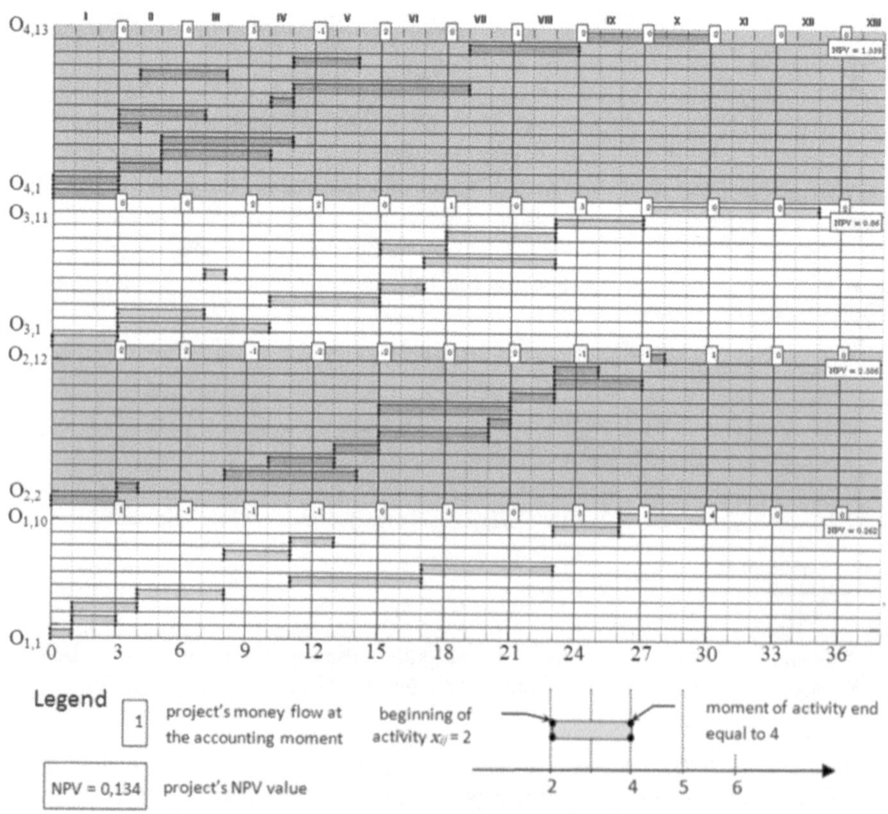

Fig. 11. The Gantt's chart of the projects portfolio execution

The considered problem CS and the question q_2 can be specified in terms of CCS:

$$CCS = ((X, D), C \cup Cy),$$

where: Cy - the set of output constraints corresponding to q_1: $Cy = \{cy_1, cy_2\}$,

$$cy_1: h \leq 36; \; cy_2: NPV > 0$$

Result of CCS examination by the consistency-checking procedure is positive, so there exists the solution Vu following the all constraints $C \cup Cy$.

The results obtained from OzMozart implemented procedure consists of the non-empty set of solutions Vu. The first admissible solution has been obtained in 250 s. So, the sequences of obtained activities operation times are as follows:

$T_1 = (1, 2, 3, 4, 6, 6, 3, 2, 3, 4)$, $T_2 = (3, 1, 6, 3, 2, 5, 1, 6, 2, 4, 2, 1)$,
$T_3 = (3, 7, 4, 5, 2, 1, 6, 3, 5, 4, 8)$, $T_4 = (3, 3, 2, 5, 6, 1, 4, 1, 8, 4, 3, 5, 6)$.

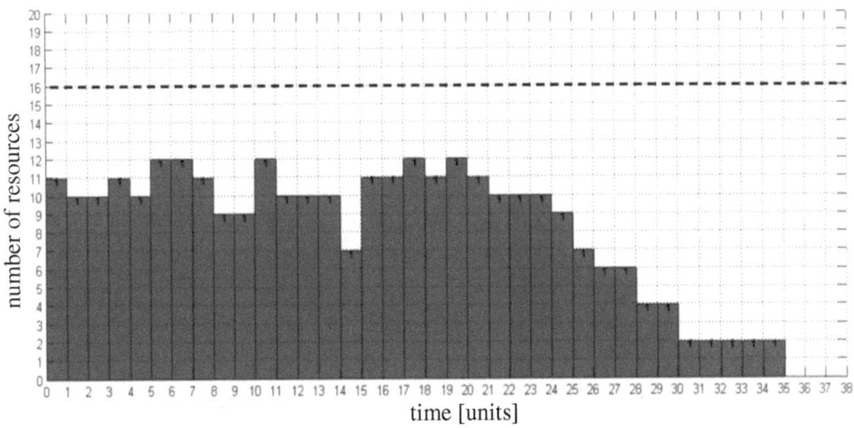

Fig. 12. Illustration of the resource zo_1 changes

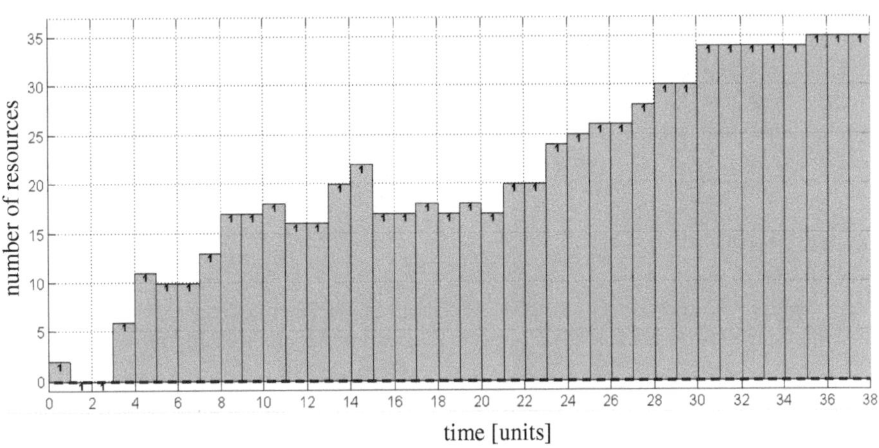

Fig. 13. Illustration of the resource zn_1 changes

In turn, the sequences of the moments of activities beginning are as follows:

$S_1 = (0,1,1,4,11,17,8,11,23,26)$, $S_2 = (0,3,8,10, 13, 15, 20, 15, 21, 23, 23, 27)$,
$S_3 = (0,3,3,10,15,7,17,15,18,23,27)$, $S_4 = (0,0,3,5,5,3,3,10,11,4,11,19,24)$.

The NPV index value calculated for projects: P_1, P_2, P_3, P_4 follow the requirement $NPV > 0$, i.e $NPV_{P1} = 0.262$, $NPV_{P2} = 2.386$, $NPV_{P3} = 0.86$, $NPV_{P4} = 1.339$.

The graphical illustration of the solution obtained is shown in Fig. 11, where on the base of the Gantt's chart an admissible activities schedule, i.e. the beginning and duration of activities following all the constraints imposed by availability as well as allocation of renewable and non-renewable resources is provided. The corresponding charts illustrating the changes of resources zo_1 and zn_2 availability are shown in Fig. 12 and Fig. 13, respectively.

5 Concluding Remarks

The introduced *CP*-based reference model provides a formal framework allowing one to formulate the projects portfolio planning problems in direct and reverse way. In other words it provides a base to an interactive task oriented decision support tools designing. That offers a possibility to respond to the questions like: What values and of what variables guarantee the production orders will complete due to assumed values of performance indexes? (besides of such standard questions as: What is the project portfolio completion time?).

The main idea standing behind of this approach lies in searching for the conditions guaranteeing the existence of responses to the standard queries as well as for conditions guaranteeing the employed search strategies can be used in on-line mode for the given size of project planning problems. Therefore, the reference model of decision problems can be seen as a knowledge base kernel of a methodology aimed at designing of dedicated and interactive decision support systems.

The routine questions can be answered in the case they are formulated in terms of CCS_P. That guarantee follows from *CCS* consistency examination, i.e. *LAM*-based consistency-check procedure. CCS_q can be treated as a knowledge base where already mentioned procedure and *LAM* implemented in *CLP*/*CP* languages play a role of an inference engine. Consequently, the direct and reverse formulations of decision problems (size of that typical for SMEs) can be easily implemented in OzMozart language and then resolved in an on-line mode.

References

[1] Bach, I., Bocewicz, G., Banaszak, Z.: Constraint programming approach to time-window and multiresource-constrained projects portfolio prototyping. In: Nguyen, N.T., Borzemski, L., Grzech, A., Ali, M., et al. (eds.) IEA/AIE 2008. LNCS (LNAI), vol. 5027, pp. 767–776. Springer, Heidelberg (2008)

[2] Bach, I., Wójcik, R., Bocewicz, G.: Projects portfolio prototyping subject to imprecise activities specification. In: Conference proceedings of 14th International Congress of Cybernetics and Systems of WOSC – ICCS 2008, Wroclaw, Poland, pp. 261–272 (2008)

[3] Banaszak, Z.: CP-based decision support for project-driven manufacturing. In: Józefowska, J., Węglarz, J. (eds.) Perspectives in Modern Project Scheduling. International Series in Operations Research and Management Science, vol. 92, pp. 409–437. Springer, New York (2006)

[4] Banaszak, Z., Bocewicz, G., Bach, I.: CP-driven Production Process Planning in Multiproject Environment. In: Decision Making in Manufacturing and Services, vol. 2(1-2), pp. 5–32 (2008)

[5] Banaszak, Z., Zaremba, M.,, Muszyński, W.: CP-based decision making for SME. In: Horacek, P., Simandl, M., Zitek, P. (eds.) Preprints of the 16th IFAC World Congress, DVD, Prague, Czech Republic (2005)

[6] Barták, R.: Incomplete Depth-First Search Techniques: A Short Survey. In: Figwer, J. (ed.) Proceedings of the 6th Workshop on Constraint Programming for Decision and Control, pp. 7–14 (2004)

[7] Beale, E.M.L.: Branch and bound methods for mathematical programming systems. In: Hammer, P.L., Johnson, E.L., Korte, B.H. (eds.) Discrete Optimization II, pp. 201–219. North Holland Publishing Co., Amsterdam (1979)

[8] Bocewicz, G., Banaszak, Z., Wójcik, R.: Design of admissible schedules for AGV systems with constraints: a logic-algebraic approach. In: Nguyen, N.T., Grzech, A., Howlett, R.J., Jain, L.C. (eds.) KES-AMSTA 2007. LNCS (LNAI), vol. 4496, pp. 578–587. Springer, Heidelberg (2007)

[9] Bubnicki, Z.: Learning processes and logic-algebraic method for the systems with knowledge representation. Systems analysis and management. PAS, Warsaw (1999)

[10] Chanas, S., Komburowski, J.: The use of fuzzy variables in PERT. Fuzzy Sets and Systems 5(1), 11–19 (1981)

[11] Dubois, D., Fargier, H., Fortemps, P.: Fuzzy scheduling: Modeling flexible constraints vs. coping with incomplete knowledge. European Journal of Operational Research 147, 231–252 (2003)

[12] Martinez, E.C., Duje, D., Perez, G.A.: On performance modeling of project-oriented production. Computers and Industrial Engineering 32, 509–527 (1997)

[13] Puget, J.-F.: A C++ Implementations of CLP. In: Proceeding of SPICS 1994 (1994)

[14] Russell, S., Norvig, P.: Artificial Intelligence: A Modern Approach, 2nd edn. Prentice Hall, Englewood Cliffs (2002)

[15] Schutle, H., Smolka, G., Wurtz, J.: Finite Domain Constraint Programming in Oz. German Research Center for Artificial Intelligence, Germany, D-66123 Saarbrucken (1998)

[16] Sung, C.S.: A Production Planning Model for Multi-Product Facilities. Journal of the Operations Research Society of Japan 28(4), 345–385 (1985)

[17] Van Hentenryck, P.: Constraint Logic Programming. Knowledge Engineering Review 6, 151–194 (1991)

[18] http://choco-solver.net

[19] http://www.sics.se/sicstus.html

[20] http://labix.org/python-constraint

CSP Techniques for Solving Combinatorial Queries within Relational Databases

Malek Mouhoub and Chang Feng

University of Regina
Wascana Parkway, Regina, SK, Canada, S4S 0A2
{mouhoubm,feng202c}@cs.uregina.ca

Abstract. A combinatorial query is a request for tuples from multiple relations that satisfy a conjunction of constraints on tuple attribute values. Managing combinatorial queries using the traditional database systems is very challenging due to the combinatorial nature of the problem. Indeed, for queries involving a large number of constraints, relations and tuples, the response time to satisfy these queries becomes an issue. To overcome this difficulty in practice we propose a new model integrating the Constraint Satisfaction Problem (CSP) framework into the database systems. Indeed, CSPs are very popular for solving combinatorial problems and have demonstrated their ability to tackle, in an efficient manner, real life large scale applications under constraints. In order to compare the performance in response time of our CSP-based model with the traditional way for handling combinatorial queries and implemented by MS SQL Server, we have conducted several experiments on large size databases. The results are very promising and show the superiority of our method comparing to the traditional one.

1 Introduction

It is generally acknowledged that relational databases have become the most significant and the main platform [1] to implement data management and query searching in practice. The efficiency and veracity for query solving is acquiring more concerns and has been improved among various database systems. However, along with the complex requirement for query from the customer, a problem is triggered: the conventional relational database system often fails to obtain answers to the combinatorial query efficiently. Indeed, current Relational Database Management Systems (RDBMS) require a lot of time effort in answering complex combinatorial queries. [2] has given an accurate definition of a combinatorial query: *a request for tuples from multiple relations that satisfy a conjunction of constraints on tuple attribute values.* The following example illustrates the above definition.

Example 1. Let us assume we have a relational database containing three tables describing computer equipments: CPU (*CPU_id*, Price, Frequency, Quality), Memory (*Memory_id*, Price, Frequency, Quality), and Motherboard (*Motherboard_id*, Price, Frequency, Quality).

N.T. Nguyen and E. Szczerbicki (Eds.): Intel. Sys. for Know. Management, SCI 252, pp. 131–151.
springerlink.com

CPU_id	Price	Freq	Quality
1	200	5	80
2	500	10	80
3	1100	8	90
4	2000	15	95

Mem_id	Price	Freq	Quality
1	1000	7	80
2	900	9	90
3	800	10	90
4	1000	15	100

Mboard_id	Price	Freq	Quality
1	100	6	50
2	200	6	50
3	300	11	80
4	400	15	70

```
Select CPU_id, Memory_id,Motherboard_id
From CPU, Memory, Motherboard  Where
 CPU.Price + Memory.Price + Motherboard.Price < 1500
 And CPU.Frequency + Memory.Frequency + Motherboard.Frequency < 30
 And CPU.Quality + Memory.Quality + Motherboard.Quality < 250
```

Fig. 1. An example of a combinatorial query.

Figure 1 illustrates the three tables and an example of a related combinatorial query. Here the combinatorial query involves a conjunction of three arithmetic constraints on attributes of the three tables.

Although a variety of query evaluation technologies such as hash-join [3], sorted-join, pair-wise join [4,1] have been implemented into the relational model attempting to optimize searching process, the effect is extremely unsatisfactory since most of them concentrate on handling simple constraints, not the complex ones [2,3,5,6]. For instance, in example 1 the arithmetic query involves a conjunction of three constraints that should be satisfied together. This requires higher complexity and selectivity [6] since the traditional model has to perform further complex combinatorial computations after searching for each sub query (corresponding to each of these 3 constraints) which will cost more time effort. In order to overcome this difficulty in practice, we propose in this paper a model that integrates the Constraint Satisfaction Problem (CSP) framework in the relational database system. Indeed, CSPs are very powerful for representing and solving discrete combinatorial problems. In the past three decades, CSPs have demonstrated their ability to efficiently tackle large size real-life applications, such as scheduling and planning problems, natural language processing, business applications and scene analysis. More precisely, a CSP consists of a finite set of variables with finite domains, and a finite set of constraints restricting the possible combinations of variable values [7,8]. A solution tuple to a CSP is a set of assigned values to variables that satisfy all the constraints. Since a CSP is known to be an NP-hard problem in general[1], a backtrack search algorithm of exponential time cost is needed to find a complete solution. In order to overcome this difficulty in practice, constraint propagation techniques have been proposed [7,9,11,10]. The goal of theses techniques is to reduce the size of the search space before and during the backtrack search. Note that some work on CSPs and databases has already been proposed in the literature. For instance, in [12] an analysis of the similarities between database theories and CSPs has been conducted. Later, [2] has proposed a general framework for modeling the combinatorial aspect, when dealing

[1] There are special cases where CSPs are solved in polynomial time, for instance, the case where a CSP network is a tree [9,10].

with relational databases, into a CSP. Query searching has then been improved by CSP techniques. Other work on using CSP search techniques to improve query searching has been reported in [13,14,15].

In this paper, our aim is to improve combinatorial query searching using the CSP framework. Experimental tests comparing our model to the MS SQL Sever demonstrate the superiority in time efficiency of our model when dealing with complex queries and large databases.

In the next section we will show how can the CSP framework be integrated within the RDBMS. Section 3 describes the details of our solving method. Section 4 provides the structure of the RDBMS with the CSP module. The experimental tests we conducted to evaluate the performance of our model are reported in section 5. We finally conclude in section 6 with some remarks and possible future work.

2 Mapping a Database with the Combinatorial Query into a CSP

There are many similarities between the structure of a database and the CSP model. Each table within the database can be considered as a relation (or a CSP constraint) where the table attributes correspond to the variables involved by the constraint. The domain of each variable corresponds to the range of values of the corresponding attribute. The combinatorial query is here an arithmetic constraint of the related CSP. For instance, the database of example 1 with the combinatorial query can be mapped into the CSP defined as follows.

- **Variables:**
  - CPU.Price,
  - CPU.Frequency,
  - CPU.Quality,
  - Memory_id.Price,
  - Memory_id.Frequency,
  - Memory_id.Quality,
  - Motherboard_id.Price,
  - Motherboard_id.Frequency
  - and Motherboard_id.Quality.
- **Domains:** Domains of the above variables.
  - CPU.Price: $\{200, 500, 1100, 2000\}$.
  - CPU.Frequency: $\{7, 9, 10, 15\}$.
  - CPU.Quality: $\{80, 90, 95\}$.
  - Memory_id.Price: $\{800, 900, 1000\}$.
  - Memory_id.Frequency: $\{7, 9, 10, 15\}$.
  - Memory_id.Quality: $\{80, 90, 100\}$.
  - Motherboard_id.Price: $\{100, 200, 300, 400\}$.
  - Motherboard_id.Frequency: $\{6, 11, 15\}$.
  - Motherboard_id.Quality: $\{100, 200, 300, 400\}$.

- **Constraints:**
  - **Semantic Constraints:**
    - ∗ CPU(*CPU_id*, Price, Frequency, Quality)
    - ∗ Memory(*Memory_id*, Price, Frequency, Quality)
    - ∗ Motherboard(*Motherboard_id*, Price,Quality)
  - **Arithmetic constraint:**

```
CPU.Price + Memory.Price + Mother.Price < 1500
And CPU.Frequency + Memory.Frequency + Motherboard.Frequency < 30
And CPU.Quality + Memory.Quality + Motherboard.Quality < 250
```

After this conversion into a CSP, we use CSP search techniques, that we will describe in the next section, in order to handle the combinatorial query in an efficient manner.

3 Solving Method

As we mentioned earlier, solving a CSP requires a backtrack search algorithm of exponential time cost. In order to overcome this difficulty in practice, local consistency techniques have been proposed [7,9,11,10]. The goal of theses techniques is to reduce the size of the search space before and during the backtrack search. More precisely, local consistency consists of checking to see if a consistency is satisfied on a subset of variables. If the local consistency is not successful then the entire CSP is not consistent. In the other case many values that are locally inconsistent will be removed which will reduce the size of the search space. Various local consistency techniques have been proposed [7]: node consistency which checks the consistency according to unary constraint of each variable, arc consistency which checks the consistency between any subset sharing one constraint, path consistency which checks the consistency between any subset sharing two constraints ..., etc.

In this paper we use arc consistency in our solving method. More formally, arc consistency [7] works as follows. Given a constraint $C(X_1,\ldots,X_p)$ then arc consistency is enforced through C on the set $X_1,\ldots,X_p$ as follows. Each value $v \in D(X_i)$ $(1 \le i \le n)$ is removed if it does not have a support (value such that C is satisfied) in at least one domain $D(X_j)$ $(1 \le j \le n$ and $j \neq i)$. $D(X_i)$ and $D(X_j)$ are the domains of X_i and X_j respectively.

Our method is described as follows.

1. First, arc consistency is applied on all the variables to reduce their domains. If a given variable domain is empty then the CSP is inconsistent (this will save us the next phase) which means that the query cannot be satisfied.
2. A backtrack search algorithm using arc consistency [7,9] is then performed as follows. We select at each time a variable and assign a value to it. Arc consistency is then performed on the domains of the non assigned variables in order to remove some inconsistent values and reduce the size of the search space. If one variable domain becomes empty then assign another value to the current variable or backtrack to the previously assigned variable in order to assign another value to this

latter. This backtrack search process will continue until all the variables are assigned values in which case we obtain a solution (satisfaction of the combinatorial query) or the entire search space has been explored without success. Note that during the backtrack search variables are selected following the *most constrained first* policy. This rule consists of picking the variable that appears in the largest number of constraints.

Before we present the details of the arc consistency algorithm, let us see with the following examples how arc consistency, through arithmetic and semantic constraints, is used to reduce the domains of the different attributes in both phases of our solving method.

Example 2. Let us consider the database of example 1. The domain of the variable CPU.Price is equal to $\{200, 500, 1100, 2000\}$. Using arc consistency with the subquery `CPU.Price + Memory.Price + Mother.Price < 1500`, we can remove the values 1100 and 2000 from the domain of CPU.price. Indeed there are no values in the domains of Memory.Price and Mother.Price such that the above arithmetic constraint is satisfied for 1100 and 2000.

Example 3. Let us now take another example where each attribute has a large number of values within a given range. For example let us consider the following ranges for the attributes of the table CPU.

```
200 <= Price <= 2000.
5 <= Frequency <= 15.
80 <= Quality <=95.
```

Using the above range for Price and the subquery:

```
CPU.Price + Memory.Price + Mother.Price < 1500
```

we can first reduce, the range of the attribute CPU.Price as follows:

```
200 <= CPU.Price < 1500.
```

When using the attributes Memory.Price and Motherboard.Price we reduce the upper bound of the above range as follows.

```
CPU.Price < 1500 - [Min(Memory.Price) + Min(Motherboard.Price)]
          < 1500 - [800 + 100]
          < 600
```

We will finally obtain the following range for CPU.Price:

```
200 <= CPU.Price < 600.
```

During the second phase of our solving method arc consistency is used through arithmetic and semantic constraints. The following example shows how an assignment of a value to a variable, during backtrack search, is propagated using arc consistency through the arithmetic constraint to update the domains (ranges) of the non assigned variables which will reduce the size of the search space.

Example 4. Let us consider the arithmetic constraint $A+B+C < 100$, where the domains of A, B and C are respectively: [10, 80], [20, 90] and [20, 70]. Using arc consistency we will reduce the domains of A and B as follows. $10 < A < 100-(20+20) = 60$. $20 < B < 100-(10+20) = 70$. Suppose now that, during the search, we first assign the value 50 to A. Through arc consistency and the above arithmetic constraint, the domain of B will be updated as follows: $20 < B < 100-(50+20) = 30$.

Example 5. Let us consider the previous example and assume that we have the following semantic constraints (corresponding to three tables):
Table1$(A, D, \ldots)$,
Table2$(B, E, \ldots)$
and Table3$(C, F, \ldots)$.

After assigning the value 50 to A as shown above in example 4, the domain of B will be reduced to [20,30]. Now through the semantic constraint Table1 all the tuples where $A \neq 50$ should be removed from Table1. Also, through Table2 all tuples where $B \notin$ [20,30] will be removed from Table2. This will reduce the domains of D, E and F.

Arc consistency is enforced with an arc consistency algorithm. In the past three decades several arc consistency algorithms have been developed. However most of these algorithms deal with binary constraints [11,16,17,18,19,20]. In the case of n-ary constraints there are three types of algorithms [7].

- Generalized Arc Consistency: for general non binary constraints.
- Global constraints: for constraints involving all the variables of a given CSP (such as the constraint $allDifferent(X_1,\ldots,X_n)$ which means that the values assigned to each of the variables $X_1,\ldots,X_n$ should be mutually different).
- Bounds-Consistency: which is a weaker (but less expensive) form of arc consistency. It consists of applying arc consistency to the upper and lower bounds of the variable domains. Examples 3 and 4 above use a form of bounds-consistency which motivates our choice of this type of algorithm for arithmetic constraints as shown below.

In our work we use the improved generalized arc consistency (GAC) algorithm [21] for semantic constraints (since this algorithm is dedicated for positive table constraints) and the bound consistency algorithm for discrete CSPs [22] in the case of arithmetic constraints. More precisely, bounds consistency is first used through arithmetic constraints to reduce the bounds of the different domains of variables. The improved GAC [21] is then used through the semantic constraints to reduce the domains of the attributes

Algorithm GAC

```
Given a constraint network CN = (X,C)
    (X: set of variables, C: set of constraints between variables)
    Q ← {(i,j) | i ∈ C ∧ j ∈ vars(C)}
    (vars(C) is the list of variables involved by C)
    While Q ≠ Nil Do
      Q ← Q − {(i,j)}
      If REVISE(i,j) Then
        If Domain(j) = Nil Then return false
        Q ← Q ⊔ {(k,l) | k ∈ C ∧ j ∈ vars(k)
        ∧ l ∈ vars(k) ∧ k ≠ i ∧ j ≠ l}
      End-If
    End-While
Return true
```

Function *REVISE*(*i*, *j*)
(REVISE for bound consistency)

```
domainSize ← |Domain(j)|
While |Domain(j)| > 0
    ∧¬seekSupportArc(i, j, min(j)) Do
  remove min(j) from Domain(j)
End-While
While |Domain(j)| > 1
    ∧¬seekSupportArc(i, j, max(j)) Do
  remove max(j) from Domain(j)
End-While
Return domainSize ≠ |Domain(j)|
```

Function *REVISE*(*i*, *j*)
(REVISE for handling semantic constraints)

```
REVISE ← false
nbElts ← |Domain(j)|
For each value a ∈ Domain(j) Do
  If ¬seekSupport(i, j, a) Then
    remove a from Domain(j)
    REVISE ← true
  End-If
End-For
Return Revise
```

Fig. 2. GAC algorithm and Revise for bound consistency (bottom left) and for handling semantic constraints (bottom right).

even more. Let us describe now the details of our method. The basic GAC algorithm [23,21] is described in figure 2. This algorithm enforces the arc consistency on all variables domains. GAC starts with all possible pairs (i, j) where j is a variable involved by the constraint i. Each pair is then processed, through the function *REVISE* as follows. Each value v of the domain of j should have a value supporting it (such that the constraint j is satisfied) on the domain on every variable involved by i otherwise v will be removed. If there is a change in the domain of j (after removing values without support) after calling the function *REVISE* then this change should be propagated to all the other variables sharing a constraint with j.

When used with arithmetic constraints (as a bound consistency algorithm) C contains the list of arithmetic constraints and the *REVISE* function (the function that does the actual revision of the domains) is defined as shown in figure 2 [22]. In the other case where GAC is used with semantic constraints C contains these arithmetic constraints and the *REVISE* function is defined as shown in the bottom right of figure 2 [21]. In the function *REVISE* (for bound consistency) of figure 2, the function *seekSupportArc* (respectively thefunction *seekSupport* of *REVISE* for semantic constraints in figure 2) is called to find a support for a given variable with a particular value. For instance when called in line 2 of the function *REVISE* for bound consistency, the function *seekSupportArc* looks, starting from the lower bound of *j*'s domain, for the first value that has a support in *i*'s domain. When doing so, any value not supported will be removed.

4 Structure of the RDBMS with the CSP Module

Figure 3 presents the architecture of the module handling combinatorial queries. In the CSP module, the CSP converter translates the query and the other information obtained

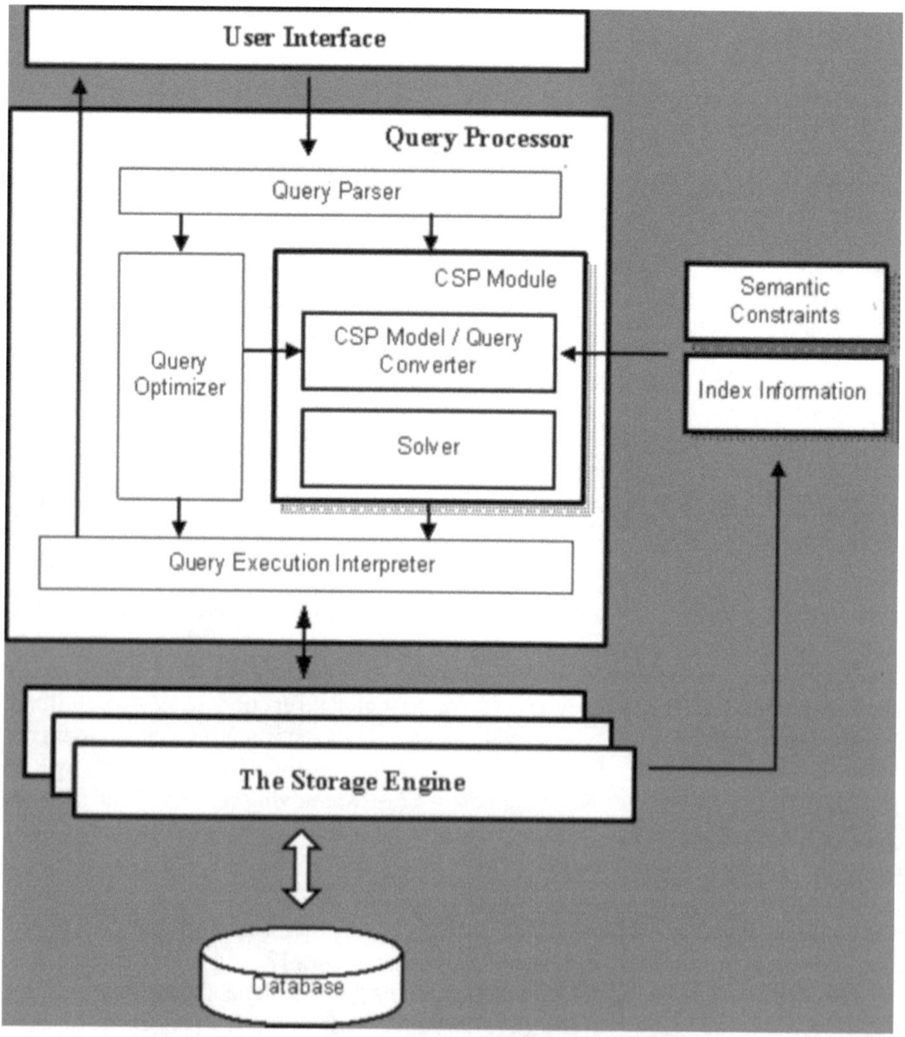

Fig. 3. Relational database model with the CSP module.

for the database such as the semantic constraints into a CSP. The solver contains the solving techniques we described in the previous section.

More precisely, in the traditional RDBMS model the query submitted by the user through the user interface is first received by the query parser which will transform the string form of the query into query objects that will be handled by the query processor. Regularly, the query optimizer will choose the appropriate query solving plan (query execution order with the least estimated cost) and send the execution request to the interpreter. The interpreter will then handle the request and implements the input/output operations by driving storage engine.

We address the function and working process of every component indicated in Figure 3 as follows.

User Interface. This is the interface being responsible to communicate with "front stage" such as website or other applications which have needs for data operations or retrievals. For instance, RDBMS like Microsoft SQL server provides a common user interface by setting connection string and corresponding parameters for .Net, PHP or Java platform.

Query Processor. The query processor is the most significant part in RDBMS that will handle query construction and optimization once the query is approached from the User Interface. It has to engender query execution plan sending to the next part of the system. The traditional model includes query parser, query optimizer and query execution interpreter. For our improved RDBMS model, the CSP module will be added into the system to deal with combinational query solving.

Query Parser. The query parser is the component in the query processor which will create the query to other solvers of RDBMS according to the requirement of query from User.

Query Optimizer. The query optimizer is one of the key parts in the query processor that attempts to determine the more efficient query execution plan of the given query. Usually, the query optimizer is called as cost-based because it mainly considers system resource like memory and CPU cost as well as I/O operations to estimate the cost for the query execution and then select the best one to implement. The join algorithm [63] (pair-wise join, hash join or sorted merge join) will be formed in the query optimizer supported with indexing.

Query Execution Interpreter. This part takes responsibility of translating the query plan into the language which can be understood by the Storage Engine. Meanwhile, the solution returned from the database will be posted back to the application by the query execution interpreter.

CSP Module. The new module added into RDBMS to solve combinational query has two sub components: CSP Query Converter and CSP Solver. The CSP Query Converter will convert the combinational query into the CSP which can be accessed and solved by the CSP Solver. Since some combinational query still need structural adjustment or join order optimization, for some cases the query should be optimized by the query optimizer before it is sent to the CSP query parser. Moreover, some database technologies and information will support the CSP query parser to reduce the size of the search problem such as indexing and semantic constraints.

Storage Engine. The Storage Engine implements related operations with data in database based on the order submitted by the query processor.

In the case of the RDBMS model integrating the CSP module, once the combinatorial query is parsed by the query parser, it is sent to the CSP module. The converter will convert the query (including the arithmetic constraints) and the semantic constraints obtained from the database into the CSP. The solver will then solve it and send the solution to the execution interpreter.

4.1 CSP Module Design

The class diagram in Figure 4 gives the details of how the CSP module is designed to handle combinational queries. The sequence diagram in Figure 5 indicates the working steps of the CSP module during the search.

The query parser gains the query by using the GetQuery function from the application. The system will decide to activate the query optimizer or not. Because sometimes the order of query can be optimized that will increase the efficiency of search, but sometimes it cannot. After index and semantic constraints information are obtained by the Query Converter, the optimized combinational query will be converted into a CSP. Continuously, the QueryAnalyze function is run to analyze the CSP and decide which specific consistency checking strategy will be used to prune the domains of the CSP. The solving method we presented in the previous section will then be applied to find a solution to the query. The Solver will then call the query interpreter to send request to the lower component to carry out the compilation.

The sequence diagram in Figure 5 shows how CSP module solves the combinational query step by step based on the main functions made in the Figure 4. Sometimes the

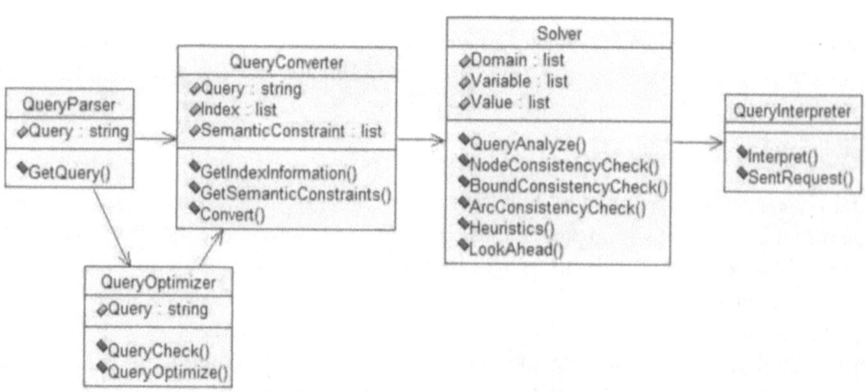

Fig. 4. Class Diagram for the CSP module.

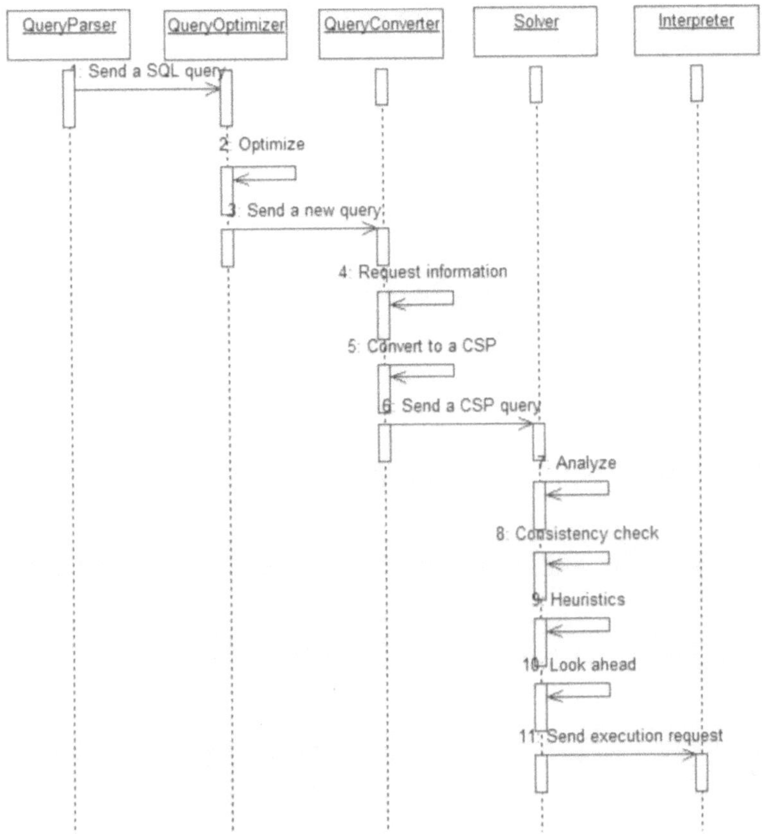

Fig. 5. Sequence Diagram for the CSP module.

query optimizer will not be activated during the search process, since it is not a necessary part for solving combinational query. The relations among it and other components have been described in Figures 4 and 5. In the following we will provide two concrete examples to show the whole solving process.

4.2 Examples

4.2.1 Computer Components Selection

We take a combinational query as an example to introduce the solving process in the CSP module as shown in Figure 4. Suppose the data source for the combinational query in Figure 6 is the one shown in Figure 1.

After optimization by the query optimizer, the second sub query is removed from the combinational query by using the merge sorted join algorithm [26]. The reason is although the constraint for the first and second sub query is all about the sum of computer components price, the arithmetic constraint condition of the first one is smaller than the second, which means the first sub query requires the smaller variable domain

```
Select CPU_id, Memory_id, Motherboard_id
  From CPU, Memory, Motherboard Where
    CPU.Price + Memory.Price + Motherboard.Price <1500 And
    CPU.Price + Memory.Price + Motherboard.Price <1600 And
    CPU.Frequency + Memory.Frequency + Motherboard.Frequency <32
    And CPU.Quality + Memory.Quality + Motherboard.Quality < 250
```

Fig. 6. A SQL combinational query.

```
Select CPU_id, Memory_id, Motherboard_id
  From CPU, Memory, Motherboard Where
    CPU.Price + Memory.Price + Motherboard.Price <1500 And
    CPU.Frequency  + Memory.Frequency + Motherboard.Frequency <32 And
    CPU.Quality + Memory.Quality + Motherboard.Quality < 250
```

Fig. 7. The optimized SQL combinational query

for solving than the second. As a result, the condition for the combinational query is optimized as shown in Figure 4.

In the next step, the query converter will implement converting from a combinational query into a CSP. The optimized SQL combinational query in Figure 5 is converted as follows.

- **Variables:**
  - CPU_id,
  - CPU.Price,
  - CPU.Frequency,
  - CPU.Quality,
  - Memory_id,
  - Memory.Price,
  - Memory.Frequency,
  - Memory.Quality,
  - Motherboard_id,
  - Motherboard.Price,
  - Motherboard.Frequency,
  - Motherboard.Quality
- **Domains:** Domains of the above variables.
  - CPU_id [1, 4],
  - CPU.Price [200, 2000],
  - CPU.Frequency [5, 15],
  - CPU.Quality [80, 95],
  - Memory_id [1, 4],
  - Memory.Price [800, 1000],
  - Memory.Frequency [7, 15],
  - Memeory.Quality [80, 100],
  - Motherboard_id [1, 4],

  - Motherboard.Price [100, 400],
  - Motherboard.Frequence [6, 15],
  - Motherboard.Quality [50, 80]
- **Constraints:**
  - **Semantic Constraints:**
    * CPU(*CPU_id*, Price, Frequency, Quality)
    * Memory(*Memory_id*, Price, Frequency, Quality)
    * Motherboard(*Motherboard_id*, Price,Frequency,Quality)
  - **Arithmetic constraint:**

```
CPU.Price + Memory.Price + Motherboard.Price < 1500
And CPU.Frequency + Memory.Frequency + Motherboard.Frequency < 32
And CPU.Quality + Memory.Quality + Motherboard.Quality < 250
```

Furthermore, index information (filtering information) needs to be collected for consistency checking of Solver. For example, the filtering information in Figure 1 is:

- CPU.Price [200, 2000],
- CPU.Frequency [5, 15],
- CPU.Quality [80, 95],
- Memory.Price [800, 1000],
- Memory.Frequency [7, 15],
- Memeory.Quality [80, 100],
- Motherboard.Price [100, 400],
- Motherboard.Frequence [6, 15],
- Motherboard.Quality [50, 80],

The above information could be applied for the node or bound consistency checking in the solving process. The semantic constraints shown above will be used for arc consistency checking.

Once all necessary information has been gained by the system, our solver will look for a possible solution. Finally, the interpreter will translate the solution found by the solver into the execution order, which can be understood by the storage engine and will be carried out by the compilation. One solution to the above example is:

- CPU (1, 200, 5, 80),
- Memory (1, 1000, 7, 80),
- Motherboard (1, 100, 6, 50).

4.2.2 Vehicle Elements Selection

Let us take another example to explain the solving process. Since the solving process is similar to the previous example, only the result of each step will be represented in this example. The data source for the combinational query are shown in Figure 8.

To look for the available combinations of vehicle elements, the user creates a combinational query as follows.

Engine_id	Price	Usage	Rate
1	4000	11	8
2	5000	10	8.5
3	2000	8	9
4	5000	15	9

Clutch_id	Price	Usage	Rate
1	2000	5	7
2	1000	7	6
3	1100	8	5.5
4	1500	10	5

Gear_id	Price	Usage	Rate
1	1000	3	7
2	500	5	8
3	800	8	7
4	1000	10	6

Fig. 8. Tables of vehicle elements

```
Select Engine_id, Clutch_id, Gear_id
From Engine, Clutch, Gear Where
Engine.Price + Clutch.Price + Gear.Price >5500 And
Engine.Usage + Clutch.Usage + Gear.Usage <20 And
Engine.Rate + Clutch.Rate + Gear.Rate < 28
```

Unlike the combinational query in the previous example, there is no sub query that can be optimized or removed in the Query Optimizer. As a result, the combinational query is directly converted into a CSP by the query converter as follows.

- **Variables**
  - Engine_id,
  - Engine.Price,
  - Engine.Usage,
  - Engine.Rate,
  - Clutch_id,
  - Clutch.Price,
  - Clutch.Usage,
  - Clutch.Rate,
  - Gear_id,
  - Gear.Price,
  - Gear.Usage,
  - Gear.Rate
- **Domains**
  - Engine_id [1, 4],
  - Engine.Price [2000, 5000],
  - Engine.Usage [8, 15],

- Engine.Rate [8, 9],
- Clutch_id [1, 4],
- Clutch.Price [1000, 2000],
- Clutch.Usage [5, 10],
- Clutch. Rate [5, 7],
- Gear_id [1, 4],
- Gear.Price [500, 1000],
- Gear.Usage [3, 10],
- Gear.Rate [6, 8]

– **Constraints**
  - **Semantic Constraints:**
    - Engine(*Engine_id*, Price, Usage, Rate)
    - Memory(*Clutch_id*, Price, Usage, Rate)
    - Motherboard(*Gear_id*, Price,Usage,Rate)
  - **Arithmetic constraint:**

```
Engine.Price + Clutch.Price + Gear.Price < 5500
And Engine.Usage + Clutch.Usage + Gear.Usage < 20
And Engine.Rate + Clutch.Rate + Gear.Rate < 28
```

The index information is collected by RDBMS: Engine.Price [2000, 5000], Engine.Usage [8, 15], Engine.Rate [8, 9], Clutch.Price [1000, 2000], Clutch.Usage [5, 10], Clutch. Rate [5, 7], Gear.Price [500, 1000], Gear.Usage [3, 10], Gear.Rate [6, 8]. This will be applied in node and bound consistency checking to prune search space in the solving process. In the final step, the system will run a backtrack search to find the solution using the algorithm we presented earlier. One solution to this problem is Engine (3, 2000, 8, 9), Clutch (1, 2000, 8, 9), Gear (1, 1000, 3, 7).

5 Experimentation

In order to compare the time performance of our query processor with one of the most advanced relational databases (MS SQL server 2005) we run several tests on randomly generated databases and take the running time in seconds needed (by our method and MS SQL) to satisfy the query. The tests are conducted on a IBM T42 with a P4 1.7 GHz processor and 512 MB RAM memory, running Windows XP.

We have built a web-based application to simulate the CSP model and run the experiments. This application has been developed using ASP.NET (*C#*) connected with SQL server (See Figure 10). In the tests, we can compare the efficiency of operations from RDBMS and the simulated CSP model. At first, three tables are set up in the RDBMS, and the data in the tables are randomly created and input by data access layer of the application for tests. We create all the data creation, manipulation and solving classes in this layer, which is responsible to connect with the relational database and implement all data operations. CSP search algorithms, constraint propagation and heuristic methods are written in this layer, all of which achieve the functions of the CSP solver as shown in Figure 3. The size and number of tables, as well as the complexity of combinational queries are changed during the testing for the performance comparison between the old and new models. Index is added for the first non-primary key column in the table, in order to offer the acceleration for traditional search and "filtering" information for consistency checking of the new CSP module.

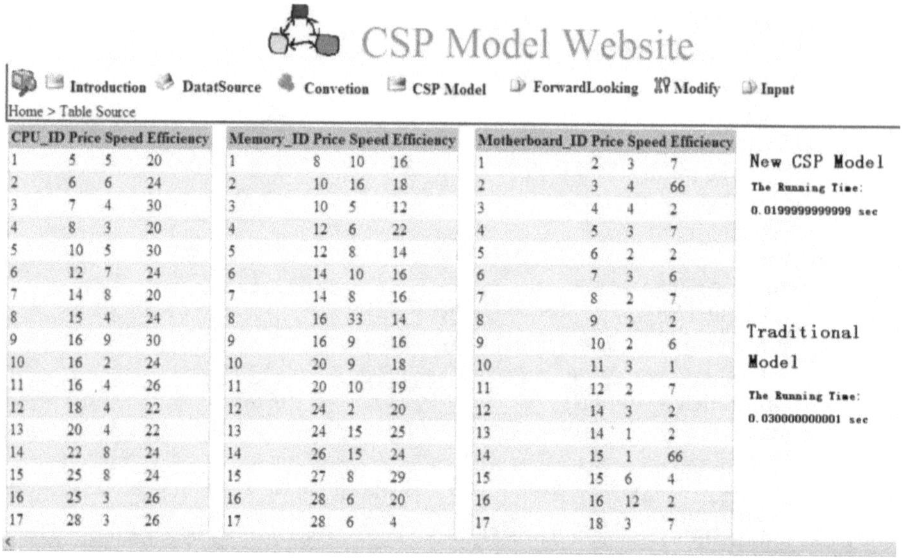

Fig. 9. The user interface of new RDBMS with CSP module.

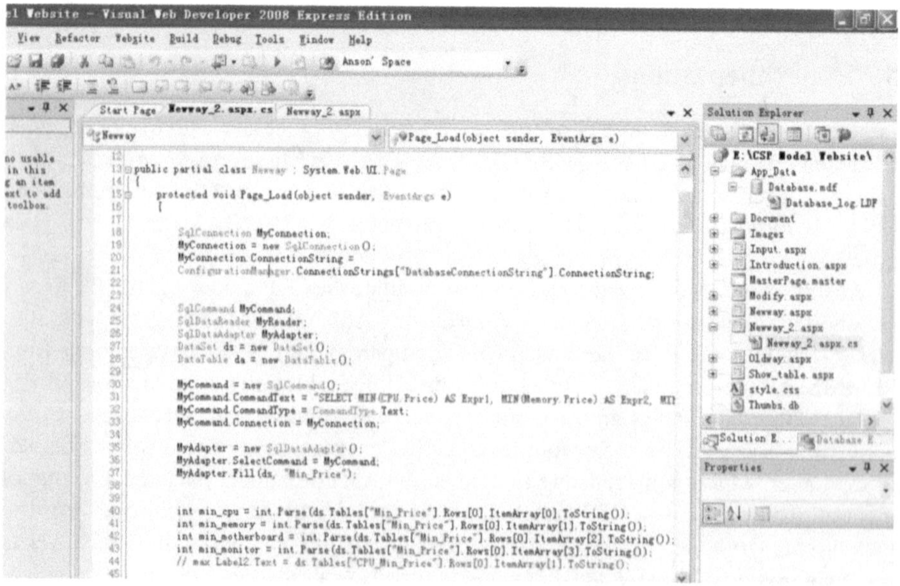

Fig. 10. The programming environment for the CSP module.

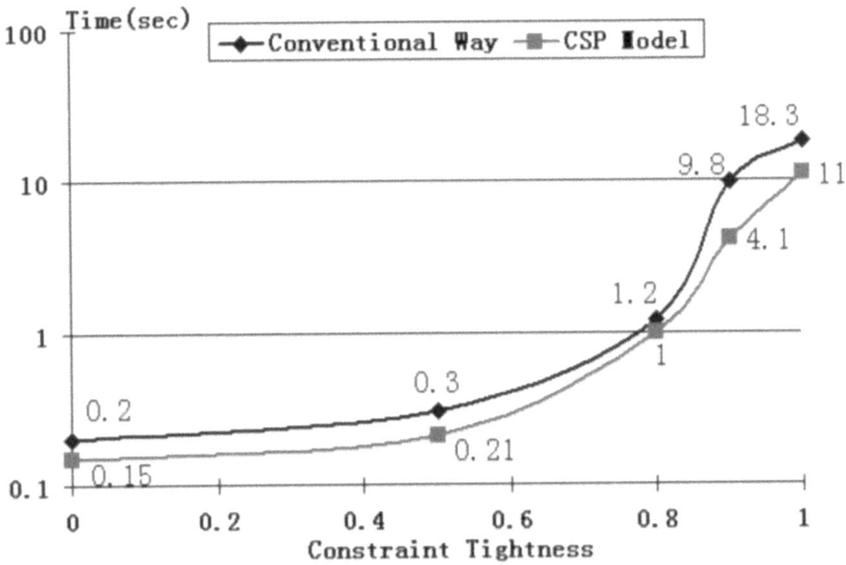

Fig. 11. Test results when varying the tightness.

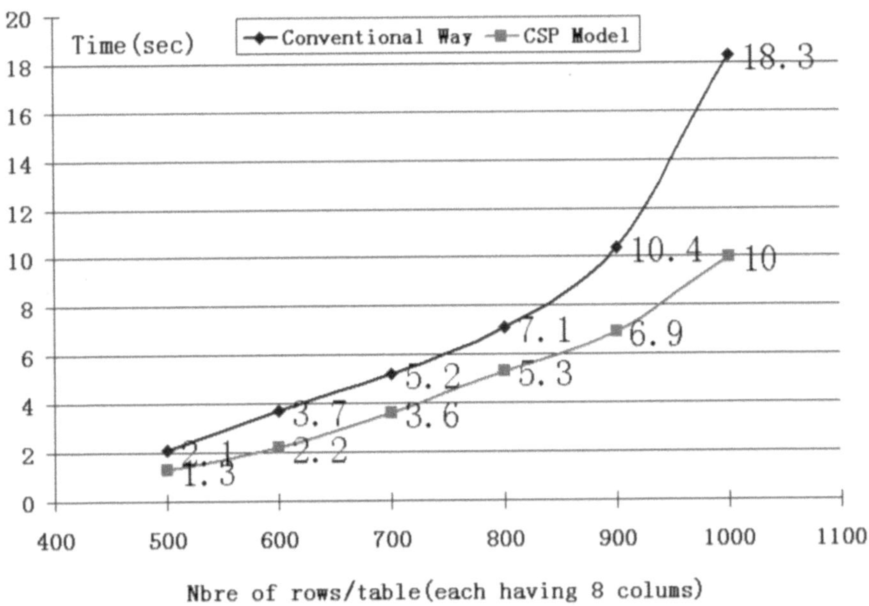

Fig. 12. Test results when varying the number of rows.

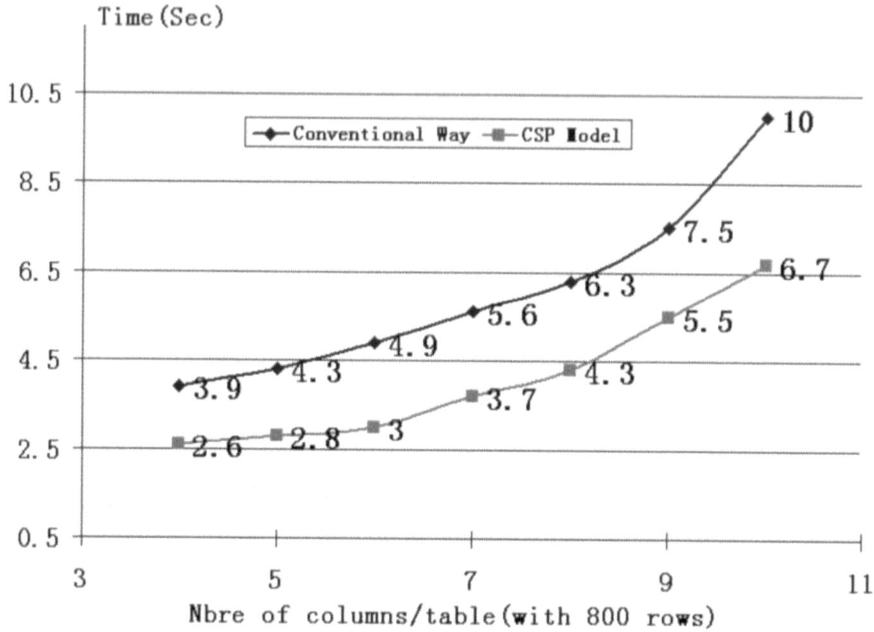

Fig. 13. Test results when varying the number of columns.

Figure 9 is the screen shot of the comparative result of the test on the combinational query of the example in Section 4.1. The three tables are displayed with the corresponding results from the traditional RDBMS model and our model as shown in the right. Figure 10 is the screen shot of the programming environment for the experiments.

Each database and its corresponding combinatorial query are randomly generated according to the following parameters: *T* (the number of tables within the database), *R* and *C* (respectively the number of rows and columns of each table within the database); and *P* the constraint tightness. This last parameter defines how easy|hard is the CSP corresponding to the generated database. More precisely, the constraint tightness *P* of a given constraint is defined as "*the fraction of all possible pairs of intervals from the domains of the variables (involved by the constraint) that are not allowed by the constraint [24].*" According to this definition, the value of *P* is between 0 and 1. Easy problems are those where the tightness is small and hard problems correspond to a high tightness value.

Figure 11 presents tests performed when varying the tightness *P*. *T*,*R* and *C* are respectively equal to 3, 800 and 8. Note that the logarithmic scale is used here for the y coordinates As we notice on the chart, when *T* is below 0.8 (the case where %80 of the possibilities are not solutions) the 2 methods have similar running time. However when *T* is more than 0.8 (which corresponds to hard problems where only few possibilities are solutions) we can see the superiority of our method over the traditional model. The other figures 14, 12, and 13 correspond to the situation where we vary *T*, *R* and *C* respectively. In all these 3 cases we can easily see that our method outperforms MS SQL (especially when *R* or *T* increases). Note that in the case where *T* is greater than 3

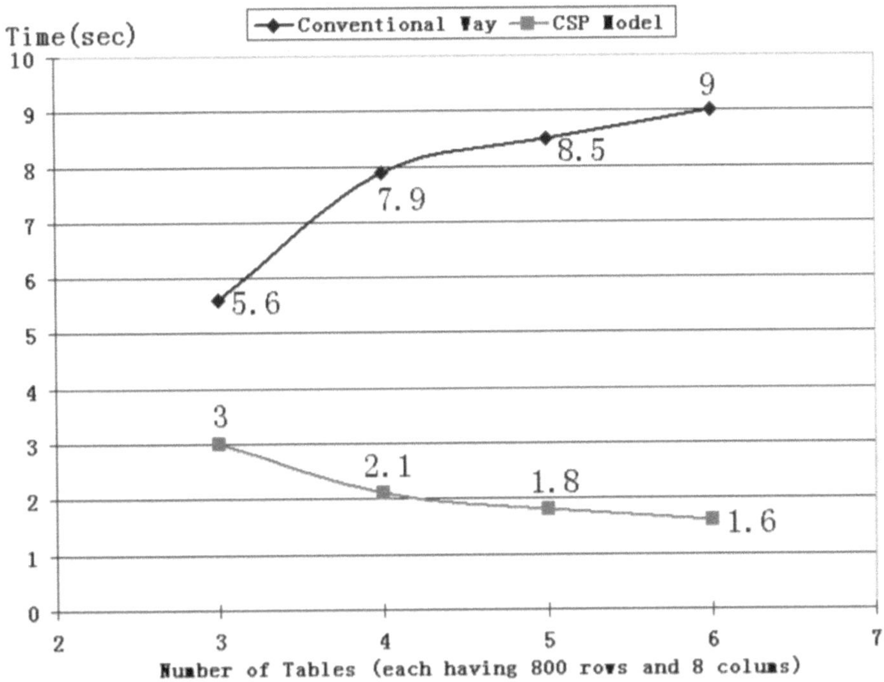

Fig. 14. Test results when varying the number of tables.

(in figure 14), the problem becomes inconsistent (the query cannot be satisfied). In this particular case, our method detects the inconsistency before the backtrack search (see step 1 of our solving method in Section 3). Since the backtrack search phase is saved in this case, the total running time is even better than the case where T is equal to 3.

6 Conclusion and Future Work

CSPs are a very powerful framework to deal with discrete combinatorial problems. In this paper we apply CSPs in order to solve a particular case of combinatorial applications consisting of satisfying a combinatorial query. More precisely our method translates the combinatorial query with the database information into a CSP. CSP techniques including constraint propagation and backtrack search are then used to satisfy the query by looking for a possible solution to the CSP. In order to demonstrate the efficiency in practice of our method, we conducted different tests on large databases and compared the running time needed by our method and the well known SQL server 2000 in order to satisfy combinatorial queries. The results of the tests demonstrate the superiority of our method for all the cases.

In the near future, we intend to integrate other search methods such as local search techniques and genetic algorithms in order to speed up the query search process. Indeed, we believe that while these approximation algorithms do not guarantee a complete solution they can be very useful in case we want an answer within a short time.

Another problem we intent to explore is handling several combinatorial queries at the same time. The obvious way is to process them one by one using the method we have proposed in this paper. However, this might not be the best idea. May be it is better to first pre-process the set of queries using arc consistency and then (if the queries are arc consistent) we can proceed with solving these queries together. An experimental comparative study of both ways needs to be carried out in order to find out which one is better in terms of processing time.

References

1. Atzeni, P., Antonellis, V.D.: Relational database theory. Benjamin-Cummings Publishing Co. (1993)
2. Liu, C., Foster, I.T.: A framework and algorithms for applying constraint solving within relational databases. In: W(C)LP 2005, pp. 147–158 (2005)
3. Revesz, P.: Introduction to Constraint Databases. Springer, New York (2002)
4. Chuang, L., Yang, L., Foster, I.T.: Efficient relational joins with arithmetic constraints on multiple attributes. In: IDEAS, pp. 210–220 (2005)
5. Vardi, M.Y.: The complexity of relational query languages. In: Annual ACM Symposium on Theory of Computing (1982)
6. Chandra, A.: Structure and complexity of relational queries. In: The 21st IEEE Symposium, pp. 333–347 (1980)
7. Dechter, R.: Constraint Processing. Morgan Kaufmann, San Francisco (2003)
8. Hentenryck, P.V.: Constraint Satisfaction in Logic Programming. MIT Press, Cambridge (1989)
9. Haralick, R., Elliott, G.: Increasing tree search efficiency for Constraint Satisfaction Problems. Artificial Intelligence 14, 263–313 (1980)
10. Mackworth, A.K., Freuder, E.: The complexity of some polynomial network-consistency algorithms for constraint satisfaction problems. Artificial Intelligence 25, 65–74 (1985)
11. Mackworth, A.K.: Consistency in networks of relations. Artificial Intelligence 8, 99–118 (1977)
12. Vardi, M.Y.: Constraint sastisfaction and database theory: A tutorial. In: PODS 2000 (2000)
13. Swami, A.: Optimization of large join queries: combining heuristics and combinatorial techniques. In: The 1989 ACM SIGMOD international conference on Management of data, pp. 367–376 (1989)
14. Jarke, M., Koch, J.: Query optimization in database systems. ACM Computing Surveys (CSUR), 111–152 (1984)
15. Miguel, I., Shen, Q.: Solution techniques for constraint satisfaction problems: Foundations. Artificial Intelligence Review 15(4) (2001)
16. Mohr, R., Henderson, T.: Arc and path consistency revisited. Artificial Intelligence 28, 225–233 (1986)
17. Bessière, C.: Arc-consistency and arc-consistency again. Artificial Intelligence 65, 179–190 (1994)
18. Bessière, C., Freuder, E., Regin, J.: Using inference to reduce arc consistency computation. In: IJCAI 1995, Montréal, Canada, pp. 592–598 (1995)
19. Zhang, Y., Yap, R.H.C.: Making ac-3 an optimal algorithm. In: Seventeenth International Joint Conference on Artificial Intelligence (IJCAI 2001), Seattle, WA, pp. 316–321 (2001)
20. Bessière, C., Régin, J.C.: Refining the basic constraint propagation algorithm. In: Seventeenth International Joint Conference on Artificial Intelligence (IJCAI 2001), Seattle, WA, pp. 309–315 (2001)

21. Lecoutre, C., Radoslaw, S.: Generalized arc consistency for positive table constraints. In: Benhamou, F. (ed.) CP 2006. LNCS, vol. 4204, pp. 284–298. Springer, Heidelberg (2006)
22. Lecoutre, C., Vion, J.: Bound consistencies for the csp. In: Proceeding of the second international workshop Constraint Propagation And Implementation (CPAI 2005) held with the 10th International Conference on Principles and Practice of Constraint Programming (CP 2005), Sitges, Spain (September 2005)
23. Mackworth, A.K.: On reading sketch maps. In: IJCAI 1977, pp. 598–606 (1977)
24. Sabin, D., Freuder, E.C.: Contradicting conventional wisdom in constraint satisfaction. In: Proc. 11th ECAI, Amsterdam, Holland, pp. 125–129 (1994)
25. Ribeiro, C.C., Ribeiro, C.D., Lanzelotte, R.S.: Query optimization in distributed relational databases. Journal of Heuristics 65, 5–23 (1997)
26. Sagiv, Y.C., Arbor, A.: Optimization of queries in relational databases, pp. 1–10. UMI Research Press (1981)

Predictive Capabilities of Adaptive and Evolutionary Fuzzy Cognitive Maps - A Comparative Study

Wojciech Froelich[1] and Przemyslaw Juszczuk[2]

[1] Institute of Computer Science, University of Silesia, ul.Bedzinska 39, Sosnowiec, Poland
wojciech.froelich@us.edu.pl
[2] Institute of Computer Science, University of Silesia, ul.Bedzinska 39, Sosnowiec, Poland
przemyslaw.juszczuk@gmail.com

Abstract. Fuzzy cognitive maps (FCMs) represent the decision process in the form of a graph that is usually easy to interpret and, therefore, can be applied as a convenient decision-support tool. In the first part of this chapter, we explain the motivations for the research on FCMs and provide a review of the research in this area. Then, as stated in the title of the chapter, we concentrate our attention on the comparative study of adaptive and evolutionary FCMs. The terms adaptive and evolutionary refer to the type of learning applied to obtain a particular FCM. Despite many existing works on FCMs, most of them concentrate on one type of learning method. The purpose of our research is to learn FCMs using diverse methods on the basis of the same dataset and apply them to the same prediction problem. We assume the effectiveness of prediction to be one of the quality measures used to evaluate the trained FCMs. The contribution of this chapter is the theoretical and experimental comparison of adaptive and evolutionary FCMs. The final goal of our research is to determine which of the analyzed learning methods should be recommended for use with respect to the considered prediction problem. To illustrate the predictive capabilities of FCMs, we present an example of their application to the prediction of weather conditions.

1 An Introduction to the Theory of Cognitive Maps

The inspiration for the development of cognitive maps originated from biological experiments [9,24]. The initial model of a cognitive map, represented by a graph, was proposed and applied in the social sciences [3]. The nodes within the graph were annotated in natural language and represented intuitively-understood concepts. The arcs represented a binary relationship labeled with "+" or "-" signs, denoting intuitively-understood, positive or negative causality (cause-and-effect relationship), respectively. The example of such dependency is the positive causal impact of concept "Number of people in the city" on concept "Modernization." The entire graph was prepared by domain experts. The intuitively-understood

N.T. Nguyen and E. Szczerbicki (Eds.): Intel. Sys. for Know. Management, SCI 252, pp. 153–174.
springerlink.com

notions of concepts and the cause-and-effect relationship among them could not be used for the construction of any computational decision support system, therefore were attempts to formalize this idea. For the purpose of this chapter, we define a cognitive map as an order pair:

$$CM =< C, W >, \tag{1}$$

where C is the set of concept labels, and W is the connection matrix representing the binary relation between concepts. The matrix W stores the weights assigned to the pairs of concepts. We assume that the nodes are indexed by subscripts i(cause node) and j(effect node). In simplest case, it is possible to distinguish binary cognitive maps (BCM) for which the concept labels are mapped to binary states denoted as $a_i \in \{0, 1\}$, where the value 1 means that the concept is activated. The weights of BCM are usually mapped to the crisp set, i.e., $w_{ij} \in \{-1, 0, 1\}$. The value 1 represents, positive causality, understood e.g. such way, that the activation (change from 0 to 1) of concept c_i occurs concurrently with the same activation of concept c_j or that deactivation (change from 1 to 0) c_i occurs concurrently with the same deactivation of concept c_j. The value -1 represents the opposite situation, in which the activation of c_i deactivates the concepts c_j or vice versa. The $w_{ij} = 0$ means that there are no concurrently occurring changes of the states of the concepts. Some researchers [22,11] assume that that the elements on the diagonal of matrix W are not considered.

The extension of CM in the form of fuzzy cognitive maps was proposed by Kosko [8]. In FCM, the concepts are represented by fuzzy sets, and the concepts labels are mapped to a real-valued activation level from the closed continuous interval $a_i \in [0, 1]$, where 0 means no activation and 1 means full activation. The arcs between concepts are also labeled with real-value weights $w_{ij} \in [-1, 1]$. The state of the FCM at time t is fully described by the state vector $A(t)$.

There have been numerous enhancements to the initial model of FCM. The addition of memory to the concept nodes within the HO-FCM [19] enables the representation of high-order dynamics of the modeled decision process. The rule-based fuzzy cognitive maps RBFCM [4] assigned rules to the arcs of FCMs, significantly enhancing their reasoning capabilities. There are also the extensions that enable the incorporation of the variable time delays between concept activation in the FCMs [15]. FCMs can be applied in economics (e.g., stock market prediction), medical science (e.g., medical diagnosis) [23], or even intrusion detection systems [1].

In most known approaches to learning FCMs, the set of concept labels C is provided a-priori by expert, and only the matrix W is learned from raw data. Diverse adaptive learning methods [10] have been applied by many authors, and the most recent proposals involve Hebbian learning [11] and non-linear Hebbian learning [14]. Evolutionary learning methods have been also investigated by many researchers [25]. A successful application of an evolutionary FCM to the prediction problem is presented in [18]. A survey of research papers on FCMs can be found in [2].

2 Problem Formulation

For the comparative analysis of adaptive and evolutionary FCMs, we will consider a prediction problem formulated in the following way. Let $T = \{t_0, t_1, \ldots, t_n\}$ be an ordered set of time labels, where $\forall i. t_i - t_{i-1} = \Delta t$ is a constant parameter that will be referred to as "base time." Let C represent the set of symbolic labels pointing to the subsets of observations (concepts) acquired from the raw, temporal data. Every concept label $c_i \in C$ is uniquely identified by its subscript. For every concept label, we define a mapping $a : T \times C \rightarrow [0, 1]$ that assigns the state (activation level) of a concept at time t. The state of the ith concept at time step t will be denoted as $a_i(t) = a(t, c_i)$. The concepts for which $a_i(t) > 0$ will be referred to as "active concepts at time t." The $A(t)$ denotes the state vector for all concepts. Let us assume now that the set T is divided into the known historical sequence (pointing to learning data), $T_H = \{t_0, t_1, \ldots, t_{s-1}\}$, and the unknown sequence representing the unknown future, $T_F = \{t_s, t_{s+1}, \ldots, t_e\}$ (pointing to testing data), where $card(T_H) = card(T_F)$. The sequence T_H constitutes the time horizon that will be used to learn a model. Let us assume that the state of concepts at time $A(t_s)$ can be observed. We would like to forecast the state of concepts $A(t)$ at some time $t_e > t_s$ by the reconstruction of the unknown sequence $A(t_s), A(t_{s+1}), \ldots, A(t_e)$, where the length of the prediction horizon $t_e - t_s = t_{pred}$ is a constant parameter. The prediction will be produced using a model based on FCM. On the basis of the prediction errors, we will estimate the quality of the obtained FCMs and, indirectly, the effectiveness of the applied learning method. For the purposes of this research, we make the simplification (as is usually done by researchers [25,11]) of assuming that the concept labels are simply the identifiers of attributes (columns) within the data table stored in the relational database.

3 Prediction with the Use of FCM

The standard exploitation form of FCM is usually the same for most applications and is independent of the applied learning method. The goal is to simulate the temporal behavior of FCM in order to predict the unknown sequence of the state vectors $A(t_s), A(t_{s+1}), \ldots, A(t_e)$. The exploitation of a given FCM starts from the acquisition of a new observation of concept activations $A(t_s)$ and then performing a simulation (that can be understood also as a kind of numerical forward reasoning) of the modeled process.

There are many papers that describe the reasoning process in FCMs, however, to our knowledge, there is no one that explicitly specifies their diverse variants. The justification of the applied reasoning method is usually done experimentally. In almost all cases, the researchers use the scaled summation of factors, each representing the linear dependency of concept activations. For all computational experiments described in this chapter, we chose to use for reasoning in FCM the Equation (2):

$$a_j(t+1) = \gamma(\sum_{i=1, i\neq j}^{n} w_{ij} a_i(t)), \tag{2}$$

where $\gamma(x) = 1/(1 + e^{-cx})$ is the threshold function that serves to confine unbounded values to a strict range, c is a constant (we assume $c = 5$), and $n = card(C)$. Some of the researchers report the use of Equation (3), involving a full impact of every concept on itself:

$$a_j(t+1) = \gamma(a_j(t) + \sum_{i=1, i\neq j}^{n} w_{ij} a_i(t)). \tag{3}$$

The other possibility is to use an arbitrarily assumed scaling coefficient, $k < 1$, as is done in Equation (4):

$$a_j(t+1) = \gamma(k a_j(t) + \sum_{i=1, i\neq j}^{n} w_{ij} a_i(t)), \tag{4}$$

or use the non-zero diagonal values of matrix W and apply Equation (5):

$$a_j(t+1) = \gamma(\sum_{i=1}^{n} w_{ij} a_i(t)). \tag{5}$$

The simulation of the FCM-based reasoning process can lead to three types [17] of behavior of the state vector $A(t)$:

1. fixed-point attractor, known also as hidden pattern (the state vector becomes fixed after some simulation steps);
2. limit cycle (the state vector keeps cycling); and
3. chaotic attractor (the state vector changes in a chaotic way).

The type of behavior of state vector depends on the characteristics of the raw data and on the algorithm applied during the learning phase of the FCM. Due to the assumed reasoning method Equation 2, the quality of prediction may depend on the type of data used during learning. One of the goals of this research is to evaluate whether knowledge of the type of learning data influences the quality of prediction or the choice of the learning algorithm.

4 Learning FCM

As mentioned before, it is possible to distinguish two general approaches to learning FCMs, i.e., adaptive and evolutionary.

4.1 Adaptive Algorithms

The main idea of the most adaptive learning methods of FCMs is to modify its weights incrementally as the new learning data in T_H become available. For the purposes of this chapter, we recall and use the two best-known adaptive algorithms. Historically, the first approach was inspired by the Hebbian algorithm used for learning artificial neural networks. It was adapted by Kosko [8] to FCMs in the form of the differential Hebbian learning (DHL) algorithm. Assuming that the ith causal concept changes, i.e., $\Delta a_i \neq 0$. The DHL algorithm modifies the weight between concepts i(cause) and j(effect), using the Equation (6):

$$w_{ij}(t+1) = w_{ij}(t) + c(t)[\Delta a_i \Delta a_j - w_{ij}(t)]. \tag{6}$$

If $\Delta a_i = 0$, then the corresponding weight is not changed, i.e., $w_{ij}(t+1) = w_{ij}(t)$. The coefficient $c(t)$ used in Equation (6) plays a key role during adaptive learning. It can be assumed that:

$$c(t) = 0.1[1 - t/1.1q] \tag{7}$$

changes over time [6]. The constant parameter $q \in \aleph$ should ensure that the value of the weight holds in interval $[-1, 1]$. Therefore, $q = card(T_H)$ can be assigned to the number of steps of the learning period. Notice that using Equation (6) updates the weight between two concepts only on the basis of changes in their activation values. The drawback of such an assumption, i.e., its negative influence on predictive capabilities of BCM, was shown in [6].

The balanced differential learning algorithm (BDA) [6] takes into account the changes of activation values of other nodes (not directly connected by the considered arc). The Equations given in [6] adapted to our notation (with the exchange of cause and effect indices) look as follows. Let us assume that $n = card(T_H)$ is the number of temporal steps in the training set. The diagonal of matrix W, which reflects the self-impact of concepts, is computed as in Equation (8):

$$w_{jj}(t+1) = w_{jj}(t) + \frac{a_j(t)}{n}. \tag{8}$$

For two different concepts, the update equations look as follows.
For $\triangle a_i \triangle a_j > 0$:

$$w_{ij}\,(t+1) = w_{ij}\,(t) + c\,(t)\,[\frac{\triangle a_j / \triangle a_i}{\sum_{k=1, \triangle a_i \triangle a_j > 0}^{n} \triangle a_j / \triangle a_k} - w_{ij}\,(t)]. \tag{9}$$

For $\triangle a_i \triangle a_j < 0$:

$$w_{ij}\,(t+1) = w_{ij}\,(t) + c\,(t)\,[\frac{-\triangle a_j / \triangle a_i}{\sum_{k=1, \triangle a_i \triangle a_j < 0}^{n} \triangle a_j / \triangle a_k} - w_{ij}\,(t)]. \tag{10}$$

The process of learning is thus extended to the scope of the entire FCM. In our opinion, one of the other important enhancements of the BDA algorithm is that it involves the factors in form of division $\Delta a_j / \Delta a_i$ instead of multiplication, $\Delta a_i \Delta a_j$, as used in Equation (6). The explanation of this opinion follows later in this chapter. As opposed to Huerga [6], who used the BDA algorithm for binary cognitive maps, we decided to try it for learning FCMs.

4.2 Evolutionary Algorithms

The other branch of computational methods for learning FCMs involves the application of evolutionary algorithms. For our purposes in this chapter, we decided to use the real coded genetic algorithm (RCGA) [25] and differential evolution (DE) [21] algorithms. Both require the transformation of connection matrix W to the chromosome vector, which can be done by taking the following rows from W, as shown in Fig. 1. The number of genes in a single genotype can be calculated in accordance with the formula: $Dim(W) = n \cdot (n-1)$, where: n is the number of concepts in FCM. According to the definition of FCM, the values of each individual gene are limited to the range $x \in [0, 1]$.

The RCGA algorithm applies a real-coded genetic algorithm to develop FCM from the set of historical data. In each iteration of the algorithm, the historical data are compared with the data generated by the FCM model. In RCGA, each chromosome is a vector of floating point values that correspond to FCM weights. The core of this algorithm is the fitness function, which can be assumed as in Equation (11), given in [25]. This is the sum of the differences between the activations of concepts generated from the FCM, and the values taken from historical data:

$$f = \frac{1}{(t_e - 1) \cdot n} \cdot \sum_{t=t_s}^{t_e} \sum_{i=1}^{n} |a_i(t) - a_i'(t)|^p, \tag{11}$$

where:

- t_s is the lower bound of the time window and specifies the initial index of the learning data,
- t_e is the upper bound of the time window and determines the final index of the learning data,
- $n = card(C)$ is the number of concepts,
- p is the constant parameter ; for our experiments we assumed $p = 1$,
- $a_i(t)$ is the observed value of the ith concept activation at the time moment t, and
- $a_i'(t)$ is the acivation value of the ith concept at the time moment t - predicted by the FCM.

The other evolutionary algorithm investigated in this chapter is the differential evolution method [20]. The main advantage of it is its ability to designate the optimal direction and speed of convergence on the basis of the position of individuals in the current population. It is necessary to initialize the initial population P_0, providing an even distribution of individuals in exploration space. Mutation used in the DE is a much more complex process than is the use of traditional genetic algorithms, but, thanks to the use of a mechanism known as differential vectors, it is possible to perform a directed search.

Let us denote $\boldsymbol{X}_{ig}$ as the ith genotype within the population of the gth generation. Let the constant n represent the number of genotypes, and m is the number of genes in the genotype. The following algorithm describes the basic steps of the differential evolution [20] algorithm.

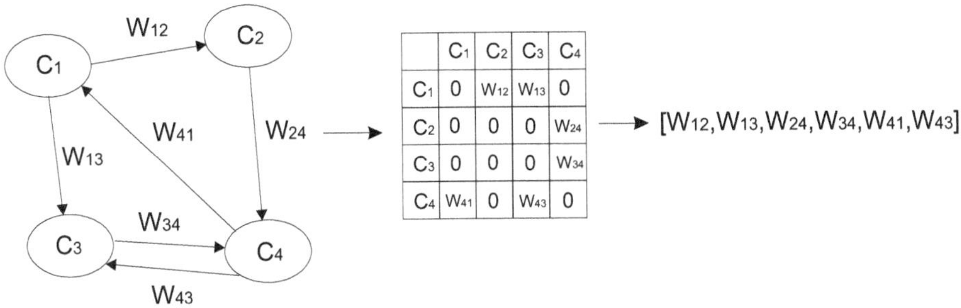

Fig. 1. Creation of the genotype on the basis of connection matrix

1. Create the initial population of genotypes $P_0 = \{\boldsymbol{X}_{1,0}, \boldsymbol{X}_{2,0}, ..., \boldsymbol{X}_{n,0}\}$,
2. Set the generation number $g = 0$
3. When the stop criterion is not met:
   (a) Compute the fitness function for every genotype in the population $\{f(\boldsymbol{X}_{1,g}), f(\boldsymbol{X}_{2,g}), ..., f(\boldsymbol{X}_{n,g})\}$
   (b) Create the population of trail genotypes (algorithm 2)V_g based on P_g
   (c) Make crossover of genotypes from the population P_g and V_g to create population U_g
   (d) Choose the genotypes with the highest fitness function from population U_g and P_g for the next population
   (e) If $g = g + 1$, go to step a.

The subroutine called in step 3b refers to the creation of trial genotypes, and it is described as follows:

1. Set the β parameter
2. For i to n do
   (a) Generate two different random numbers r_1, r_2 between range $(1, ..., n)$
   (b) for j to m do $v_{ij} = v_{ij} + \beta \cdot (v_{r_{1j}} - v_{r_{2j}})$.

The parameter $\beta \in [0, \infty]$ specifies the strength of impact of the difference vector (between the two genotypes from the population P_g) on the value of the newly-created genotype belonging to population V_g, and the numbers r_1, and r_2 determine the indices of the genotypes; j is the index of the selected gene; and v_{ij} is a new value of the ith gene of the jth genotype.

We would like also to mention here two other approaches to evolutionary learning of FCMs: the particle swarm optimization algorithm [12,13]. and the algorithm proposed by Khan and Chong [7].

5 Theoretical Comparison of Adaptive and Evolutionary FCMs

In this section, we present some discussion of the theoretical aspects of adaptive and evolutionary FCMs.

Due to diverse possibilities of modeling causality [16], the interpretation of an arc within FCM as cause and effect dependency may raise doubts. The computational interpretation of FCM weights is usually done in two phases:

1. during the exploitation of FCM - the weights substantially influence the reasoning process,
2. during the learning process - the weights are interpreted differently, depending on the method applied to automatically obtain the FCM.

Let us notice here a trivial case that can lead to better understanding of equations (2) through (5). Let us temporarily assume that the scaling function $\gamma(x) = 1$ is trivial and analyze the mutual impact of only two different concepts without self-dependency (i.e., the weights on the diagonal of W are equal to 0). Thus, independent of whether we apply equations (2), (3), (4), or (5), we use a simple linear dependency for reasoning between concept activations: $a_j(t+1) = w_{ij}a_i(t)$. Let assume, we would like to compute the weight w_{ij} on the basis of raw data available in two consecutive time steps, $< t_1, t_2 >$. In the case of linear dependency, we can do this easily by using the Equation (12):

$$w_{ij} = \frac{\Delta a_j}{\Delta a_i}, \tag{12}$$

where: $\Delta a_j = a_j(t_2) - a_j(t_1)$, $\Delta a_i = a_i(t_2) - a_i(t_1)$. For the causal concept with i^{th} subscript, we assume $\Delta a_i \neq 0$. This way, we achieve a simple FCM - trivial but perfect from the point of view of one-step prediction at time step t_1. Of course, the model can be valid for the next time steps $t_2, t_3, \ldots, t_n$ only when assuming the existence of the same linear dependency between the activations of the two concepts. Another notice is such that, if the activation of ith concept grows and $w_{ij} > 0$, then the the value of activation of jth concept also increases. Moreover, the application of scaling function $\gamma(x) = \frac{1}{(1+e^{-cx})}$ does not change this fact. The conclusion is that, if the effect concept should drop and we want to achieve this using any one of equations 2 through 5, we need a negative impact of other concepts (with negative value of corresponding weights). Now, you can see how the weights of FCM are mutually dependent and that their values can be difficult to determine considering only two concepts.

5.1 The Temporal Aspects of Adaptive and Evolutioary Types of Learning

The temporal aspects of adaptive and evolutionary types of learning are briefly presented in Fig. 2. In evolutionary algorithms, the detection of concept activation changes is made within two successive time steps. The length of the moving time-window is constant and equal to 2. In opposite, the evolutionary algorithms use the fixed time-window of the length $t_e - t_s$, see Equation (11) to evaluate every candidate FCM. Moreover, the fitness function that makes the evaluation of FCMs can be constructed in different ways and adjusted to the considered application.

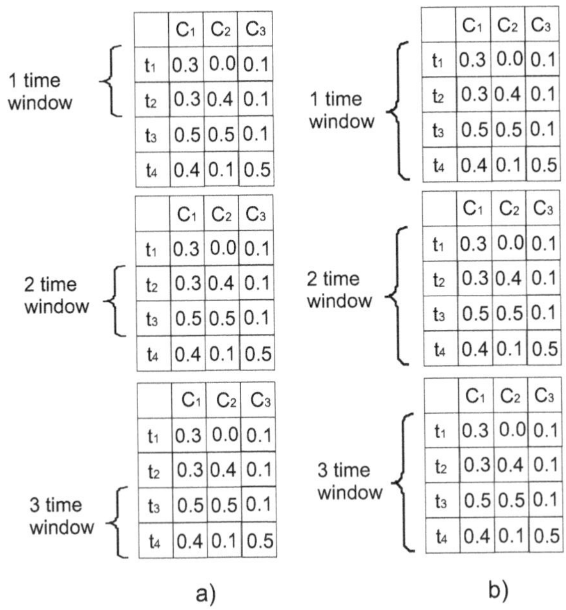

Fig. 2. Successive time-windows: a) adaptive algorithms and b) evolutionary algorithms

Let us look once more at Equation (6) used in the DHL learning algorithm, and let us consider another trivial case. If the learning coefficient $c(t) = 1$, Equation (6) would assume the form as in Equation (13):

$$w_{ij}(t+1) = \Delta a_i \Delta a_j. \tag{13}$$

The weight w_{ij} assumes a value that is dependent only on the last two time steps from the learning sequence. If we assume that the activation values of both concepts change equally and quite strongly, e.g., $\Delta a_i = 0.5$ and $\Delta a_j = 0.5$, the maximal impact (assuming $c(t) = 1$) of this single change $\Delta a_i \Delta a_j = 0.25$ on w_{ij} is only a quarter as high as the theoretical weight computed using Equation (12), namely $\frac{\Delta a_j}{\Delta a_i} = \frac{0.5}{0.5} = 1$. The application of $c(t) < 1$ will make this influence even weaker.

Now, you can see clearly that the intention of using 6 and the value of $c(t) < 1$ is to perform a gradual modification of the weight w_{ij} as the new data are observed. The lower the $c(t)$ coefficient is, the weaker is the influence of a singular change of concepts' activation on the weight w_{ij}. According to (6), the modification of FCM weight always is proportional to the $c(t)$ fraction of the product of the detected changes.

5.2 The Dynamics of the Learning Process

Due to the changing learning parameter $c(t)$, it can be noticed that, within the first few steps of the adaptive learning algorithm, large changes in FCM weights are more likely to occur. Further iterations are used to gradually stabilize the values of the weights of FCM. The changes in concept activations that occurred earlier have greater influence on the FCM weights than the changes in activations during the final stages of the learning period.

In comparison, evolutionary algorithms use the fitness function to evaluate the quality of the entire FCM. In the following iterations, the value of the fitness function decreases to zero, which means that the solution approaches the global optimum. However, a small improvement of the evaluated FCM very often means large fluctuations of the FCM weights. The evolutionary method allows the assessment of the quality threshold and makes possible the cessation of the learning process when the desired threshold is reached. In adaptive methods, the evaluation of FCM is possible only after the entire process of learning is finished. This has to do with the fact that, in adaptive methods, there is no function that is used for the evaluation of the FCM that can examine the difference between the solution obtained and the expected optimum solution.

While learning large FCMs with only a few concepts changing at some time step, the adaptive algorithms theoretically have some advantage over the evolutionary technique. The adaptive algorithm modifies the mutual weight for the arc between two concepts that have changed their values. In contrast, when evolutionary algorithms are applied, all the weights of the FCM are processed (even if only one concept value has changed). This is due to the application of crossover and mutation operators (used in genetic and differential evolution algorithms), which operate on the entire population of genotypes. As a result of genetic operators, completely new individuals are produced - each of which may have different values of genes in the genotype.

5.3 Sensitivity to Incompleteness of Data

For both types of learning methods, the data must be complete. Especially in case of adaptive algorithms, note that the lack of information on concept activation value at a particular time step does do not imply that the concept value has not changed.

One of the simplest possible solutions to this situation is to supplement the learning data. The concepts that are not observed at a given moment are transcribed from the previous or next time step. The situation is shown in Fig. 3 ,

	C_1	C_2	C_3
t_1	0.3	X	0.1
t_2	X	0.4	X
t_3	0.5	0.5	X
t_4	0.4	0.1	0.5

a)

	C_1	C_2	C_3
t_1	0.3	0.0	0.1
t_2	0.3	0.4	0.1
t_3	0.5	0.5	0.1
t_4	0.4	0.1	0.5

b)

Fig. 3. Raw data tables: a) missing data and b) supplemented data.

and the missing values are denoted by the letter "X." However, the intuitively understood confidence in such "repair" can be limited and can cause a decrease in the quality of the obtained FCM.

5.4 Complexity and Scalability

The complexity of learning algorithms can be reduced by providing the values of some weights in matrix W by an expert. It allows the reduction of the dimensions of the search space. In the case in which we know nothing about FCM, we must assume that increasing the number of concepts significantly increases the number of connections between them. the FCM is initially represented by a full graph. In this case, the number of edges is n^2, where $n = card(C)$ is the number of concepts. The fitness function evaluates the entire FCM; therefore, the scalability with respect to the number of concepts is the main weakness of evolutionary learning. The adaptive methods are not so sensitive to the amount of concepts within the FCM. The DHL and BDA iterative algorithms work locally, modifying only the weights between currently changing concepts. Therefore, the scalability of adaptive method is quite good. The main drawback of the evolutionary methods is the time required to generate the solution. This is due to the fact that, in each iteration, the algorithm evaluates m number of FCMs (where m is the number of genotypes in the population). The total number of cognitive maps created by the algorithm is $m \cdot g$ (where g is number of generations). The result of a smaller size population or reducing the size of the time-window is the learning time is shortened. Unfortunately, this reduces the quality of the obtained FCM.

Using adaptive methods, only one FCM is generated, and its weights are then modified in subsequent learning steps. The function used to modify weights assigned to the edges is less complicated than the fitness function used in evolutionary methods. In addition, there are no operations, such as crossover, mutation, or selection, that increase the time needed to find the satisfactory solution.

5.5 Intuitive Interpretation of the Relationship between Concepts

Adaptive methods allow much more intuitive understanding of how the process of learning FCM weights takes place. Recall that a change in the activation of the cause concept occurs concurrently with a change of the activation value of the effect concept. These changes are reflected in the value of weight of the edge between two concepts. In our opinion, the ease of such interpretation is highly desirable in an FCM-based decision support system. The FCM learned using adaptive DHL method is easier for a human expert to interpret. Even when the automatic reasoning process based on FCM (e.g., using Equation (2)) is not performed, the learned FCM can be helpful for the expert as the illustration of the possible changes of concept activations within the concept space.

6 Experimental Comparison of Adaptive and Evolutionary FCMs

Due to the specific inductive features of FCMs, we were interested in knowing whether the predictive capabilities of the obtained FCM-based models depend on the type of raw data used to learn the model. For the experimental evaluation of adaptive and evolutionary learning algorithms, we decided to use three classes of learning data, according to the dynamic features of the FCMs, i.e., 1) fixed point attractor, 2) limit cycle, and 3) chaotic attractor. The exemplary raw data will always be shown in the chart before presenting the results of the experiment.

In many practical applications, we have raw data available in the form of observations stored in a relational database. The construction of complex concepts on the basis of such data is not trivial and usually requires sophisticated machine-learning techniques, e.g., learning of fuzzy membership functions for all concepts. The substantial support of a domain expert may also be necessary. Therefore, for the purposes of this chapter, we decided to simplify the construction of concepts. We assumed that the concept labels correspond to the attributes of the data table, as is often done by researchers [25,11]. Instead of the complex construction of fuzzy concepts, we simply normalized the value sets of the attributes to the $[0, 1]$ range. The minima and maxima of value sets are computed as a data processing step.

6.1 Experimental Settings

For all experiments described in this section we assume the following settings:

- the base time is $\Delta t = 1$;
- $card(T_H) = card(T_F) = t_{pred} = 10$;
- the fitness function (11) is used in all our experiments, the parameter $p = 1$;
- we use the Equation (2) for reasoning.

We define the prediction error for the ith concept at time step t according to Equation (14):

$$err(t) = |a_i(t) - a_i'(t)|, \tag{14}$$

where $a_i(t)$ denotes the activation of concept observed from real data, and $a_i'(t)$ is the predicted activation of concept using the FCM-based model. The cumulative prediction error for n time steps is defined as in Equation (15):

$$err(n) = \sum_{t=1}^{n} err(t). \tag{15}$$

In the following experiments we test the ability of FCMs to perform two particular prediction tasks:

1. reproducing the original learning data assuming: $A(t_s) = A(t_0)$,
2. predicting the unknown future data assuming: $A(t_s) \neq A(t_0)$.

6.2 Quality of Data Reproduction

In this section we would like to test experimentally the ability of the FCMs to mimic the historical data used for learning. In Fig. 4 we show the raw data used to learn FCM in the first experiment.

The matrix W learned using four different algorithms is shown in Table 1. Due to the assumed reasoning method (2), the values of weights on the diagonal of W are equal to 0.

For the fixed-point attractor data, in the case of large fluctuations in the concept activations, the adaptive algorithms are not able to reflect changes within the full learning sequence. In Fig. 5a and 6a, we can see that the DHL algorithm cannot learn weights properly, and the FCM cannot duplicate historical data well. The DHL algorithm assumes that, in any given time step, only two concepts influence each other. Therefore, it is not possible to properly identify relationships between all concepts. In the subsequent time moments, the

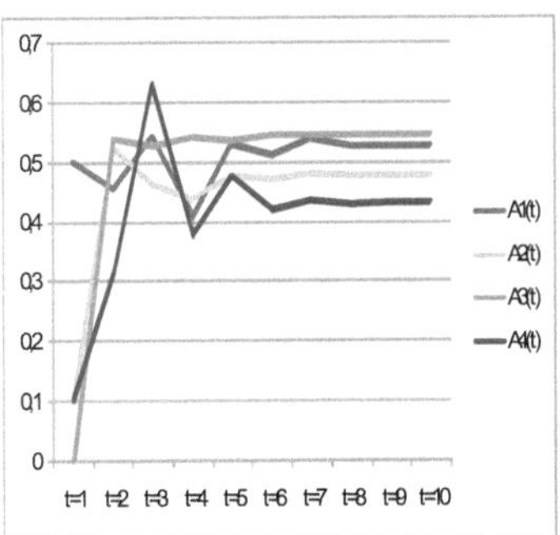

	A_1	A_2	A_3	A_4
t=1	0.5	0.1	0	0.1
t=2	0.455	0.524979	0.539915	0.310026
t=3	0.54312	0.464262	0.527132	0.629644
t=4	0.408378	0.439597	0.542621	0.375921
t=5	0.531948	0.479105	0.538987	0.474907
t=6	0.512085	0.474444	0.545696	0.421097
t=7	0.540683	0.481264	0.546106	0.435934
t=8	0.52738	0.479386	0.546894	0.430208
t=9	0.528659	0.479986	0.546892	0.431707
t=10	0.528275	0.479721	0.546911	0.431404

Fig. 4. Learning data of the "fixed-point attractor" type

Table 1. The weights of the learned FCMs for "fixed-point attractor" data

	DHL	BDA	RCGA	DE
W_{12}	-0.00068	0.075172	-0.02245	-0.34378
W_{13}	-0.00172	-0.12873	0.178	0.647781
W_{14}	0.005257	0.012696	-0.15158	-0.38127
W_{21}	-0.00068	0.113095	0.110021	0.135948
W_{23}	0.013431	-0.10934	-0.12272	-0.08979
W_{24}	0.004659	0.14378	-0.01398	-0.09226
W_{31}	-0.00172	-0.2221	0.06246	0.062624
W_{32}	0.013431	-0.10535	0.027094	0.026622
W_{34}	0.006162	-0.19794	-0.02007	-0.01977
W_{41}	0.005257	0.013787	-0.51977	-0.52
W_{42}	0.004659	0.051657	0.999514	1
W_{43}	0.006162	-0.01206	-0.46926	-0.46932

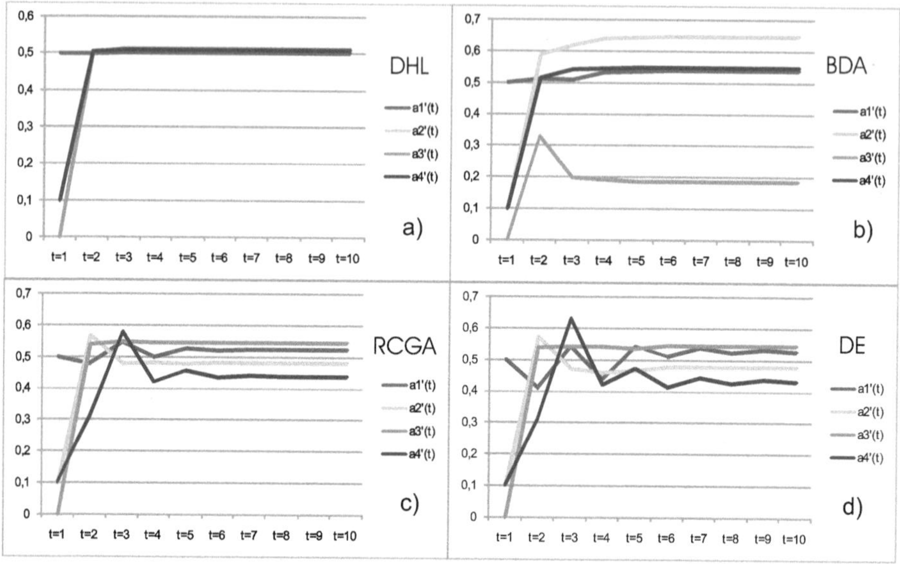

Fig. 5. The results of prediction for "fixed-point attractor" data

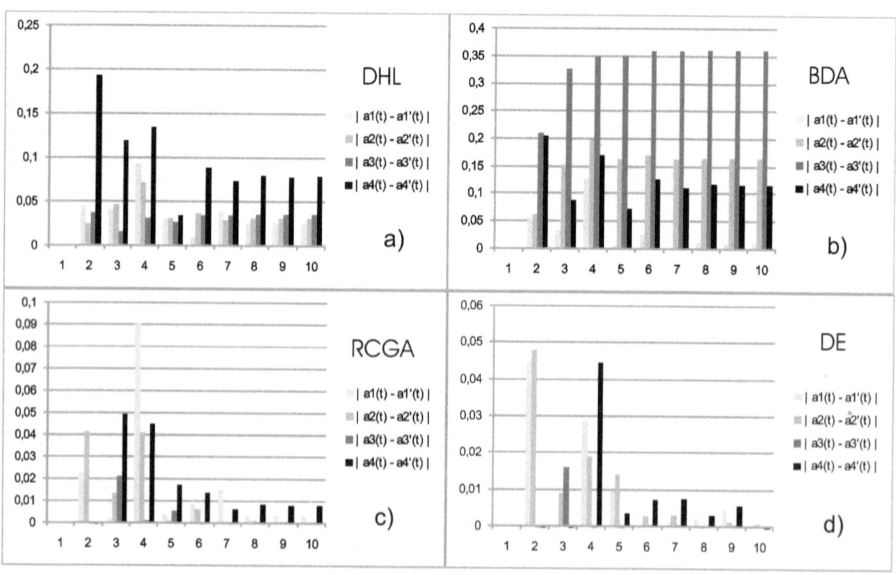

Fig. 6. Prediction errors for "fixed-point attractor" data

multiplication of two concept changes with values lower than 1.0 causes the weight values to gradually approach zero. At the same time, during the prediction, concept values that are close to zero scaled with the sigmoidal function are

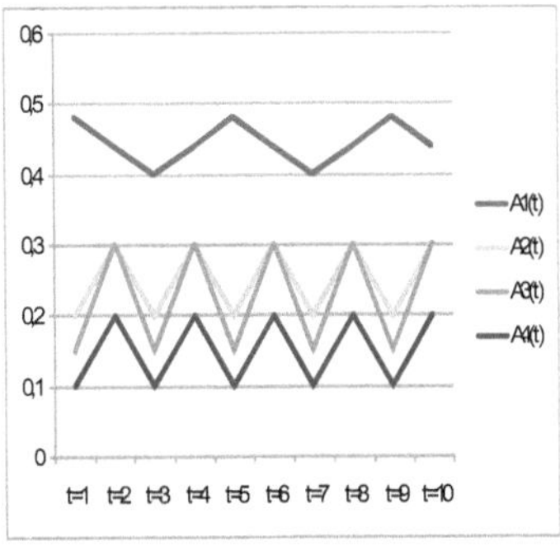

	A_1	A_2	A_3	A_4
t=1	0.48	0.2	0.15	0.1
t=2	0.44	0.3	0.3	0.2
t=3	0.4	0.2	0.15	0.1
t=4	0.44	0.3	0.3	0.2
t=5	0.48	0.2	0.15	0.1
t=6	0.44	0.3	0.3	0.2
t=7	0.4	0.2	0.15	0.1
t=8	0.44	0.3	0.3	0.2
t=9	0.48	0.2	0.15	0.1
t=10	0.44	0.3	0.3	0.2

Fig. 7. Learning data of the "limit cycle" type

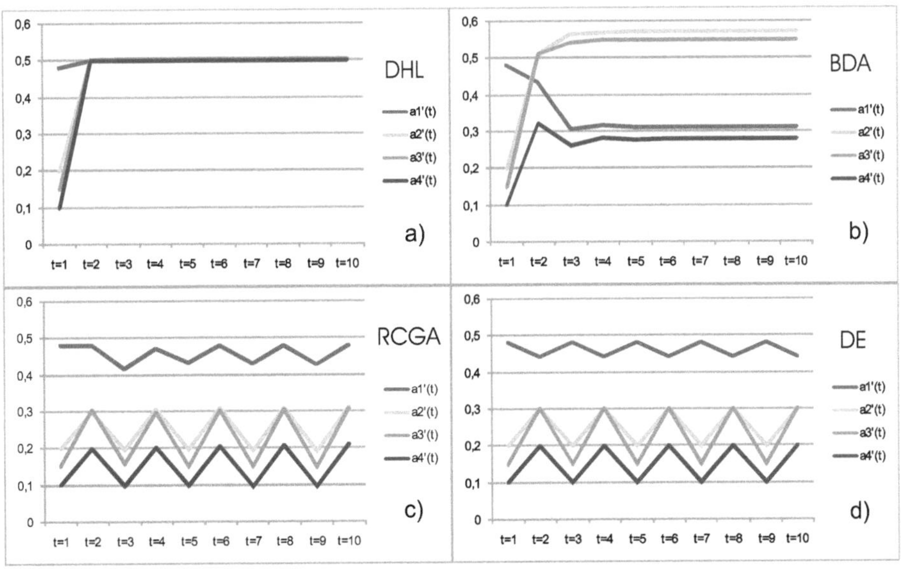

Fig. 8. Prediction results for "limit cycle" data

set to the value 0.5. Therefore after a few iterations, almost all concepts reach steady state.

The BDA adaptive algorithm deals with learning FCMs much better than DHL. Due to the use of more concepts in each time step, the values of the weights are not close to zero. In Fig. 5b and 6b it can be seen that, for fixed data, the value of concepts c_1 is predicted quite well and that the prediction of the activation value of concept c_2 contains a relatively small error. Unfortunately, the other two concepts reach an erroneous steady state.

Another considered algorithm, real coded genetic algorithm RCGA, deals quite well with the prediction problem as you can see in Fig. 5c and 6c. Only

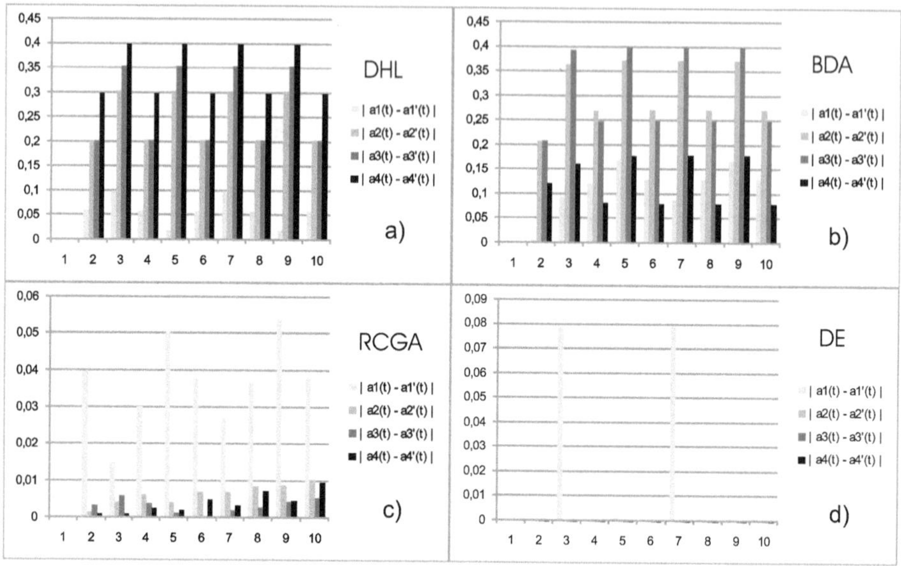

Fig. 9. Prediction errors for "limit cycle" data

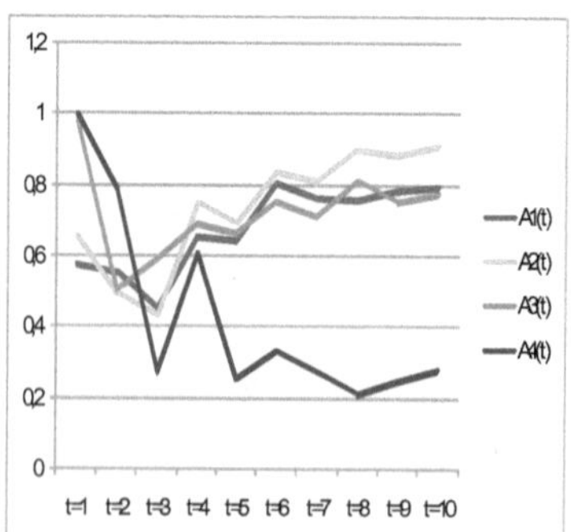

	A_1	A_2	A_3	A_4
t=1	0.57609	0.65286	0.98	1
t=2	0.5563	0.49524	0.5	0.78788
t=3	0.45	0.43	0.59	0.27273
t=4	0.65434	0.75056	0.69	0.60607
t=5	0.6413	0.69	0.66	0.25212
t=6	0.80435	0.83928	0.75	0.33333
t=7	0.76304	0.81	0.71	0.27273
t=8	0.756	0.9	0.81	0.21212
t=9	0.78319	0.88381	0.75	0.24697
t=10	0.7963	0.9112	0.77	0.28234

Fig. 10. Learning data of the "chaotic attractor" type

slight errors are seen in earlier iterations. In subsequent time steps, when the data reach steady state, the error rate becomes very small (lower than 0.01).

The differential evolution algorithm is the second evolutionary method that is able to give very good results in the process of FCM learning (Fig. 5d and 6d). The prediction errors are comparable to those achieved using the RCGA algorithm. Initially, the prediction error was about 0.05, and, in the next iterations, it was reduced to a value of 0.01.

A similar situation occurs in the case of using cycle data for learning. The DHL and BDA algorithms cannot mimic cycle data (Fig. 8a, Fig. 8b, Fig. 9a and Fig. 9b). The application of BDA leads to a situation in which concept

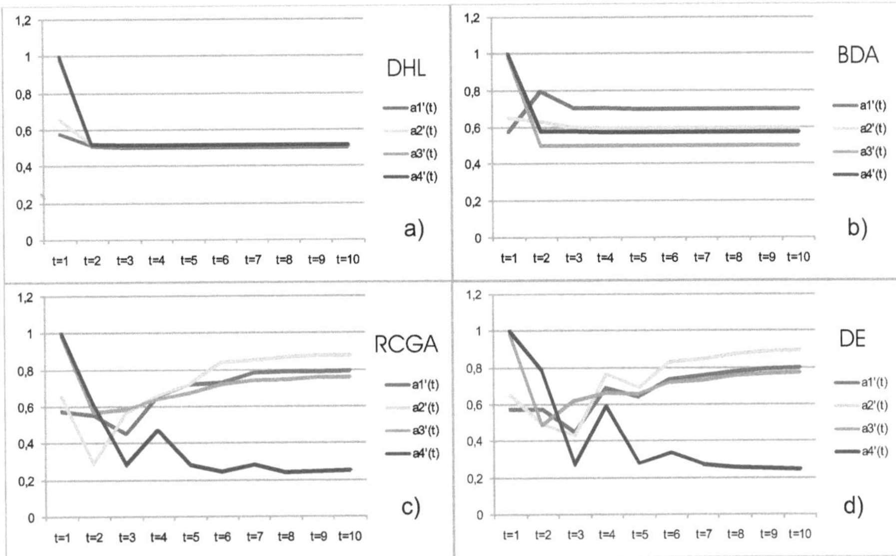

Fig. 11. Prediction results for "chaotic attractor" data

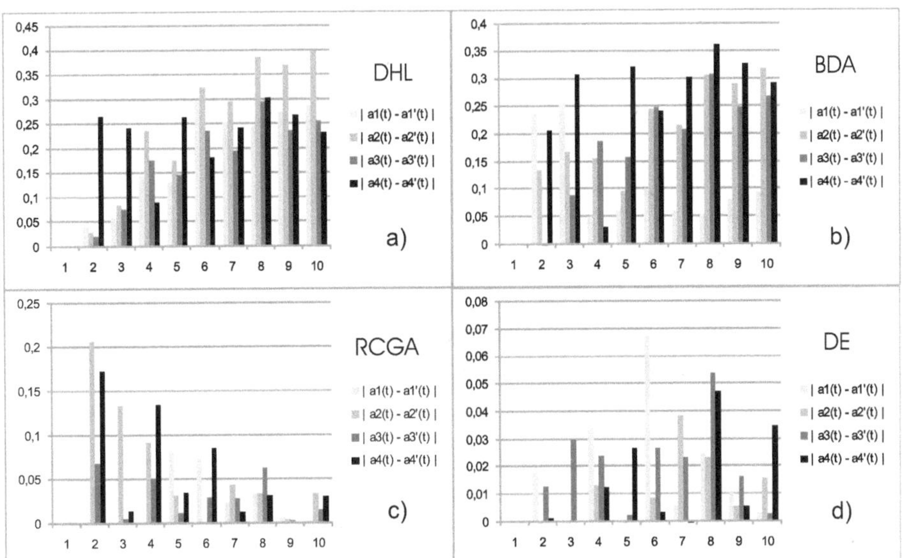

Fig. 12. Prediction errors for "chaotic attractor" data

activations reach steady state in subsequent time steps. The application of the RCGA algorithm effectively facilitates learning the FCM (Fig. 8c). The sum of the prediction errors is quite small (Fig. 9c).

Chaotic data cannot be learned by DHL and BDA either, during the reasoning phase every concept activation reaches steady state (Fig. 11a, Fig. 11b, Fig. 12a and Fig. 12b).

It may be seen in Fig. 11c and 12c that, for chaotic data, the application of the RCGA algorithm during learning also leads to reaching the equilibrium state during reasoning. In this case, it may be assumed that concept values can be duplicated with some reasonable approximation, but, at the end of the reasoning process, the generated data have the tendency to reach steady state. Chaotic data were also difficult to duplicate by an FCM learned by the DE algorithm (Fig. 11d), but the prediction errors were much less than those for the RCGA algorithm (Fig. 12d). Note that the scales on the vertical axes of Fig. 12c and Fig. 13d differ.

In any case, the evolutionary methods of learning FCMs seem to be much better than adaptive. Therefore, we decided to present the cumulative error, defined in Equation (15), only for the evolutionary learning methods. The cumulative error $err(n)$ is summed for all three types of data, where the number of time steps $n = 10$. Every test, understood as learning-and-testing cycle, was run 10 times. The first column in Table 2 denotes the number of the particular test.

Table 2. Cumulative prediction errors

Test	Chaotic		Cycle		Fixed	
	DE	RCGA	DE	RCGA	DE	RCGA
1	0.595297	5.823462	0.173247	0.173247	0.279861	1.682506
2	1.31491	1.924132	0.17417	0.17417	0.279845	4.513751
3	2.110496	1.59704	0.1766	0.1766	0.279843	0.45272
4	1.513919	2.126377	0.166013	0.166013	0.27989	2.382618
5	2.315513	2.918895	0.171755	0.171755	0.279894	2.394499
6	2.213198	1.878982	0.168795	0.168795	0.279847	3.684865
7	1.6112	2.942499	0.17147	0.17147	0.279856	1.019418
8	1.455	3.888239	0.165302	0.165302	0.279871	2.620143
9	1.617157	2.210706	0.167254	0.167254	0.279872	2.227744
10	1.345905	3.923134	0.1708	0.1708	0.279887	4.89989

Evolutionary algorithms allow the learning of FCMs of quite good quality. Even for cases in which there are large variations in the concept activation values, problems are not created for the RCGA or the differential evolution algorithms.

6.3 Quality of Data Prediction

In his section we would like to test the behavior of the achieved FCMs assuming that the initial state vector $A(t_s)$ is changed i.e. $A(t_s) \neq A(t_0)$. For the following experiments we assume: $a_1(t_s) = 0.3$, $a_2(t_s) = 0$, $a_3(t_s) = 0.1$, $a_4(t_s) = 0.3$. We decided to compare the predicted sequence of state vectors with respect to the original learning data. We show the prediction errors only for "fixed point

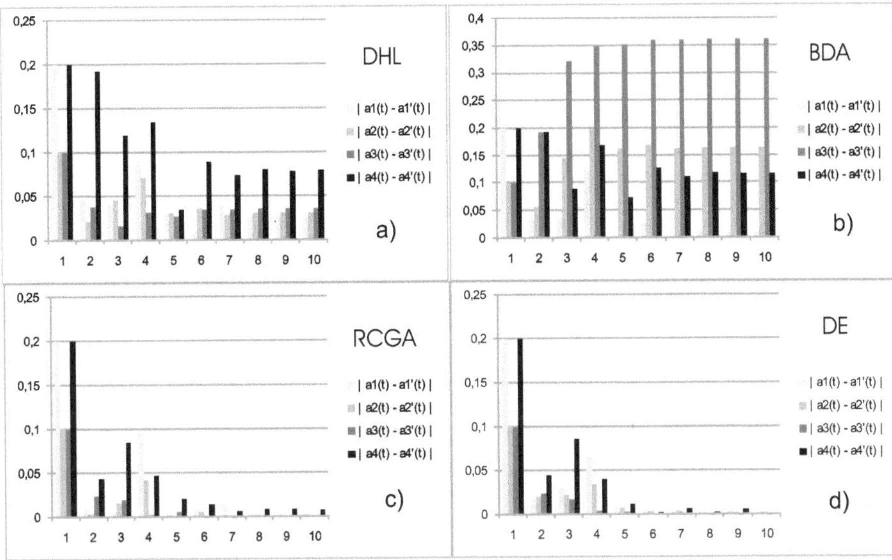

Fig. 13. Prediction errors for "fixed-point attractor" data

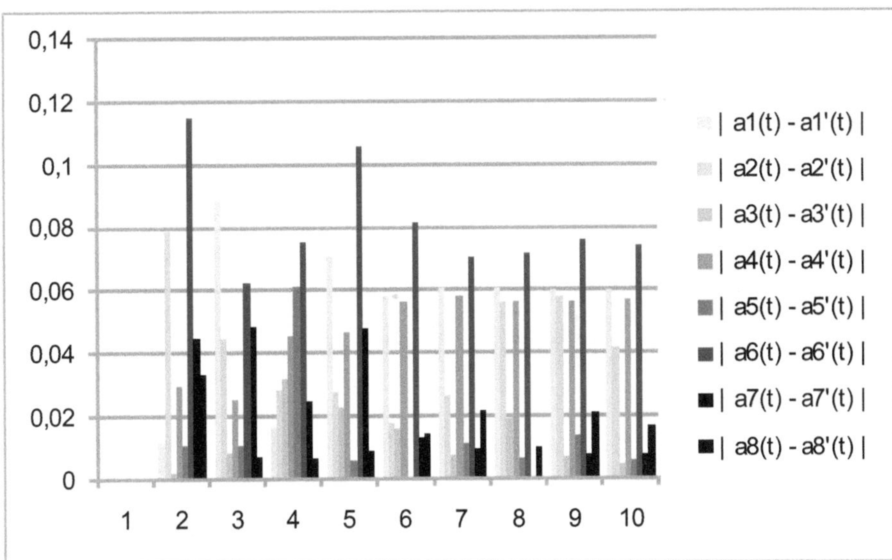

Fig. 14. Prediction errors for "fixed-point attractor" data - 8 concepts

attractor" data (Fig. 13). Also this time, there were problems with the prediction of state vector by the adaptive FCMs. In case of using the BDA algorithm the stability of the state vector was not properly achieved. The activation value of concept c_3 differs significantly from its original stability value. In case of evolutionary method you can notice that the large initial error (due to the forced initial state) is quickly reduced. The values of all concepts in stability state are also properly achieved. Unfortunately, in case of cyclic and chaotic data the problems with adaptive algorithms were similar. Also in this case the DE algorithm was the best.

6.4 Scaling up the Evolutionary Learning Methods

On the basis of our previous experiments we have decided to choose the DE algorithm as the best for learning small FCMs consisting of only 4 concepts. It was interesting for us to check experimentally the possibility to scale up the learning algorithms for larger FCMs. In the next step of the analysis we decided to use 8 concepts. In this case, the dimension of the optimization problem (the length of chromosome) is equal to 56. The obtained FCM is of quite good quality. The prediction errors are shown in Fig 14.

7 Example of Application

The prediction of weather conditions is a good illustration of the application of FCM to a practical problem. The goal is to predict the weather in the following days on the basis of current measurements and the available FCM model. The set of five concepts is as follows: $C = \{c_1$ - temperature, c_2 - dew point temperature, c_3 - humidity, c_4 - wind speed, c_5 - atmospheric pressure$\}$. To learn the FCM,

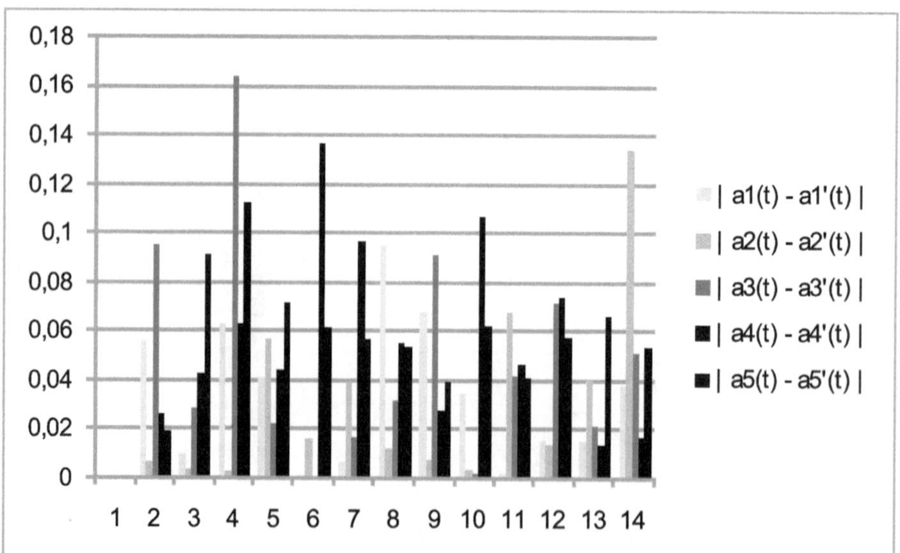

Fig. 15. Prediction errors for weather data - 5 concepts

we used the short-term meteorological data [5] collected for the central part of Poland within the period 01-15 January 2008. We used the differential evolution algorithm and assumed the following constraints and parameters: the cardinality of the initial population was 100, the number of learning iterations was 10,000. On the basis of the achieved FCM, we made a one-day prediction of weather conditions. The prediction results were collected for the next 14 days. The results were in accordance with our intuitive expectations. Fig. 15 presents the achieved prediction errors, which show that the FCM can be used quite successfully to predict weather conditions in a short time horizon.

8 Conclusions and Recommendations

In this chapter, we conducted a comparative analysis of adaptive and evolutionary learning methods of FCMs. In our opinion, the main advantage of evolutionary methods is the relative high quality of the obtained FCM with respect to the considered prediction task. This seems to be valid for any considered types of raw data used for learning. Using the standard prediction method for FCM, the predictive capabilities of the adaptive FCMs are completely unsatisfactory. In our opinion, the cumulative and linear features of the standard reasoning process seem to be inappropriate for use with adaptive learning methods. On the other hand, the intuitive interpretation of adaptive FCMs by humans seems to be better. In conclusion, we definitely recommend evolutionary methods for learning FCMs that are intended for use in making predictions. In some cases in which the quality of prediction is not crucial, the existing adaptive methods of learning FCMs can be also useful.

Acknowledgments

The research has been partially supported by the grant N N516 368334 from the Polish Ministry of Science and Higher Education.

Finally, we would like to thank the anonymous reviewers for their helpful comments.

References

1. Vaughn, R.B., Siraj, A., Bridges, S.M.: Fuzzy cognitive maps for decision support in an intelligent intrusion detection system. In: IFSA World Congress and 20th NAFIPS International Conference, vol. 4, pp. 2165–2170 (2001)
2. Aguilar, J.: Survey about fuzzy cognitive maps papers. International Journal of Computational Cognition 3(2) (2005)
3. Axelrod, R.: Structure of Decision–The Cognitive Maps of Political Elites. Princeton University Press, Princeton (1976)
4. Carvalho, J.P., Paulo, J., Jose, C., Tome, A.B.: Rule based fuzzy cognitive maps - qualitative systems dynamics (2000)
5. Air club of 'Ziemia Lubuska' (2008), `http://www.meteo.azl.pl/historia.php`

6. Huerga, A.V.: A balanced differential learning algorithm in fuzzy cognitive maps. In: Proceedings of the 16th International Workshop on Qualitative Reasoning (2002)
7. Khan, M.S., Chong, A., Quaddus, M.: Fuzzy cognitive maps in intelligent decision support - a review, int. Journal of Systems Research and Information Science (1999)
8. Kosko, B.: Fuzzy cognitive maps. International Journal of Man-Machine Studies 24(1), 65–75 (1986)
9. Okeef'e, J., Nadel, L.: The Hippocampus as Cognitive Map. Calerendon Press, Oxford (1978)
10. Papageorgiou, E., Stylios, C.D., Groumpos, P.P.: Learning algorithms for fuzzy cognitive maps. In: EUSFLAT Conf., pp. 83–88 (2001)
11. Papageorgiou, E., Stylios, C.D., Groumpos, P.P.: Active hebbian learning algorithm to train fuzzy cognitive maps. Int. J. Approx. Reasoning 37(3), 219–249 (2004)
12. Papageorgiou, E.I.: Fuzzy cognitive maps learning using particle swarm optimization. Journal of Intelligent Information Systems (2005)
13. Papageorgiou, E.I., Parsopoulos, K.E., Groumpos, P.P., Vrahatis, M.N.: Fuzzy cognitive maps learning through swarm intelligence. In: Rutkowski, L., Siekmann, J.H., Tadeusiewicz, R., Zadeh, L.A. (eds.) ICAISC 2004. LNCS (LNAI), vol. 3070, pp. 344–349. Springer, Heidelberg (2004)
14. Papageorgiou, E.I., Stylios, C.D., Groumpos, P.P.: Fuzzy cognitive map learning based on nonlinear hebbian rule. In: Australian Conference on Artificial Intelligence, pp. 256–268 (2003)
15. Park, K.S., Kim, S.H.: Fuzzy cognitive maps considering time relationships. International Journal Human-Computer Studies 42, 157–168 (1995)
16. Pearl, J.: Causality, Models Reasoning and Inference. Cambridge University Press, Cambridge (2000)
17. Stach, W., Kurgan, L.A., Pedrycz, W.: A survey of fuzzy cognitive map learning methods. In: Grzegorzewski, P., Krawczak, M., Zadrozny, S. (eds.) Issues in Soft Computing: Theory and Applications (2005)
18. Stach, W., Kurgan, L.A., Pedrycz, W.: Numerical and linguistic prediction of time series with the use of fuzzy cognitive maps. IEEE T. Fuzzy Systems 16(1), 61–72 (2008)
19. Stach, W., Kurgan, L.A., Pedrycz, W., Reformat, M.: Higher-order fuzzy cognitive maps. In: Proceedings of NAFIPS 2006 (2006)
20. Storn, R., Price, K.: On the usage of differential evolution for function optimization. In: IEEE NAFIPS 1996, pp. 519–523 (1996)
21. Storn, R., Price, K.: Differential evolution - a simple and eficient heuristic for global optimization over continuous spaces. Springer Journal of Global Optimization 11(4), 341–359 (1997)
22. Stylios, C.D., Georgopoulos, V.C., Groumpos, P.P.: The use of fuzzy cognitive maps in modeling systems. In: Proceeding of 5th IEEE Mediterranean Conference on Control and Systems, Paphos (1997)
23. Stylios, C.D., Georgopoulos, V.C.: Fuzzy cognitive maps structure for medical decision support systems. Studies in Fuzziness and Soft Computing 218, 151–174 (2008)
24. Tolman, E.C.: Cognitive maps in rats and men. The Psychological Review 55(4), 189–208 (1948)
25. Stach, W., Kurgan, L.A., Pedrycz, W., Reformat, M.: Genetic learning of fuzzy cognitive maps. Fuzzy Sets and Systems 153(3), 371–401 (2005)

An Architecture for the Semantic Enhancement of Virtual Engineering Applications

Carlos Toro[1], Manuel Graña[3], Jorge Posada[1], Cesar Sanín[2], and Edward Szczerbicki[2]

[1] VICOMTech Research Centre, Spain
[2] Faculty of Engineering and Built Environment, University of Newcastle, Australia
[3] University of the Basque Country - Facultad de Informática de San Sebastián, Spain

Abstract. Virtual Engineering (VE) is defined as the integration of geometric models and related engineering tools (such as analysis, simulation, optimization and decision-making, etc), within a computerized environment that facilitates multidisciplinary and collaborative product development [1]. The focus of Virtual Engineering is to engage the human capacity for evaluation of complex systems and situations [2]. In this article, we present an architecture for the Semantic enhancement of Virtual Engineering Applications (VEA) through their embedded Virtual Engineering Tools. Our architecture follows a User-Intention-Domain schema and makes use of state of the art technologies like the Set of Experience Knowledge Structure and the Reflexive Ontologies in order to offer a better User experience with VEA. The main advantages of using Semantics in VEA can be summarized are: (*i*) an improved information and embedded Knowledge management, (*ii*) Enhancements in the search, knowledge and information sharing processes during the use of the VEA, (*iii*) The use of the intrinsic Knowledge embedded in the elements being described by the VEA and (*iv*) the empowerment of the User's knowledge and embedding of such knowledge in a structured and explicit conceptualization.

Keywords: Domain Modeling, Knowledge Based Systems, ontologies.

1 Introduction

Virtual Engineering (VE) is defined as the integration of geometric models and related engineering tools (such as analysis, simulation, optimization and decision-making, etc), within a computerized environment that facilitates multidisciplinary and collaborative product development [1]. According to McCorkle [2] "a key aim of Virtual Engineering is to engage the human capacity for evaluation of complex systems and situations". In the scope of this work, we consider a Virtual Engineering Application (VEA), any engineering-focused software able to perform the aforementioned task. Also, a Virtual Engineering Tool (VET) as any of the components of a VEA that offer a user or machine interaction capability. In this paper, we will introduce an architecture for the semantic enhancement of VEA through their embedded VET. This paper is presented as follows: In section 2, we present some relevant concepts for this paper. In section 3 we introduce a concept model for VEA, discussing some of its properties. In section 4, we present our

N.T. Nguyen and E. Szczerbicki (Eds.): Intel. Sys. for Know. Management, SCI 252, pp. 175–195.
springerlink.com

architecture for the semantic enhancement of VEA. In section 5 we introduce an example case where our methodology was successfully applied and lastly in section 6 we present some conclusions and future work.

2 Relevant Concepts

In his section, we introduce some of the core concepts that will be used along the paper. We do not intend to give an extensive overview of such concepts as any of them could easily cover extensive reviews. We propose instead brief descriptions and relevant works from which an interested reader could extend this content.

2.1 Knowledge Engineering

Knowledge is considered an invaluable resource of great benefit for most purposes in life. For this reason, mankind has always attempted to make it part of their assets. Knowledge itself seems to be an attribute of human beings; it may be defined [3] as: (*i*) the expertise and skills acquired by a person through experience or education via a theoretical or practical understanding of a subject, (*ii*) what is known in a particular field or in total related to facts and information or (*iii*) the awareness or familiarity gained by experience of a fact or situation. The acquisition of Knowledge involves complex cognitive processes such as: perception, learning, communication, association and reasoning. The processes involved are the result of the development of the person and closely related with his intelligence. The word Knowledge is also used to mean the confident understanding of a subject with the ability to use it for a specific purpose if appropriate. Philosophical debates in general start with Plato's formulation of Knowledge as "justified true belief". There appears however, that no single agreed definition of Knowledge has been reached at the present, and one of the main reasons for this is the fact that continuously new competing theories are presented by the scientific community. There are theories that come from the medical, the psychological and even the sociological points of view. However in this paper we will address a very specific application of the term for the fields of Engineering and Computer Science. According to Feigenbaum [4]: "Knowledge Engineering (KE) is an engineering discipline that involves integrating Knowledge into computer systems in order to solve complex problems normally requiring a high level of human expertise". [4] Knowledge Engineering, hence, relies on instructional methodologies and Computer Science in general, trying to mimic Knowledge and behaviors that are intrinsically human in a certain Domain and within the scope of an artificial system. This broad definition implies not only the need of specific technologies, but also the need to overcome related implementation issues. Recent developments in ontology theory have been introduced to the engineering world through Computer Science. Specifically the Knowledge Engineering community is giving a lot of attention to ontologies, thanks to the fact that this theory offers an interesting tool to describe objects and the relations amongst them within the scope of a given Domain. In Computer Science, and hence within the scope of this paper, ontologies are used as an information support schema for Knowledge representation of a Domain. They provide a vocabulary for representing and communicating Knowledge about some topic and a set of relations holding amongst the terms in that vocabulary.

2.2 Ontologies

There are many possible definitions to describe what ontology is. In the Computer Science domain, the widely accepted definition states that "an ontology, is the explicit specification of a conceptualization" [5], or in other words an ontology is the description of the concepts and relationships in a domain of study. Some of the motivations to model a domain with ontologies are (*i*) to share common understanding of the structure of information among people or software agents, (*ii*) to enable reuse of domain knowledge, to make domain assumptions explicit, (*iii*) to separate domain knowledge from the operational knowledge, and (*iv*) to analyze the domain's modelled knowledge. Ontologies can be modelled using different languages, for instance, RDF, RDFS and OWL, the later is a new standard from the W3C consortium available in three flavours, OWL-Lite, OWL-DL and OWL-Full, depending on the desired level of semantic load [6]. The main characteristic of an ontology-based solution is its capacity to semantically infer newly derived information. The user does not explicitly specify such information and in order to obtain it modern inference engines and reasoners, like Racer or Pellet [7], are used.

2.3 The Reflexive Ontologies (RO)

RO are a computational methodology for the enhancement of the ontology query process introduced by Toro et.al in [8]. A RO is a description of the concepts and relations of such concepts with a set of self-contained queries over instances in the domain of study. RO can be used whenever an ontology query is needed. The aforementioned technique enhances a Knowledge Base (KB) by (*i*) the speeding up of the query process (*ii*) giving to the ontology the possibility of adding new queries on individuals with the corresponding answers to such queries (a feature that adds knowledge about the domain); and (*iii*) the self containment of the Knowledge Structure in a single file; including the model, the relations between the elements of the model, the individuals (instances) and queries over such individuals. Fig 1 depicts the RO concept.

2.4 The Set of Experience Knowledge Structure (SOEKS)

The Set of Experience Knowledge Structure is an experience tool able to collect and manage explicit knowledge of different forms of formal decision events [9]. The SOEKS has been developed as part of a platform for transforming information into knowledge named Knowledge Supply Chain System. In this methodology, there are four basic components: variables, functions, constraints and rules associated and stored in a combined dynamic structure. The combination of the four components of the SOEKS offers distinctiveness due to the elements of the structure that are connected among themselves, imitating part of a long strand of DNA. A SOEKS produces a value of decision, called efficiency. Besides, it is possible to group sets of experience by category, that is, by objectives. These groups could be store a "strategy" for such category of decisions.

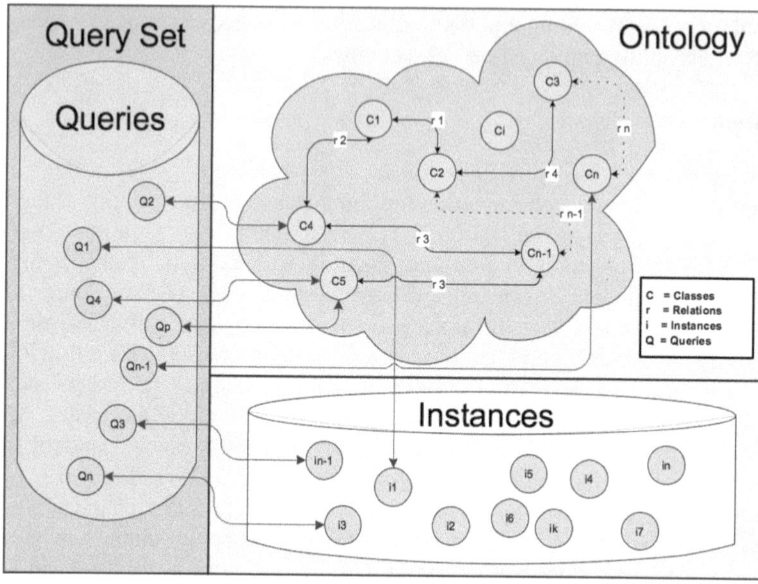

Fig. 1. Reflexive Ontology concept

2.5 Toward Semantically Enabled Virtual Engineering Applications

The ideas and technologies that support the Semantic Web provide an interesting platform to support the Virtual Engineering Applications. This can aid in the development, use and re-use of knowledge from a networked environment in which the information is meaningfully distributed. According to McCorkle the underlying question that must be answered today is: "How will information be integrated in a manner that will allow commercial and proprietary software tools to remain separate while also being integrated so that the end user can control and query these tools with little to no knowledge of the tools implementation or inner-working details?" [10] The answer to this question will depend largely on the ability to implement open interfaces and schemes that can evolve over time as well as open source toolkits that enable development teams to collaborate at a high level. The enhancement of VEA by Knowledge-oriented (or Knowledge-supported) technologies has been arguably tangentially reached by some independent efforts. In [11], a discussion about potential applications of the Semantic Web to explore the above question is given, presenting some specific capabilities that should be fulfilled by Semantically oriented Virtual Engineering Tools. Huang [12] presents a decision support platform for the interactive design that integrates mathematical optimizations with human interactions based on VET. In this approach, the designer's interaction causes the optimization process to dynamically change by adding, deleting, and modifying objectives, constraints, and other parameters that govern processes.

As an illustration, a coal pipe design case was used to demonstrate the platform's capabilities. Huang's case study demonstrated that adding user interaction into the design process has the potential to improve design efficiency and quality. Kuntz's [13] research was focused on the application of computational simulation models used in other branches of engineering to project planning, the authors propose a model, called the Virtual Design Team (VDT), that represents the structure and capabilities of organizational entities, such as teams (called actors in the model), activities, and both direct and coordination work.

The model links the organization chart and the activity diagram of projects, and can be used to predict the effect of different organizational structures or of the use of different project constraints. In this respect, the proposed method outperforms planning tools based on the Critical Path Method, as it explicitly incorporates the necessary communication and coordination among the actors assigned to different project activities. As mentioned previously, some research into the topic has been carried out, however, at the present time, the scientific community has not reached consensus in how to extend Virtual Engineering Tools with Semantics. An architecture to solve the mentioned scenario is an open problem.

3 Conceptual Model of a Virtual Engineering Application

We model a VEA in terms of the different components depicted in Fig 2. The question that we want to address is: How can Semantic Technologies work with the Virtual Engineering Application components? To find answers to such a question, we need first to describe the VEA elements: (*i*) characteristics, (*ii*) requirements and (*iii*) interaction.

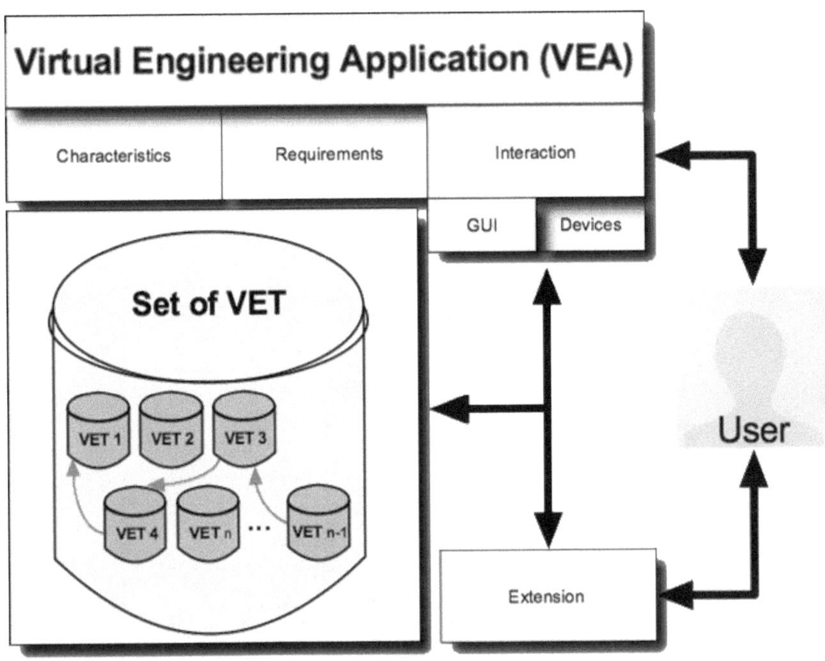

Fig. 2. Components of a VEA

Set of Characteristics

Usually the set of characteristics describes the expected benefits of the VEA in terms of capabilities, features, compatibility, accepted formats for input/output interactions and the overall benefits of using the VEA. The set of characteristics defines a general overview of the competences of the VEA, and the comparison of such competences against the requirements, is in most of the cases the main reason to use the VEA. However, characteristics are usually presented with a marketing focus. For the aforementioned reason, it is wise to perform an evaluation of the VEA before the making the purchase decision. Aware of such a scenario, some VEA providers offer the possibility to test before buying. Fig 3 presents a typical set of characteristics extracted from an advertising pamphlet

Set of Requirements

The set of requisites are the minimum requirements by the computer system where the VEA will be used. Usually the requisites are given as the specification of a minimal configuration. The requirements are a good way to balance the VEA needs as they relate to hardware and software requisites. Fig 4 depicts a typical requisite sheet excerpt.

Set of Interaction Paradigms

The set of interaction paradigms, include different GUI's, and input/output device characteristics and requirements e.g. a mouse, an output screen, etc. The interaction

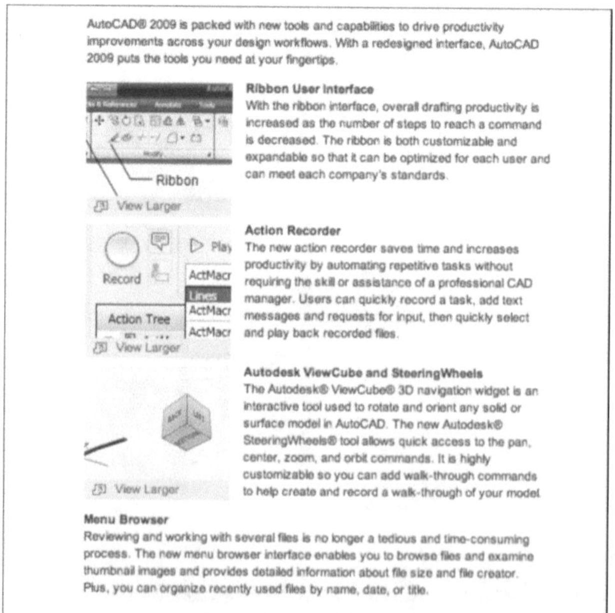

Fig. 3. VEA characterization and advertising

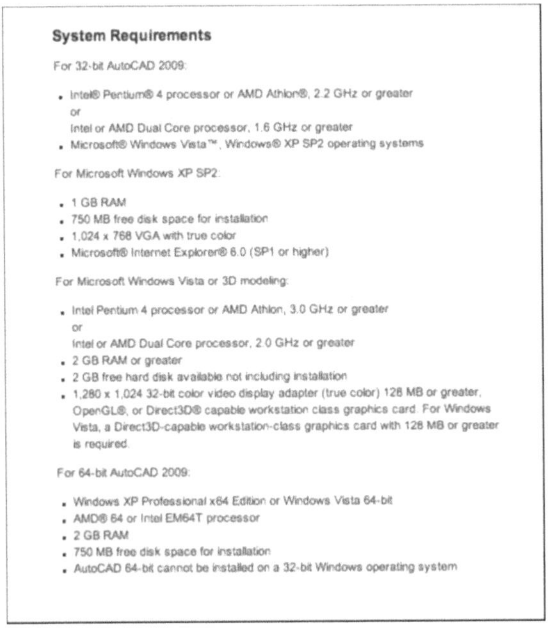

System Requirements

For 32-bit AutoCAD 2009:

- Intel® Pentium® 4 processor or AMD Athlon®, 2.2 GHz or greater
  or
  Intel or AMD Dual Core processor, 1.6 GHz or greater
- Microsoft® Windows Vista™, Windows® XP SP2 operating systems

For Microsoft Windows XP SP2:

- 1 GB RAM
- 750 MB free disk space for installation
- 1,024 x 768 VGA with true color
- Microsoft® Internet Explorer® 6.0 (SP1 or higher)

For Microsoft Windows Vista or 3D modeling:

- Intel Pentium 4 processor or AMD Athlon, 3.0 GHz or greater
  or
  Intel or AMD Dual Core processor, 2.0 GHz or greater
- 2 GB RAM or greater
- 2 GB free hard disk available not including installation
- 1,280 x 1,024 32-bit color video display adapter (true color) 128 MB or greater, OpenGL®, or Direct3D® capable workstation class graphics card. For Windows Vista, a Direct3D-capable workstation-class graphics card with 128 MB or greater is required.

For 64-bit AutoCAD 2009:

- Windows XP Professional x64 Edition or Windows Vista 64-bit
- AMD® 64 or Intel EM64T processor
- 2 GB RAM
- 750 MB free disk space for installation
- AutoCAD 64-bit cannot be installed on a 32-bit Windows operating system

Fig. 4. VEA System Requirements

paradigms are the places where the user executes the input and output interface actions. Some example interaction paradigms possible are: (*i*) the use of multiple views in the same display, (*ii*) the capability to accept not traditional GUI interactions e.g. voice commands, and (*iii*) the use of alternative input devices. In Friedell's work [14], input devices and display devices for the CAD Domain are presented. Friedell's focuses on the question of what is the fundamental nature of the interaction between a human and a computer in the design process supported by the CAD system, and presents a discussion on the best approaches.

Set of Virtual Engineering Tools

The set of VET provided is arguably the most important argument to use the VEA. We base our approach to Semantically enhancing a VEA in the fact that a given VET belonging to the set of Virtual Engineering Tools can be Semantically enhanced by means of considering a User-Domain-Intention approach. Fig 5 depicts an excerpt that characterizes a VET from a VEA intended for Steel Detailing. The set of Virtual Engineering Tools can be composed of complementary tools that are used in conjunction with other tool or isolated tools that can be used upon necessity. When we introduced the GOMS methodology for User Modeling, we described the Operators, the Set of available VET are equivalent to such Operators in the sense that they represent the means for accomplishing a goal or sub-goal.

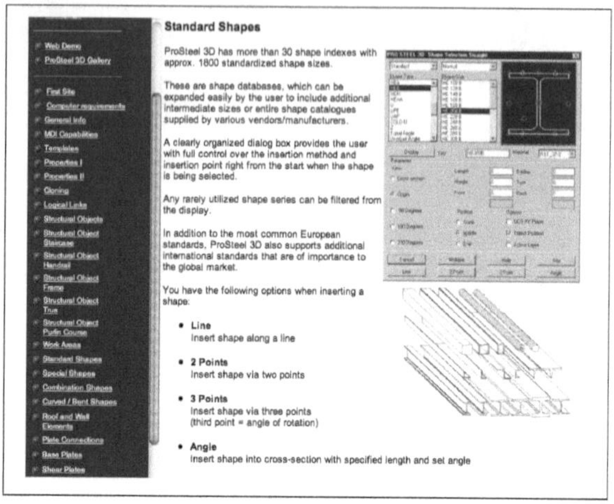

Fig. 5. VEA Tools (excerpt from Prosteel 3D manual)

Extension Capabilities

We propose our VET enhancements by means of using the programming capabilities of the VEA. Such programming capabilities allow the extension of the functionality of the VEA and its VET. Generally, the capability to extend a VEA is provided via an Application Programming Interface (API) or by means of scripting languages. As a general rule, the more open the API, the easier it is to enhance it. The most common computer languages used for API interfaces are C/C++, Java and Visual Basic. The extension capabilities of the VEA, also allow the ease of repetitive actions. Fig 6 depicts a sample code excerpt from the AutoCAD C++ extension API (called ObjectARX). Undoubtedly the two ways of User-VEA interaction are (i) the direct manipulation of the component Virtual Engineering Tools that belong to the VEA via the GUI and Devices and (ii) the use of the extension capabilities of such VEA. As we are interested in semi-automatic features, our approach focuses on the second interaction method, being the Extension Capabilities of the VEA a fundamental stepping-stone for their Semantic enhancement. Of course we do not categorically say that a VEA without extension capabilities cannot be semantically enhanced. However such scenario is not considered in our approach and it would be a possible extension to our work as a future development.

```
// create a new line
AcDbObjectPointer<AcDbLine> line = new AcDbLine();

// set the properties for it
line->setStartPoint(AcGePoint3d(10,10,0));
line->setEndPoint(AcGePoint3d(20,30,0));
```

Fig. 6. Extending a VEA using an API

4 Methodology for the Semantic Enhancement of VEA

In this section we will present a methodology for the Semantic Enhancement of Virtual Engineering Applications. Our methodology uses the recommendations previously introduced in previous papers [15] e.g. User Modeling, Domain Modeling

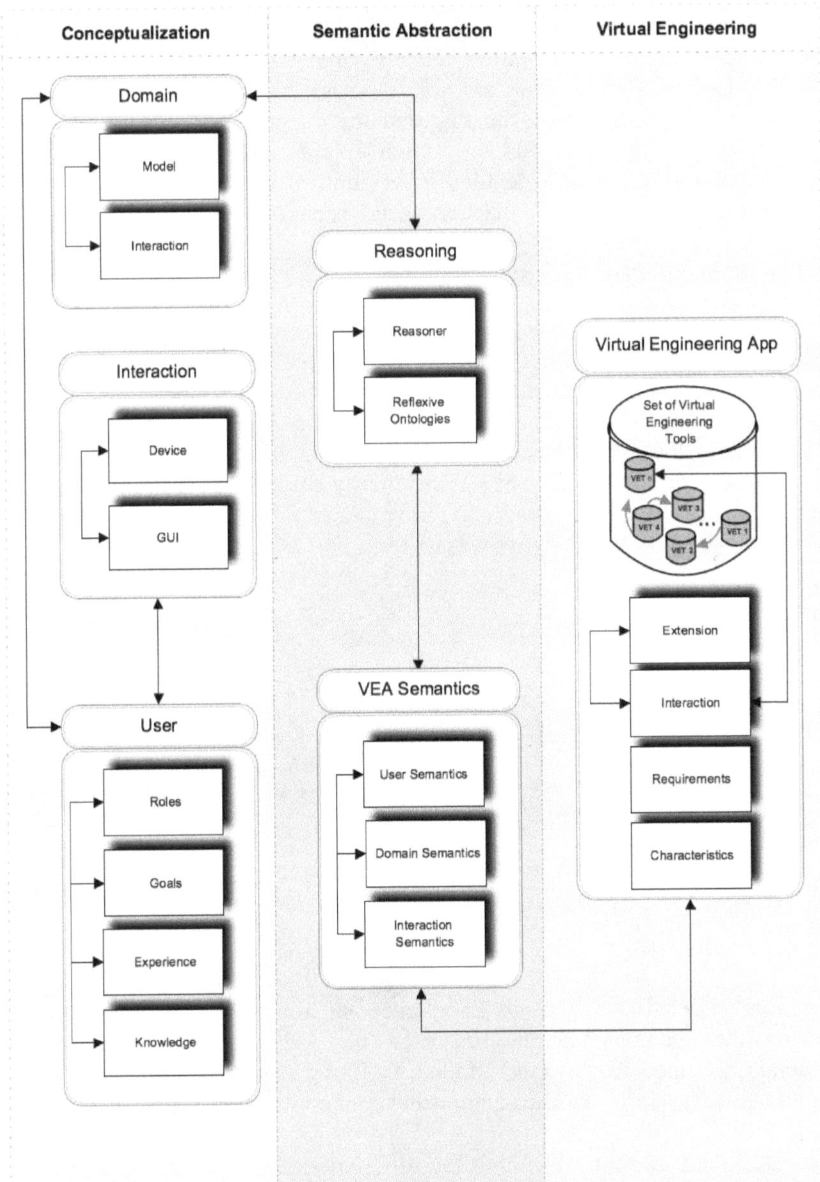

Fig. 7. Architecture for the Semantic Enhancement of VEA

and Interaction Modeling. We consider a simple, yet efficient procedure to enhance any API extensible VEA by dividing our methodology in three layers: (i) the Conceptualization layer, (ii) the Semantic Abstraction layer and (iii) the Virtual Engineering Application layer. In our methodology, different modules and sub-modules compose the layers following a bottom-up approach where the sub-modules are bottom concepts to be modeled and the layers are the top level of conceptualization. Our approach makes strong use of ontology modeling for diverse reasons pointed by research works in the filed [15] the most important being the benefits related to the subsumption relations that allow us to take advantage of implicit Semantics. Sub-modules are able to share information between each other when they belong to the same module and the connections between the different layers are made through module-to-module bidirectional connections. In the following sections we give a detailed description of the different elements of our methodology. A general view of the architecture is depicted in Fig 7.

4.1 The Conceptualization Layer

The general idea of this layer is to provide a conceptual model of the User goals, his expertise, how the user interacts with the VEA and the scenario where the interaction is taking place. The conceptualization layer comprises three modules: The domain Module, the User Module and in the Interaction Module. The importance of conceptualizing the characteristics of the User and the Domain recommendations were discussed in [15]. In this part we will briefly describe the overall characteristics of such modules. To help the reader to follow our explanations, Figure 8 depicts only the Conceptualization Layer components.

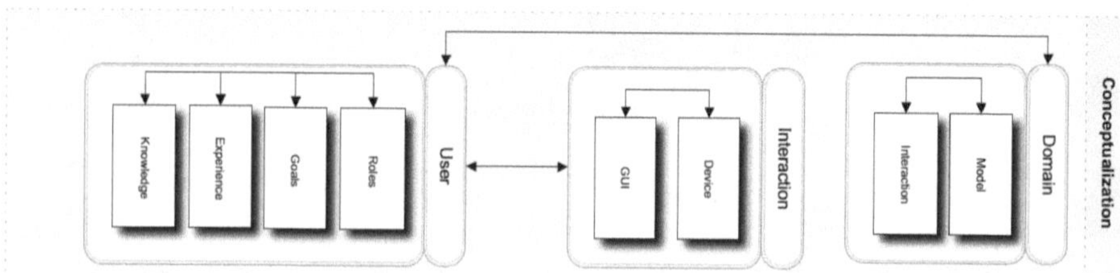

Fig. 8. The Conceptualization Layer

The User Module

The user is modeled from a Goals-Experience and Knowledge-Role points of view, following the recommendations presented in [15] for User Modeling. Each consideration is modeled as a sub-module featuring Knowledge Bases or each one using an ontology approach. The component sub-modules are:

- **Role sub-module:** Defining the different Roles. The user role sub-module can be modeled as a sub-ontology or as a full ontology.

- Goals sub-module: This sub-module considers high and low level goals, being the over-all describing what a user wants to accomplish (or is needed to accomplish). User goals are lists of requirements on the domain.
- **Experience sub-module:** This sub-module is very important in order to determine recommendations based on previous users experiences. This User Experience sub-module has been modeled in various presented approaches using the Set of Experience Knowledge (SOEKS), described by Sanin et. al in [9].
- **Knowledge sub-module:** This sub-module is equivalent to the definition of a User Type. It contains the Knowledge about the Domain that defines the basic differences between Users. We intentionally make User Knowledge and the User Type synonymous because our approach is oriented towards the functional aspect of the User. If the User "has a particular Knowledge of the Domain" which means that such a User can be categorized in the Domain. The User Role sub-module described before will help tighten such User Type categorization by extending the Knowledge model in the sub-module. Another tightening factor is the Experience, that enriches the Knowledge sub-module with previous experiences that are not necessary from the User being modeled, but from sets of Users in similar Domain restricted situations.

The Interaction Module

This module is in charge of conceptualization of the user-VEA interface handling. Two sub-modules compose it, the device sub-module that is a list of hardware interaction features and the GUI that contains Graphical User Interface paradigms (software). The main function of the interaction module is to serve as interface between the conceptualized model of the interaction and the VET, which is where the interaction finally takes place. The different action-reaction situations between the user (programmer or final user) with the rest of the model and the VEA through its VET are modeled from a User Perspective in terms of usability and applicability.

- **Device sub-module:** As already mentioned previously, this sub-module handles the modeling of different hardware that is capable of offer User-VET interactions. Not only typical interaction devices are considered (e.g. screen, mice, etc.) but also the device as a whole is modeled here (e.g. PDA, sub-laptop, etc).
- **GUI sub-module:** This sub-module contains the modeling of the GUI paradigms that allow a better recommendation of a VET from the set of VEA, or an enhancement to the VEA GUI itself by means of enhancements of its embedded VET GUI.
- **The Domain module:** This module was discussed in depth in [15] where we used Engineering Standards as the base for our Domain Modeling. As can be seen in Fig 7, this module has two sub-modules, the Domain Model sub-module and the Domain Interaction sub-module. The Domain module relates to the remaining of the architecture trough the User Module. The reason why

we do this is because in our approach we consider a User Oriented concept for the VEA interaction.

- **Domain model sub-module:** In this sub-module, an identification of the purpose and requirements of the Domain is made. Such identification and purpose tasks are modeled based on a suitable Engineering Standard that complies with such requirements.
- **Domain interaction sub-module:** This sub-module contains the model of the different interactions with the Domain. Such interactions are important for the User Roles correct determination and use.

4.2 The Semantic Abstraction Layer

The general idea of this layer is to use all the models gathered at the conceptualization layer and performs a reasoning process over them in order to enhance the VEA using the knowledge obtained directly, or by means of semantic reasoning from the modeling of the User, the Domain and the Interaction. Fig 9 depicts the Semantic Abstraction Layer components. The Semantic Abstraction module is the place where the Semantic Enhancement takes place. This layer has two modules: The Reasoning Module and the VEA Semantics module.

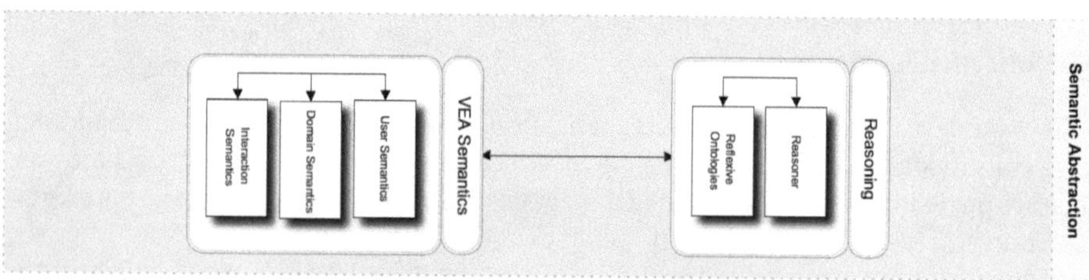

Fig. 9. The Semantic Abstraction Layer

The Reasoning Module

This module is in charge of performing the Semantic Reasoning of the gathered Knowledge from the Conceptualization Layer and also from the VEA semantics sub-module. The reasoning process is performed by a third party reasoner tool handled by the reasoner sub-module and a query engine, implemented using Reflexive Ontologies, which is needed to maintain the Knowledge reasoning in a sort of cyclic status. As can bee seen the Knowledge is gathered also from the Virtual Engineering Layer through the VEA semantics sub-module in order to feed such Knowledge from the VEA being enhanced.

- **Reasoner sub-module:** As explained before, reasoners allow the discovery of non-direct patterns (e.g subsumptions), in other words, this module is in charge of the indirect queries and their handling. The reasoner module can be implemented using the ontology editor API if the reasoning is not complex, but it is advisable to use more complex tools such as those already mentioned in the introduction (e.g. Pellet, FaCT++) as they cover more complex semantic scenarios.

- **Reflexive ontologies sub-module:** This uses the Reflexive Ontologies (RO) concept as a way to perform better and faster query processes to the KB. The generation of the RO idea was directly related to the fact that we needed a fast response tactic in order to maintain the cyclic nature of our architecture. As can be seen, it is quite important to have a good flow of information between all the modules and sub-modules of our architecture and since all the KBs are modeled as ontologies, the most natural approach in order to enhance the massive use of queries over such KB was to develop a technique that provides us with a mechanism that save time by accessing the KB only when needed. The RO concept was introduced in [8] and the concept was also used by Cobos et.al in [16].

The VEA Semantics Module

This module serves as a logical link between the conceptual modeling and the real world actions in the VEA. The module is composed by three sub-modules each one capable of handling the semantic information reasoned in the Reasoner module and the real world feeds provided by the VEA. The three components handle the three different aspects we consider in our Semantic Enhancement: User, Domain and Interaction.

4.3 The Virtual Engineering Layer

This layer contains the Virtual Engineering Application as described earlier in this section. Fig 10 depicts the Virtual Engineering Layer components.

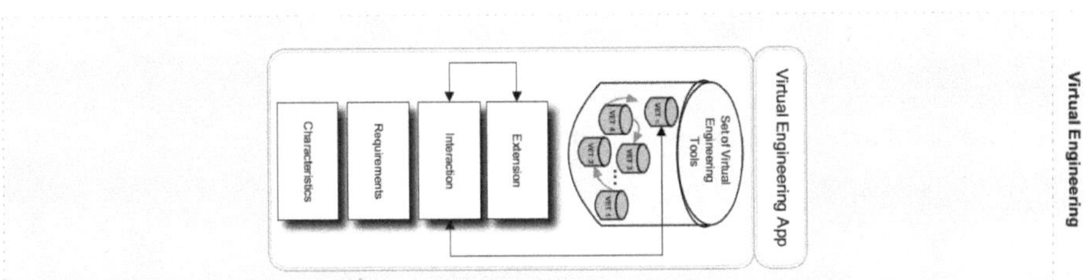

Fig. 10. The Virtual Engineering Layer

The Virtual Engineering Application Module

This module is divided in four sub-modules, those being the Extension and the Interaction sub-modules in a cyclic schema that allows enhancements from the programmatic point of view provided by the Extension sub-module, reachable by parts of the interaction sub-module whose main function is to communicate the interactions to the VET closing the logic chain between the first conceptual layer and the VEA. Interactions are not only considered at a VET level, however, in most cases, the interactions happen here and not in a VEA. Three sub-modules compose this layer: (*i*) The Extension, (*ii*) The interaction, (*iii*) the requirements and (*iv*) the Characteristics.

5 Test Case – Semantic Enhancement of Industrial Maintenance (IM) Tasks

In a previous work [17] we presented a system that considering some detected challenges in the Industrial Maintenance field, proposes an architecture and its correspondent implementation for the enhancement of the Industrial Maintenance management using Knowledge Based techniques.

In the aforementioned system we used Virtual and Augmented Reality in portable devices that in conjunction provided an enhanced experience for the user who is performing the maintenance tasks. The architecture and system implementation presented, fulfilled the following characteristics:

- It is centered in Plant Management and Industrial Maintenance Domain.
- It uses the Standard Ontology for Ubiquitous and Pervasive Applications (SOUPA) [citeulike:1282356] for the Ambient Intelligence modeling and the Set of Experience Knowledge Structure (SOEKS) [9] as the underlying Knowledge Structure for the User Experience modeling.
- We presented an architecture that we call UDKE, which uses Augmented Reality technologies in order to provide a better user interaction and is designed with portable devices in mind.

In this section we will review briefly the presented system in [17] showing in fact that such development although older than this work, follows the same directives and recommendations while at the same time proves that the conceptualization proposed here is applicable even to legacy architectures.

We will perform a comparison between the original proposed architecture and the more conceptual approach introduced in this paper.

5.1 Some Challenges in IM That Could Be Benefited from Semantics

From an application point of view, many different research initiatives have been presented by the scientific community, ranging from seminal work presented in 1929 by Wilson [18] and up to IM techniques (predictive, TPM, etc) [19] and the use of modern computer-based technologies such as AR/VR [arvika-paper].

To our knowledge, most of these approaches however, miss out on the potential of using domain specific knowledge-based theories that might enhance the user's experience. This user, in our case, is the maintenance worker in a typical industrial facility whose special needs include mobility, fast response and immediate access to the relevant data, like specifications, historical records, etc. We show in this VET that the use of Semantics and AR techniques provides additional support to maintenance tasks, by improving the user understanding of the elements under maintenance.

The enhancement is realized when the knowledge and the user experience related to the maintenance system is embedded in the AR environment as an important aid for the user, providing him with a sense of immersion. Some of the identified challenges in IM are:

Challenge 1: Domain Modeling

Since IM takes place in many differing environments and for diverse elements, the modeling of the domain represents a difficult task to accomplish. The set of actions needed to perform IM in a Chemical Plant and the actions needed in a Nuclear Plant, although similar, are at the end different (the later requires even more careful inspection). For this reason, we believe that the environment itself should provide the information for IM. Moreover, the domain modeling must be consistent and easily adaptable and extensible for new requirements (reasons that strengthen the use of ontology modeling using available maintenance databases as data providers). To our acquaintance, there is no agreement reached on the modeling of the IM domain, specially taking into account Ambient Intelligence scenarios to model Knowledge Bases for industrial Plant maintenance. In our approach we use Ambient Intelligence modeling for our Knowledge Base.

Challenge 2: User Immersion and Use of Portable Devices

This challenge has been previously addressed by various R&D projects. There have been difficulties reported with RFID tags and further difficulties in employing optical devices. The former relate to interference and the need for close proximity to a tag, and the later due to certain characteristics of the environment where the IM is taking place (darkness, contrast, etc). We do not intend to solve any of these issues; instead we concentrate on a simple VR/AR output that can be easily and cheaply integrated into any facility. We deem the benefits of portable devices in this specific VET. For this reason we intentionally included their use within our architecture. However we believe strongly that some technical problems must be solved in order to make their use beneficial.

Challenge 3: User Expertise

This challenge is related to the fact that the user experience is not typically taken directly into account with the IM tasks. At times the IM databases consider some statistical measurements for a given element lifespan, but the reasons why the element was changed before the end of the cycle are not stored anywhere. We believe that these factors are as important as the fact that the element was changed (or maintained). The possibility to store and handle User expertise is therefore desirable.

5.2 The UDKE Architecture as Presented in [17]

UDKE (User, Device, Knowledge and Experience) architecture was introduced in our paper "Knowledge based Industrial Maintenance using portable devices and Augmented Reality". Our approach takes into consideration the four challenges found

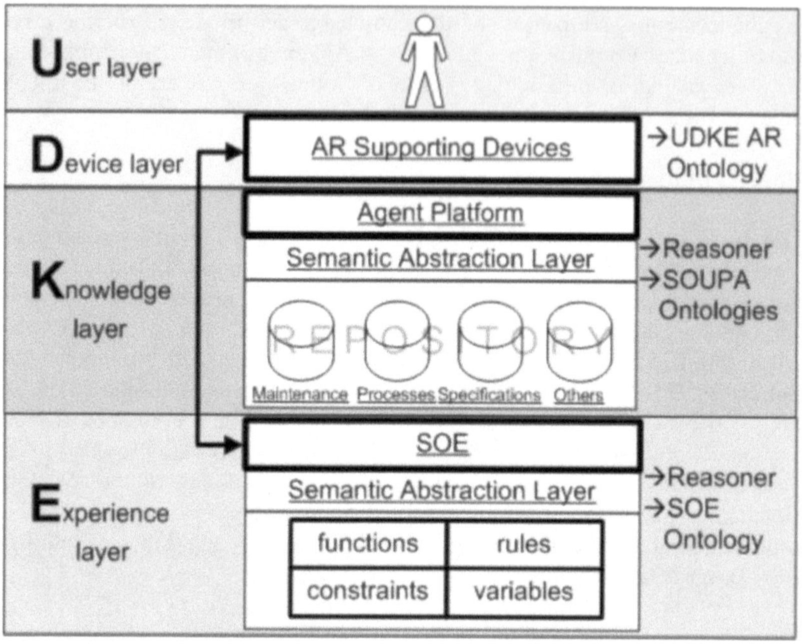

Fig. 11. Overview of the UDKE architecture

in the IM supporting problem that we discussed in the introduction of this section. It can be said that UDKE provides a possible conceptual model for a Maintenance System that combines Knowledge, User experience and AR/VR techniques and with the capability of running in portable devices like PDA's or sub-laptops, and in general in a vast majority of devices with Java support. As can be seen in Fig 11, UDKE is divided into four layers, each one describing a fundamental view from which the Maintenance task accomplishment could acquire an advantage.

In our scenario, the user during his maintenance patrol uses a portable device (PDA, UMPC or Tablet PC) with a camera connected. For every object to be maintained exists a VR marker (following the sensor concept in ambient intelligence).

Every marker is an unequivocal gray-scale pattern that can be easily printed on white paper by a regular PC printer. When the camera recognizes a marker, a matching element to be maintained is identified according to the context (user, task, priority) and a set of information is extracted from the repositories.

The output video stream of the camera is mixed with 3D objects and other relevant information and is displayed to the user in his portable device screen. As can be seen in Fig 12, when the user is in front of an element (in this case a fire extinguisher) the system recognizes the matching marker. The user receives on the device's display (shown in figure), information such as the name of the element, the next programmed change, the maintenance procedure etc. All information is obtained from the repositories in the Knowledge layer and is maintained by the Experience layer.

Fig. 12. AR Enhanced user view (UDKE Platform)

The system was tested using different portable devices, our implementation uses JAVA as the core language for the prototype implementation. The AR engine used was the JAR Toolkit library. All the ontology modeling was done in Protégé and the API used was the Protégé OWL API.

The agent platform used in our implementation was JADE, and for reasoning over the ontologies we used RACER. When a marker is detected, the system calculates the matrices necessary to place the augmented information via JARToolkit calls. Following the application flow, the Agent platform begins its work starting a majordomo service whose main function is to serve as an intermediary between the user and the rest of the architecture.

The majordomo handles the events in the knowledge layer databases through reasoning over the SOUPA ontologies. The majordomo also handles the Experience layer through reasoning over the SOE (Set of experience Knowledge Structure) ontology in order to obtain knowledge from past experiences or similar devices being maintained. Once all of the information is obtained/inferred and possible experiences are acquired from the SOE using the reasoning system, a final step is performed by returning the information to the device (UMPC, Pocket PC, etc) and displayed (streamed) in its graphical output.

5.3 Generalization of the UDKE Architecture

As pointed before, UDKE architecture was an application of Semantics to a specific VET intended for Industrial Maintenance. Fig 13 depicts a comparison between the aforementioned approach and the contribution presented in this paper.

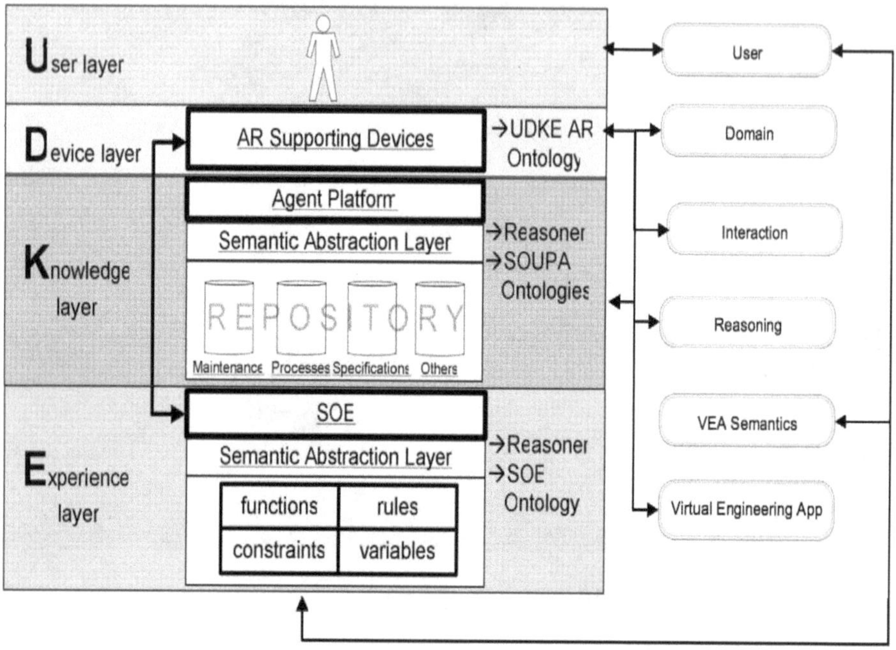

Fig. 13. UDKE and General Architecture relation

As can be seen, matching of both approaches at a concept level is quite similar, however the following highlights deserve to be mentioned:

- The Architecture presented in this article follows a more generalized approach where its component layers and modules do not necessary have to be implemented in a one-to-one matching module, meaning that an implementation could cover diverse parts of the general architecture.
- It can be argued that the UDKE architecture was an implementation of the more generalized approach presented here, showing that even generalization efforts evolve.
- When we implemented UDKE for the first time, we encountered that the Knowledge layer was above the Experience layer. By examining the outcome of the mentioned implementation on a real world test case, we found that experience should be placed above knowledge in order to filter such Knowledge for producing the best recommendation for the situation. We learnt from that observation and reflected it in the generalization shown in this paper.
- Another interesting conclusion of the comparison of both architectures is the fact that technologies (SOE/SOEKS, Reflexive Ontologies, etc) have been differentiated from conceptual models in the generalized architecture. The reason why we propose such separation is merely implementation, as with

this new point of view. The Knowledge Engineer who is designing a system that follows our recommendation can choose freely the tools for his implementation.
- We intentionally stick to the ontologies approach when comes to Knowledge Modeling for many different reasons, between those reasons a very important one to mention is the easiness of ontology modeling for overhauling. The above idea directs in other words to the fact that is a simple task to take a legacy ontology and evolve it when new concepts arise (which in fact happened when we overhauled UDKE redesigning it from the generalized perspective shown in this paper).

6 Conclusions and Future Work

Semantic technologies can indeed improve the current state of the art of Virtual Engineering Applications from diverse perspectives. In particular in this work we proposed a theoretical framework with specific recommendations, methodologies and practical tools that allows the integration of Semantics in VEA trough the Semantic enhancement of their embedded VET.

A novel modular architecture has been proposed which allows the implementation of the different conceptual steps in the methodological part as connected modules. Each module helps in having a more explicit representation of the semantics inherent to the model, but also to the user and the domain. Thus, the methodology starts from a Conceptualization Layer and passes the gathered Knowledge about the User, the Domain and the requirements to a Semantic abstraction Layer that performs the Reasoning processes that reflect the Semantic enhancement in the VEA. This architecture will served as the base for a test case implementation, proving its value as a generic and extensible way to semantically enhance a VEA through the enhancement of VET.

In the presented example, our proposed architecture enhanced the maintenance process as the knowledge coming from the user and his experience can be re-used. The aforementioned fact is one of the most important issues when one is involved in maintenance tasks as no matter how clear is the process, it must be taken into account that form a mere mechanical point of view, two elements of the same kind do not behave identically. Sometimes a fire extinguisher (for example) is changed before reaching its end-of-life due to environmental factors as humidity or excessive heat that may produce inconvenient fractures in the seals that little by little could lower the pressure of the extinguishing fluid.

An experienced maintainer could take the choice to change the extinguisher and a system like the one we present will be aware of the reasons why the change was made and eventually could advice another maintainer in a similar situation to change the extinguishers that share similar environments.

The aforementioned situation clearly produces a better and safer environment for the industrial plant and its workers.

We believe that the involvement of semantic enhanced environments will be a stepping stone in the plant of the future concept. A greener industry which at the same time holds a safe-for-the-users production will be possible if the industries involved include technologies as the ones presented in this work.

The benefits of using a semantically enhanced VET for Industrial Maintenance tasks are evident; however we believe that some challenges are still open. One of such challenges is that the discussed approach depends on the full capability of viewing all the regions on the marker (for identification purposes and calculation of the camera parameters). The problem arises when such marker cannot be seen partially or in total, as the information that relates such marker to the actual Maintenance Data-sets and hence with the Domain embedded Knowledge will not be reached. Such problem is due to poor light quality or deterioration of the marker and even because of the quality of the camera's CMOS sensor. We believe that a logical next step in our methodology would be to consider marker-less approaches such as projects like ARVIKA [20] or in the work presented by Comport [21].

References

[1] Jian, C.Q., McCorkle, D., Lorra, M.A., Bryden, K.M.: Applications of virtual engineering in combustion equipment development and engineering. In: ASME (ed.) ASME International Mechanical Engineering Congress and Expo, IMECE 2006–14362, Chicago, November 2006. ASME. ASME Publishers (2006)

[2] McCorkle, D.S., Bryden, K.M., Swensen, D.A.: Using virtual engineering tools to reduce nox emissions. In: Proceedings of ASME Power 2004. POWER2004-52021, March 2004, pp. 441–446. ASME Publishers (2004)

[3] Press, O.U.: Oxford Dictionary of English, 2nd revised edn. Oxford University Press, Oxford (2003)

[4] Feigenbaum, E., McCorduck, P.: The Fifth Generation. Addison-Wesley, Reading (1983)

[5] Gruber, T.: Toward principles for the design of ontologies used for knowledge sharing. International Journal of Human-Computer Studies 43(5-6), 907–928 (1995)

[6] Antoniou, G., Harmelen, F.V.: Web ontology language: OWL. In: Handbook on Ontologies in Information Systems, pp. 67–92. Springer, Heidelberg (2003)

[7] Haarslev, V., Muller, R.: Racer: An OWL reasoning agent for the semantic web. In: Proc. of the International Workshop on Applications, Products and Services of Web-based Support Systems, in conjunction with 2003 IEEE/WIC International Conference on Web Intel ligence, pp. 91–95. Society Press (2003)

[8] Toro, C., Sanín, C., Szczerbicki, E., Posada, J.: Reflexive Ontologies: Enhancing Ontologies with self-contained queries. Cybernetics and Systems: An International Journal 39(1-19) (2008)

[9] Sanin, C., Szczerbicki, E., Toro, C.: An owl ontology of set of experience knowledge structure. J. UCS 13(2), 209–223 (2007)

[10] McCorkle, D.S., Bryden, K.M.: Using the Semantic Web to Enable Integration with Virtual Engineering Tools. In: London, S. (ed.) Proceedings of the 1st International Virtual Manufacturing Workshop (March 2006)

[11] McCorkle, D.S., Bryden, K.M.: Using the semantic web technologies in virtual engineering tools to create extensible interfaces. Virtual Real 11(4), 253–260 (2007)

[12] Huang, G., Bryden, K.M.: Introducing virtual engineering technology into interactive design process with high-fidelity models. In: WSC 2005: Proceedings of the 37th conference on Winter simulation, Winter Simulation Conference, pp. 1958–1967 (2005)

[13] Kunz, J.C., Christiansen, T.R., Cohen, G.P., Jin, Y., Levitt, R.E.: The virtual design team. Commun. ACM 41(11), 84–91 (1998)

[14] Friedell, M.M., Corporation, L.D.: Interaction paradigms for human-computer cooperation in design

[15] Toro, C.: Semantic Enhancement of Virtual Engineering Applications. PhD thesis, University of the Basque Country (March 2009)

[16] Cobos, Y., Toro, C., Sarasua, C., Vaquero, J., Linaza, M.T., Posada, J.: An architecture for fast semantic retrieval in the film heritage domain. In: 6th International Workshop on Content-Based Multimedia Indexing (CBMI), June 2008, pp. 272–279. IEEE, Los Alamitos (2008)

[17] Toro, C., Sanín, C., Vaquero, J., Posada, J., Szczerbicki, E.: Knowledge Based Industrial Maintenance Using Portable Devices and Augmented Reality. In: Apolloni, B., Howlett, R.J., Jain, L. (eds.) KES 2007, Part I. LNCS (LNAI), vol. 4692, pp. 295–302. Springer, Heidelberg (2007)

[18] Wilson, J.: The management of the maintenance department in industrial plants. Master's thesis, Massachusetts Institute of Technology. Dept. of Business and Engineering Administration, May 17 (1929), `http://dspace.mit.edu/handle/1721.1/16521`

[19] Naka jima, S.: Introduction to TPM. Technical report, Japan Institute for Plant Maintenance (1991)

[20] Weidenhausen, J., Knoepfle, C., Stricker, D.: Lessons learned on the way to industrial augmented Reality applications, a retrospective on ARVIKA. Computers and Graphics 27(6), 887–891 (2003)

[21] Comport, A.I., March, E., Chaumette, F.: A real-time tracker for markerless augmented reality. In: ACM/IEEE Int. Symp. on Mixed and Augmented Reality, ISMAR 2003, pp. 36–45 (2003)

Effective Backbone Techniques for Ontology Integration

Trong Hai Duong[1], Ngoc Thanh Nguyen[2], and Geun Sik Jo[1]

[1] School of Computer and Information Engineering,
Inha University, Korea
haiduongtrong@gmail.com,gsjo@inha.ac.kr
[2] Institute of Informatics and Engineering,
Wroclaw University of Technology, Poland
thanh@pwr.wroc.pl

Abstract. Most previous research on ontology integration has focused on similarity measurements between ontological *entities*, e.g., *lexicons*, *instances*, *schemas* and *taxonomies*, resulting in high computational costs due to the need to consider all possible pairs between two given ontologies. In this chapter, we present a novel approach to reduce computational complexity in ontology integration. Thereby, we present an enriched ontology model for a formal ontological analysis that enables us to build ontologies with a clean and untangled taxonomic structure. The *importance* and *types of concepts*, for priority matching and direct matching between concepts, respectively are proposed. *Identity-based similarity* is computed, which enables us to avoid comparisons of all properties related to each concept, while matching between concepts. The problem of conflict in ontology integration has initially been explored on the *instance-level* and *concept-level*. This is useful to avoid many cases of mismatching.

Keywords: Ontology integration, Importance concepts, Types of concepts, Conflict, Identity-based similarity.

1 Introduction

Ontology has been important not only in the semantic web and semantic data processing, but also in various research fields and application areas, e.g., knowledge engineering, database design and integration. So these types of ontologies play a central role in facilitating knowledge exchange between several heterogeneous sources. To make this process efficient, distributed ontologies have to be integrated.

In general, the problem of ontology integration can be described as follows: *Given a set of ontologies* $\{O_1, ..., O_n\}$, *a unified ontology* O *capable of replacing them must be found* [15,30]. The ontology reflects the creator's own understanding of knowledge- this is similar to the relation of a literary work to its author. The best explanation of the phenomenon of human consciousness is William

N.T. Nguyen and E. Szczerbicki (Eds.): Intel. Sys. for Know. Management, SCI 252, pp. 197–227.
springerlink.com

James' famous stream of consciousness theory. He observed that human consciousness has a composite structure including *substantive parts* (thought or idea) and *transitive parts* (fringe or penumbra), and drifts from thought-to-thought. Thus, ontology integration is a complex task, since the ontologies have heterogeneous characteristics and forms, e.g., languages, structures, etc. Therefore, [20] suggested an ontology architecture providing a solid basis for studies about the task of ontology integration and [30] identified the activities that must be performed in ontology integration. Various tools supporting ontology integration based on different techniques have been introduced:

Schema-based Similarity. Cupid [23] implements an algorithm comprising linguistic and structural schema matching techniques, and computation of similarity coefficients using domain-specific thesauri. OntoMerge [4] is a system for ontology translation of the semantic web. Ontology translation refers to such tasks as (i) dataset translation, that is, translating a set of facts expressed in one ontology to those in another; (ii) generating ontology extensions, that is, given two ontologies o and o' and an extension (sub- ontology) o_s of the former, build the corresponding extension o'_s, and (iii) query answering from multiple ontologies. The main principle of this approach is ontology translation via ontology merging and automated reasoning. H-Match [2] is an automated ontology matching system. H- Match inputs two ontologies and outputs (one-to-one or one-to-many) the correspondences between their concepts with the same or closest intended meaning. The approach is based on a similarity analysis via affinity metrics, e.g., term-to-term affinity, data type compatibility, and thresholds. H-Match computes two types of affinities (in the [0,1] range), viz., linguistic and contextual. These are then combined via weighting schemas, thus yielding a final measure, viz., semantic affinity. Linguistic affinity builds on the thesaurus-based approach of the Artemis system.

Instance-based Similarity. FCA-Merge is a method for merging ontologies, which is a bottom-up approach supporting a global structural description of the merging process. For the source ontologies, it extracts instances from a given set of domain-specific text documents by applying natural language processing techniques. Based on the extracted instances, mathematical techniques from formal concept analysis are applied. The output is explored and transformed into the merged ontology by the ontology engineer. GLUE [5,6,7,8] is a system that employs a multi-strategy machine learning technique with joint probability distributions. This tool can identify the similarities of instances. Also, it can compare the relations between ontologies based on the similarities between instances. .GLUE uses two types of base learners: a name learner and a number of content learners.

Schema-based and Instance-based Similarity. RiMOM (Risk Minimisation based Ontology Mapping) [32] is an approach inspired by Bayesian decision theory, which formalizes ontology matching as a decision making problem. Given two ontologies, it aims for the optimal and automatic discovery of alignments, which can be complex (including concatenation operators). The approach first searches for concept-to-concept correspondences, and then searches for property-to-property correspondences.

Lexical-based and Schema-based Similarity. MAFRA [25] is an ontology mapping frame-work using Semantic Bridge Ontology (SBO). In MAFRA, The similarity between two concepts is mainly based on lexical analysis via WordNet, domain glossaries, bilingual dictionaries, and corpuses. Madhavan et al. [24] proposed an approach to schema matching; besides the input information available from the schemas under consideration, it also exploits some do- main specific knowledge via an external corpus of schemas and mappings.

Taxonomy-based and Mixed Similarity. The hybrid method [11] involves integrating multiple ontologies on several levels, such as the element level, internal structure, and relational structure. *OnConceptSNet* is a semantic network serving to reconcile multiple ontologies. The relations between the concepts of *OnConceptSNet* are derived from a semantic support environment combining special domain (e.g. Wordnet) and text corpus. *OnConceptSNet* is enhanced by the taxonomy-based rules, where the similarity between two concepts is determined by analyzing not only their subsumption relationships such as generalization, specialization, and siblings, but also their binary relations (properties). Hybrid-based matching is a heuristic that is advantageous due to the initial reduction of the complexity based on direct matching between the same types of concepts.

Most of the aforementioned works merely involve blind or exhaustive matching among all concepts in different ontologies and all properties belonging to each concept. Therefore, the computational complexity increases rapidly during the integration of large ontologies. Additionally, they have not yet explored the conflicts between ontologies on different levels such as the *instance level* (e.g. the same instance but different concepts) and the *concept level* (e.g. multiple forms of the same concept, overlapping but different concepts, the same concepts but different names.). Thus, our work focused on reducing complexity during ontology integration. The term *complexity* has two senses: First, *complexity* means the heterogeneous characteristics and forms of ontologies, e.g., languages, structures, etc. This may cause semantic conflicts during ontology integration. Second, *complexity* means the time of matching between concepts belonging to different ontologies. In this chapter, we focused on these problems, and an outline of our main contribution is as follows:

- *Enriched Ontology Model* presents an extended ontology model that enriches the standard conceptual model with semantic information which precisely characterises the conceptfs properties such as *key identity, local identity,. . .* and types of concepts including *defined concept, partition concept, inherited concept, primitive concept* and *individual concept.* The identities that hold for a concept and the type of concept are considered the main tools used for a formal ontological analysis that enables us to build ontologies with a clean and untangled taxonomic structure.

- *Similarity Measure Techniques* During ontology integration, it is important to find relations between entities belonging to different ontologies. Normally, the relations are equivalence relations that are discovered via similarity measure techniques. In this section, we present simple similarity techniques such as *string-based, language-based* and *linguistic-based similarity* and show how to combine these techniques. The methods of wordnet-based similarity measure are analyzed and a novel method of *synset-based similarity* measure is proposed. Moreover, the lexico-syntactic patterns methodis applied to discover relationships between entities from a text corpus. A collaborative method of combining a text corpus and WordNet to prove the semantic relations used for ontology integration tasks is proposed.
- *Importance Concept-based Similarity* the *importance of concepts* has been proposed for priority matching between concepts. The importance measurement of a concept takes into account the contributions from all the other concepts in the ontology, based on all types of relations, including not only sub- sumption and non-subsumption, but also concepts with associated attributes exhibiting a mutually reinforcing relationship. We can identify a concept's possible position in the hierarchy via its importance measurement. Thus, we agree that *assuming two ontologies O_i and O_j must be integrated, the concept c_i belonging to ontology O_i should be priority matched to the concept c_j or neighbors of c_j belonging to ontology O_j, where the distance between c_i's importance measurement and c_j's is minimal.*
- *Identity-based Similarity* the *types of concepts* have been expanded from [11] and combined with the importance concepts in this proposal, for direct matching between concepts of the same type, instead of using blind or exhaustive matching among all concepts. This proposal involves the following steps: First, we classify all concepts belonging to different ontologies into five disjoint groups based on their identities. Each group consists of the same type of concept. Only concepts in the same disjoint groups can be directly matched. Second, we assign a weight (importance vector) to each disjoint group in ascending order of the importance weight. Each pair of the same disjoint groups belonging to different ontologies must be directly matched. Moreover, while calculating similarities between concepts, we simply focus on the identity of the concept, instead of comparing all properties related to each concept.
- *Conflict in Ontology Integration* the problem of conflict in ontology integration has been explored on the *instance-level* and *concept-level*. It is useful for avoiding many cases of mismatching.
- In the experiment, we have applied the aforementioned methods to design an effective algorithm for ontology matching. We have also compared our method with the previous studies.

2 Enriched Ontology Model

We assume a real world $(\boldsymbol{A}, \boldsymbol{V})$ where $\boldsymbol{A}$ is a finite set of attributes and $\boldsymbol{V}$ is the domain of $\boldsymbol{A}$. Also, $\boldsymbol{V}$ can be expressed as a set of attribute values, and

$\boldsymbol{V} = \bigcup_{a \in \boldsymbol{A}} V_a$ where V_a is the domain of attribute a. In this chapter, we make the following assumptions:

Definition 1 (Ontology). *An ontology is a quadruple:*

$$O = (C, \sum, R, Z) \tag{1}$$

where,

- **C**: *set of concepts (the classes);*
- **R**: *set of binary relations between the concepts from C, or between the concepts from C and the values defined in a standard or user-defined data type;*
- **Z**: *set of axioms, which can be interpreted as integrity constraints or relationships between instances and concepts. This means that Z is a set of restrictions or conditions (necessary & sufficient) to define the concepts in C;*
- $< \mathbf{C}, \sum >$: *is the taxonomic structure of the concepts from C where $\sum$ is the collection of subsumption relationship ($\sqsubseteq$) between any two concepts from C. For two concepts c_1 and $c_2 \in C, c_2 \sqsubseteq c_1$ if and only if any instances that are members of concept c_2 are also members of concept c_1, and the converse is not true.*

R is known as the set of properties. For every $p \in R$, there is a specific domain D and range R such that $p : D \rightarrow R$, where $D \subset C$ and if $R \subset C$ then p is called an object property, otherwise if R is a set of standard or user-defined data types then p is called a data type property. We assume that concepts c and c'correspond to the domain and range of property p respectively, where p is also known as an attribute of concept c. There are two given instances v and v' that belong to the corresponding concepts c and c' respectively. We denote vR^pv' as the relation from instance v to v' based on the property (attribute) p and the relation from instance v' to v based on the property p is denoted as $vR^{-p}v'$.

Definition 2 (Concept). *A concept c of an $(\mathbf{A}, \mathbf{V})$-based ontology is defined as a quadruple:*

$$c = (z_c, A^c, V^c) \tag{2}$$

where c is the unique identifier for instances of the concept. $A^c \subseteq \mathbf{A}$ is a set of attributes describing the concept and $V^c \subseteq \mathbf{V}$ is the attributes' domain: $V^c = \bigcup_{a \in A^c} V_a$. The $z_c \subset Z$ is the set of restrictions or conditions (necessary & sufficient) to define the concept c. The z_c can be represented as a constraint function $z_c : A^c \rightarrow \mathbf{Z}$ such that $z_c(a) \in \mathbf{Z}$ for all $a \in A^c$.

Pair (A^c, V^c) is called the possible world of concept c and A^c is called the structure of the concept c. Notice that within an ontology there may be two or more concepts with the same structure. If this is the case, the constraint function z_c is useful for expressing the associated relationships. For example, two concepts $RedWine$ and $WhiteWine$ have the same structure $\{hasMaker, hasColor\}$. But $z_{RedWine}(hasColor) = \{\exists hasColor = red\}$ and $z_{WhiteWine}(hasColor) = \{\exists hasColor = white\}$.

Definition 3 (Instance). *An instance of a concept c is described by the attributes from set A^c with the values from set V^c. Thus an instance of a concept c is defined as a pair:*

$$instance = (id, v) \tag{3}$$

where id is the unique identifier of the instance in world (A, V) and v is the value of the instance, which is a tuple of type A^c and can be expressed as a function:

$$v : A^c \rightarrow V^c \tag{4}$$

such that $v(a) \in V_a$ for a A^c.

Value v is also called a description of the instance within a concept. A concept may be interpreted as a set of all instances described by its structure. We can then write $i \in c$ to express the fact that i is an instance of concept c.

We denote $Ins(O, c)$ as the set of instances belonging to concept c in ontology O and an instance of a ontology as $I = \bigcup_{(c \in C)} Ins(O, c)$. Notice that a knowledge base can be represented by an ontology and its instances.

Definition 4 (Key Identity). *The* Key Identity *(KI) of a concept is an attribute from set A^c which provides a unique value to each individual of the concept in the real world $(\mathbf{A}, \mathbf{V})$. Formally, if ki is a KI of the concept c, it satisfies the following conditions:*
- $ki \in A^c$,
- $x \in Ins(O, c), \forall v_1, v_2 \in \mathbf{V}, xR^{ki}v_1 \wedge xR^{ki}v_2 \rightarrow v_1 = v_2$, *and*
- $x \in Ins(O, c), \forall v_1, v_2 \in \mathbf{V}, xR^{-ki}v_1 \wedge xR^{-ki}v_2 \rightarrow v_1 = v_2$.

The first two conditions mean that the KI of a concept must necessarily provide the same KI value for the same instance of it. The third condition means that it must be sufficient to recognize that two actual instances with the same KI value are the same. The three conditions imply that the KI of a concept should be globally identifiable for instances in the real world $(\boldsymbol{A}, \boldsymbol{V})$. The KI is also known as the rigid property [16] that is essential to all its instances.

Definition 5 (Local Identity). *The* Local Identity *(LI) of a concept is an attribute from set A^c, which provides a unique value to each individual of the concept in the possible world (A^c, V^c). Formally, if li is an LI of concept c, it satisfies the following conditions.*
- $li \in A^c$,
- $x \in Ins(O, c), \forall v_1, v_2 \in V^c, xR^{li}v_1 \wedge xR^{li}v_2 \rightarrow v_1 = v_2$, *and*
- $x \in Ins(O, c), \forall v_1, v_2 \in V^c, xR^{-li}v_1 \wedge xR^{-li}v_2 \rightarrow v_1 = v_2$.

The difference between KI and LI is that the LI of a concept is only locally identifiable for instances in the possible world (A^c, V^c).

Example 1. We consider the concept $Person$ owning the $hasFingerprint$, which is KI. The instance $Jean$ has $hasFingerprint$ of 000155BDC, and the instance $Peggy$ has $hasFingerprint$ of 000155BDC. Because $hasFingerprint$ is a KI, we can deduce that *Jean* and *Peggy* must be the same instance. Note that because *hasFingerprint* is a *KI*, there is always an inverse relation *isFingerprintOf*. If two instances 000155BDC and 000155BEF are *isFingerprintOf* of the instance *Jean*, 000155BDC and 000155BEF must be the same. However, notice that if 000155BDC and 000155BEF were explicitly stated to be two different instances, these statements would lead to an inconsistency.

Definition 6 (Inheritance Identity). *The* Inheritance Identity *(II) of a concept is a set of identified attributes which is inherited from its super-concepts, which provides a unique II value to each instance of the concept.*

Definition 7 (Constant Property). *The* Constant Property *of the concept c is a property within* A^c *which provides a common attribute value for all individuals belonging to the concept.*

Example 2. We consider the concept *MalePerson* with its structure $\{Person \wedge has - Gender = Male\}$. The *MalePerson* is defined as the concept *Person* which satisfies $z_{MalePerson}(hasGender) = \{\forall hasGender = Male\}$. The property of *hasGender* has the constant value of *Male* for all individuas belonging to the concept *MalePerson*. Therefore, the *hasGender* is a constant property.

We found that the same concept has certain fixed roles in similar ontologies. Thus the concept's position lies in a fixed interval of the ontologies' hierarchy. Moreover, the role of a concept is determined by its attributes with the associated restrictions. Based on the concept's characteristics, we can classify concepts belonging to an ontology into five disjoint groups as follows:

Definition 8 (Defined Concept). *The* Defined Concept *(DC) is a concept which has at least one KI. Formally, if c is a DC, its constraint function* z_c *satisfies the following conditions:*
- $\exists a \in A^c, z_c(\mathrm{a})$ *is a necessary and sufficient condition, and*
- *the attribute a is a KI.*

Example 3. Consider *Example* 1 in which the concept *Person* is an example of the *DC*. The *DC* is also known as a rigid sort [16] that supplies a principle of identity for its individuals.

Definition 9 (Partition Concept). *The* Partition Concept *(PC) is part of a DC. Formally, if c is a PC, it satisfies the following conditions:*
- $\exists a \in A^c, \forall x \in v_c : x(a)$ *is a constant value, and*
- *the concept c is a defined concept satisfying* $z_c(\mathrm{a})$.

Example 4. We consider two concepts *MalePerson* and *FemalePerson* with the same structure $\{Person, hasGender\}$. The *MalePerson* is defined as the concept *Person* that satisfies $z_{MalePerson}(hasGender) = \{\forall hasGender = Male\}$. The *FemalePerson* is defined as the concept *Person* that satisfies $z_{FemalePerson}$− $(hasGender) = \{\forall hasGender = Female\}$. Thus the concepts *MalePerson* and *FemalePerson* are *PC*s.

Definition 10 (Inherited Concept). *The* Inherited Concept *(IC) is a sub-concept of either a defined concept or a partition concept, or another inherited concept. It has at least one LI. Formally, if c is a IC, then its constraint function z_c satisfies the following conditions:*
- $\exists a \in A^c, z_c(a)$ *is the necessary and sufficient condition , and*
- *the attribute a is a LI.*

Example 5. If two concepts *Student* with *KI hasIdStudent* and *Employee* with *hasIdEmployee* are sub-concepts of the concept *Person*, we can infer that *Student* and *Employee* must be *IC*s.

Definition 11 (Primitive Concept). *The* Primitive Concept *(PvC) is a concept which has neither KI nor LI and is defined from other concepts. Formally, if a concept is PvC, its constraint function z_c does not have any set of necessary & sufficient conditions and no concept is used as a sub-concept of PvC.*

Example 6. We consider the *UndergraduateStudent, MasterStudent*, and *DoctoralStudent* defined from the concept *Student*. Since they have no set of necessary & sufficient conditions, we can infer that these concepts must be *PvC*. Notice that the concepts are never used as sub-concepts of *PvC*.

Definition 12 (Individual Concept). *The* Individual Concept *(IvC) is a concept with no object properties and is not defined from other concepts.*

Example 7. Based on *Figure* 3, the concept *Subject* is an *individual concept*. An *Individual concept* can be considered as a set of instances or data types defined by the ontology creator.

Proposition 1 (Types of Concepts). *For a given ontology O belonging to the real world (A, V), we denote the five different sets of DCs, PCs, ICs, PvCs and IvCs as C^{DC}, C^{PC}, C^{IC}, C^{PvC} and C^{IvC} respectively.*

1. $C^{DC} \cup C^{PC} \cup C^{IC} \cup C^{PvC} \cup C^{IvC} = \boldsymbol{C}$
2. $C^{DC} \cap C^{PC} \cap C^{IC} \cap C^{PvC} \cap C^{IvC} = \emptyset$
3. *In ascending order, the levels of concepts are IvC, PvC, IC, PC and DC respectively.*

Proposition 1 shows that concepts belonging to an ontology can be classified into five disjoint groups.

3 Similarity Measure Techniques

The importance matter of ontology integration is finding the relations between entities expressed in different ontologies. Very often, these relations are equivalence relations that are discovered via the similarity measure between the ontological entities. According to the aforementioned studies, the techniques of similarity analysis that have been explored for ontology integration can be classified into following four groups:

- *Instance-based similarity*: The similarity between two concepts is based on common instances.
- *Lexical-based similarity*: The similarity between two concepts is based on analysis of the linguistic meanings of associated names.
- *Schema-based similarity*: The similarity between two concepts is based on analysis of the similarity between associated properties.
- *Taxonomy-based similarity*: The similarity between two concepts is based on analysis of their structural relationships, such as subsumption.

However, the above techniques are generated from the word similarity measure. A formal definition of the similarity method is as follows: let x, y, z denote words, and *sim* (x, y) represent the semantic similarity between x and y.

- $sim(x, y) \in [0, 1]$
- if $sim(x, y) = 1$ then $y = x$ or x is semantically equivalent to y.
- $sim(x, y) = 0$: two words are disjoint, i.e., no common characteristics.
- $sim(x, y) = sim(y, x)$: similarity is symmetric
- $sim(x, x) \geq sim(y, x)$: similarity is maximality

In this section we present some methods to measure the similar degree between two words and two strings (names/labels/comments) which are often used to name or describe the entities of an ontology.

3.1 Basic Techniques

String-based techniques. String-based techniques are often used in order to match names and name descriptions of ontological entities. These techniques consider strings as sequences of letters in an alphabet. They are typically based on the following intuition: the more similar the strings, the more likely that they denote the same concepts.

Entity's name coreference determines when identified names in the text refer to the same entity. Most entities' names exist in documents that have different forms. An entity's name coreference isbased on the following criteria:

- Suffix takes as input two strings and checks whether the first string ends with the second one (e.g., *Compact Disc* = *CD*).
- Prefix takes as input two strings and checks whether the first string starts with the second one (e.g., *Net* = *Network*).
- Aliases within the document that are indicated by parentheticals (e.g., *KFOR* = *Kosovo Force*)
- Variants of a name (e.g., *W. Bush* = *George W. Bush*)

A prefix or suffix pre-similarity can be defined from the prefix and suffix tests, which test whether one string is the prefix or suffix of another. Prefixes, suffixes or aliases can be recognized from a text corpus via lexico-syntax patterns such as parentheticals. Very often, this model is examined under learning approaches.

Edit distances similarity The edit distance, $e(x, y)$, from a string x to a string y is the minimum number of simple edit operations (insert, delete, replace, transpose) required to transform one string into the other. Edit distances were designed to measure the similarity between strings that may contain spelling mistakes. The edit distance similarity can be computed as follows:

$$sim(x, y) = 1 - \frac{e(x, y)}{max(size(x), size(y))} \tag{5}$$

Common substring similarity denoting z is the longest common substring between x and y starting from the first character of both and containing at least three characters. The common substring similarity is defined as follows:

$$sim(x, y) = \frac{2.length(z)}{length(x) + length(y)} \tag{6}$$

It is easy to see that this measure is indeed a similarity. One could also consider a subsequence similarity as well. This definition can be used for building functions based on the longest common prefix or longest common suffix.

Language-based techniques. consider names or express names (label or comment) as words in some natural language, e.g., English. They are based on natural language processing techniques exploiting morphological properties of the input words. They are applied to the names of entities before running string-based or linguistics-based techniques in order to improve their results. This approach is called pre-processing, which is shown in Fig. 1:

- *Tokenization* The concept labels are segmented into sequences of tokens by a tokeniser which recognises punctuation, cases, blank characters, digits, etc. For example, *has_Hands_Free_Kits* becomes tokens{*has, Hands, Free, Kits*}.
- *Stemming* The strings underlying tokens are morphologically analyzed in order to reduce them to normalised basic forms. For example, tokens $\{has, Hands, Free, Kits\}$ becomes root form $\{have, hand, free, kit\}$.

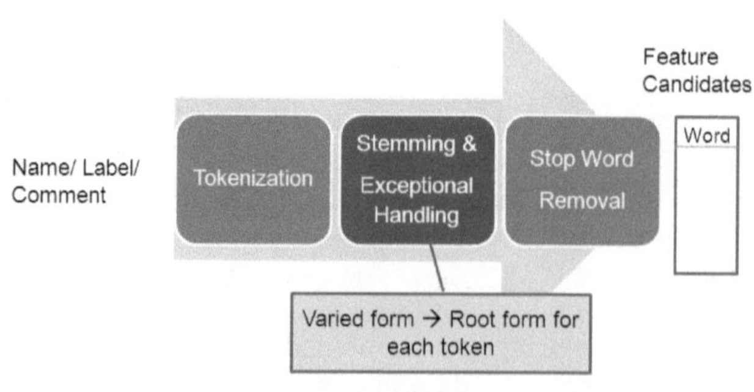

Fig. 1. Pre-Processing.

- *Stop word* is a word which is irrelevant to the content of the text, such as a pronoun, conjunction, preposition, article, and auxiliary verb. Removing stop words ensures more efficient processing by downsizing and the remaining words are relevant to the content. For example, after removing the stop words of the set $\{have, hand, free, kit\}$, it becomes $\{hand, free, kit\}$.

Linguistic resources such as lexicons or domain specific thesauri are used in order to match words based on their linguistic relations [1] , e.g., synonyms, hyponyms as several works [9,10]. A lexical resource such as the lexical database WordNet is particularly well suited for similarity measures, since it organizes nouns and verbs into hierarchies of is-a relations, and concepts can be re- lated based on the path lengths between them such as in Lch [21] and Wup [35]. In addition, WordNet has extremely fine-grained notions of word sense, which precisely capture even minor distinctions between different possible word senses, thus the similarity measures can based on the information content of the concepts [31]. Moreover, concepts can be related in many ways beyond mere similarity to each other. For example, a *wheel* is a part of a *car*, *night* is the opposite of *day*, *snow* is made up of *water*, a *knife* is used to cut *bread*, ect. As such WordNet provides additional (nonhierarchical) relations such as *has–part*, *is–made–of*, *is–an–attribute–of*, etc.

Basic Techniques for Element-level Similarity. Element-level similarity considers ontological entities independently of their relations with other entities or their instances. The similarity between entities is determined by the similarity between their names or labels. In this section, a rich semantic algorithm for the element-level is proposed, which combines the above simple techniques to measure the similarity between entities on the element-level. The algorithm is shown in *Algorithm* 1.

3.2 Wordnet-Based Similarity Measure

The lexical database WordNet is particularly well suited to similarity measures, since it organizes nouns and verbs into hierarchies of is-a relations. In version 2.0, there are nine noun hierarchies comprising 80,000 concepts, and 554 verb hierarchies comprising 13,500 concepts [29].

Many previous works focusing on finding the similarity between words have been WordNet-based. However, we can distinguish two basic approaches: (1) The similarity measures are based on the path lengths between concepts such as in Lch [21] and Wup [35]. Most of these similarity measures are subject to the *is-a* hierarchy containing the concepts. But is-a relations in WordNet do not cross part-of-speech boundaries, so these WordNet-based similarity measures are limited to distinguishing between noun pairs (e.g., cat and dog) and verb pairs (e.g., run and walk). While included in WordNet, the adjectives and adverbs

[1] In this case the names of ontology entities are considered as words of a natural language.

input : two concepts c and c'
output: The similarity between the concepts

$L = \{l_1, l_2, \ldots, l_n\}$ is tokens of labels/comments of concepts c;
$L' = \{l'_1, l'_2, \ldots, l'_m\}$ is tokens of labels/comments of concepts c';
Removing stopped words and root form of tokens in L and L';
$A_i = \{a_1, a_2, \ldots, a_k\}$ is a synset of l_i where $l_i \in L, i = 1 \ldots n$;
$A'_j = \{a'_1, a'_2, \ldots, a'_h\}$ is a synset of l'_i where $l'_i \in L', i' = 1 \ldots m$;
repeat
  foreach *pair* A_i, A'_j **do**
    if $A_i \cap A'_j \neq \emptyset$ *or* $sSim(a,b) = 1$ *or* $R(a,$ *equivalent* $,b) \mid a \in A_i$ *and* $b \in A'_j$ **then**
      count=count+1;
      removing A_i;
      removing A'_j;
    end
  end
  foreach A_i, A'_j **do**
    foreach $a \in A_i$ **do** $A_i = A_i \cup synset(a)$;
    foreach $a' \in A'_j$ **do** $A'_j = A'_j \cup synset(a')$;
  end
until *threshold* ;
return (2*count/(size(L)+size(L')));

Algorithm 1. A Rich Semantic Algorithm for the Element-Level

are not organized into is-a hierarchies. (2) The similarity measures are based on information content, which is a corpus-based measure of the specificity of a concept. These measures include Res [31], Lin [22], and Jcn [19]. Intrinsic to the calculation of information content is the use of tagged corpora. It is expected that the more often a concept appears in a corpus, the less specific it is, so the methods depend on tagged corpora. Such a strategy is not without a downside; there are two well known deep problems. Manually tagging corpora is monotonous and very time consuming; It is very difficult to obtain a statistically valid and reliable corpus that accurately reflects the word usage; many relatively common words may not appear, even in a very large corpora. This problem is usually referred to as the sparse data problem.

However, concepts can be related in many ways beyond mere similarity. For example, the words that occur together in a synset have the synonym relation. For example, the words *learner* occurs in two noun synsets {*learner, scholar, assimilator*} and {*apprentice, learner, prentice*}; *student* occurs in two noun synsets {*student, pupil, educate* } and {*scholar, s-cholarly person, bookman, student*}. Thus, *scholar* is a common word of a *student*'s and *learner*'s synset, so student and learner have a relation. If we continue finding synonyms of words in *student*'s and *learner*'s synsets, the number of similar words that occurs together with *student* and *learner* may be much larger. Thus the similarity degree between *student* and *learner* is much larger. Moreover, each word can occur in many synsets that cross-part-of-speech. For example, the word *base* occurs in seven adjective synsets, three verb synsets, and 19 noun synsets, thus the similarity crosses-part-of-speech. Thus, we proposed the following formula for measuring the semantic similarity of words:

$$sim(w_1, w_2) = \max_{level=1,\ldots,n} \left(\frac{\triangle + \sum_{w_i \in Syn_1 \cap E} (\sum_{w_j \in Syn_2 \cap E} Inc(w_i, w_j))}{min(size(Syn_1), size(Syn_2)) + size(E)} \right) \quad (7)$$

where

$$Inc = \begin{cases} 0 & \text{if } w_i \neq w_j \\ 1 & \text{if } w_i = w_j \end{cases}$$

If $Inc(w_1, w_2) = 1$ then $E = E \cup \{w_1\}$.
$\triangle$ is the total return value of *Inc* at *level*=1..*k*-1, *k* is the current *level.*
When the *level* is increased from 1 to *n*, each iteration then,
$Syn_1 = \bigcup_{w \in Syn_1 \cap E} Synonym(w)$ and $Syn_2 = \bigcup_{w \in Syn_2 \cap E} Synonym(w)$.

We experimented with the method to determine the similarities between 100 pairs of words with different similarity degree and crossing-part-of-speech. When we choose a level of three, and the limit of the size of array Syn is 1000, most of the similarities between words are determined. The higher the level, the greater the similarities between words. For example, the similarity between learner and student at level 1, 2, 3 is 0.24, 0.65, 0.84, respectively.

3.3 Lexico-Syntactic Patterns-Based Similarity

Lexico-Syntactic Patterns for Hypernym. Three syntactic phenomena commonly encode the hypernymic proposition: verbs, appositive structures, and nominal modifications. Here, we consider appositive structures [17,18,26] in which two noun phrases must be contiguous. Three types of appositive cues can then mark the second noun phrase: commas, parentheses, or lexical items (including, such as, particularly, and especially). For instance, consider a sentence (*) *We identified the activities involving students such as seminars, discussion, conferences.*, where *activity* is the hypernym of *seminar, discussion, conference.* The sentence is then transformed into the following lexico-syntactic expression:

(1a) NP_0 such as NP_1 ,$NP_2\ldots$, (and $\|$ or) NP_i $i \geq 1$, where NP_i is phrase noun i, are such that they imply
(1b) for all NP_i, $i \geq 1$, hypernym($head(NP_0)$, $head(NP_i)$) where $head(NP_i)$ is head noun of NP_i.

In [17,18,26] the authors presented most of the lexico-syntactic patterns for hyponymy. However, they had not analyzed NP to identify its head noun, so mistakes are common when finding the hypernym relation between concepts. For instance, student is the hypermyn of *seminar, discussion, conference* often acquired from sentences as (*). Therefore, in this section we analyze the grammatical noun phrase to identify its head noun, which is useful to avoid wrong relations as in the aforementioned instance.

Noun Phrase Analysis. Noun phrases normally consist of one or more in functional positions: HEAD (H:), DETERMINER (D:), POST DETERMINER (POD:), MODIFIER (M:), and POST MODIFIER (PM:), where

(1) HEAD: the noun, the personal pronoun, the demonstrative pronoun, and the possessive pronoun.

(2) DETERMINER: the definite article, indefinite articles, demonstrative articles, and possessive articles.

(3) POST DETERMINER: the cardinal numerals, ordinal numerals, and general ordinals.

(4) MODIFIER: adjective and noun.

(5) POST MODIFIER: the post modifier is a prepositional phrase which has two functional positions within it, the RELATER (R:) position and the OBJECT OF A PREPOSITION (OP:) position. The RELATER position is occupied by a preposition, for example, *from* in *from the library*, and the OBJECT OF A PREPOSITION position is typically occupied by a noun phrase, for example *the library* in *from the library*.

It must be noted that only HEAD (H:) appears in all of the 16 pat- terns (see table 1.) thus a head must appear in a noun phrase. However, all of the other functional labels appear in some patterns but not in others. This means that those functional positions are optional in the noun phrase. We showed this in (**) by placing the abbreviations for those functional positions in parentheses. Any functional position in parentheses may or may not occur in a given noun phrase. If it does occur, it must be in the order indicated by the pattern. For example, DETERMINERS are always first and POST MODIFIERS are always last; POST

Table 1. 16 patterns of noun phrase

H: *books*	D: + POD: + M: + H: + PM: *the three large books from the library*
D: + H: *the books*	POD: + M: + H: + PM: *three large books from the library*
POD: + H: *three books*	D: + M: + H: + PM: *the large books from the library*
M: + H: *large books*	D: + POD: + H: + PM: *the three books from the library*
D: + M: + H: *the large books*	M: + H: + PM: *large books from the library*
D: + POD: + H: *the three books*	POD: + H: + PM: *three books from the library*
POD: + M: + H: *three large books*	D: + H: + PM: *the books from the library*
D: + POD: + M: + H: *the three large books*	H: + PM: *books from the library*

DETERMINERS follow any co-occurring DETERMINERS and precede any co-occurring MODIFIERS.

(**) noun phrase form: (D:) + (POD:) + (M:) + H: + (PM:)
(the) (three) (large) books (from the library)

Here is a grammatical analysis of the noun phrase *the three large books from the library* based on (**):
The noun phrase (*the three large books from the library*)
DETERMINER definite article (the)
POST DETERMINER cardinal numeral (three)
MODIFIER adjective (large)
HEAD noun (books)
POST MODIFIER prepositional phrase (from the library)

Lexico-Syntactic Patterns for Similarity Measure. Most instances of a concept are the set of hyponyms of the concept. For example, when the concept *Country* has instances as *Vietnam, Korea, Poland,* it is considered as the hypernym of *Vietnam, Korea,* and *Poland.* Thus, a method to compute the similarity between two concepts via their instances has been proposed as follows:
$L_c = \{l_1, l_2, \ldots, l_n\}$, is the name/label of the instances of concept c.
$A_i = \{a_1, a_2, \ldots, a_k\}, a_j$ is the set of tokens resulting from two processes: demarcating and possible classification of sections of each string $l_i \in L_c$ and determining the root form of each token. For example, parsing the name *Hands_Free_Kits* into tokens $\{hand, free, kit\}$.
$G_i = \{g_1, g_2, \ldots, g_m\}$ is the set of more general words of $a_j \in A_i, j = 1, 2, \ldots, k$. Each of the words belonging to Gi are generated from the *lexico-syntac patterns.*

$$H = (h_1, h_2, \ldots, h_k) = \bigcup_{i=1}^{n} (G'_i) \tag{8}$$

where $G'_i \subseteq G_i$ and if $h_j \in G'_i, h_j$ exists at least $\frac{1}{2}n$ sets $G_i, i = 1, 2, \ldots, n$

The feature vector of H is denoted as follows:

$$\overrightarrow{S_H} = (w_1, w_2, \ldots, w_k) \tag{9}$$

where

$$w_i = \frac{f_i}{\sum_{j=1}^{k} f_j} \tag{10}$$

f_i is number of occurrences of h_i in the sets G_i, i=1..n.

We define the similarity between two concepts c and c' as follows:

$$sim(c, c') = sim(\overrightarrow{S_H}, \overrightarrow{S'_H}) = \frac{\sum_{(h_i, h'_i) \in K} (w_i * w'_i)}{\sqrt{\sum_{i=1}^{n} (w_i)^2} * \sqrt{\sum_{i=1}^{n} (w'_i)^2}} \tag{11}$$

where $K = \{(h_i, h'_i) | sim(h_i, h'_i) = 1\}$

3.4 Collaborative Acquisition Algorithm

While comparing a result of a relation between two concepts c_1 and c_2 denoting $R(c_1, c_2)$ for WordNet, there are three possibilities:

1. Both concepts c_1 and c_2 are in WordNet, and their relation $R(c_1, c_2)$ is already in the database of WordNet; we suggest updating *ontology integration tasks* [10].

2. Both concepts c_1 and c_2 are in WordNet, and the relation $R(c_1, c_2)$ is not; we suggest updating WordNet.

3. The concepts c_1 and c_2 are not present; we suggest adding these concepts and the corresponding $R(c_1, c_2)$ relation to the *Knowledge of Assistant WordNet.*

Here we sketch the collaborative acquisition algorithm which combines WordNet and text corpus to discover new relations between ontological entities for ontology integration tasks as follows (see *Figure* 1.):

- *Knowledge of Assistance WordNet* is a Concept Net based on the ontology with the following relations: *is kind of, is equivalent of.* It receives messages from the *Feedback component*, then updates the relations between the entities of ontologies which are not WordNet-based.
- *Mining from Text Corpus* is the procedure that is mentioned in the above section. It discovers new relations between ontological entities via the text corpus.
- *Ontology Integration Task* is presented in our previous work [10]. It receives the relation $R(c_1, c_2)$ and updates *OnConceptSNet.*

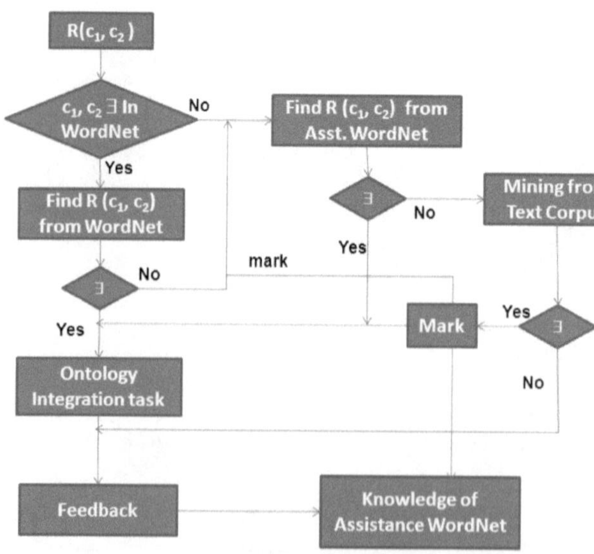

Fig. 2. The Acquisition Process.

- *Feedback* is a cache of the new relation and mark (mark is used to identify the new relation which should be updated in *Knowledge of Assistance WordNet* or *WordNet-based.*

4 Identity-Based Similarity

In this section, we present our ideas [12] on how to reduce the search space for matching between concepts belonging to different ontologies. After many experiments, we found that it is useful to classify all concepts belonging to different ontologies into more disjoint groups. We assume that *two ontologies must be integrated. Only two concepts belonging to the same disjoint groups of the ontologies can match.* It is easy to see that the assumption is a self-evident truth, because of the criteria for classifying concepts based on their attributes with associated restrictions. Naturally, two concepts that differ from their attributes should be disjoint. In *Figure* 3, we show a example of five disjoint concept groups belonging to the ontology E_1.

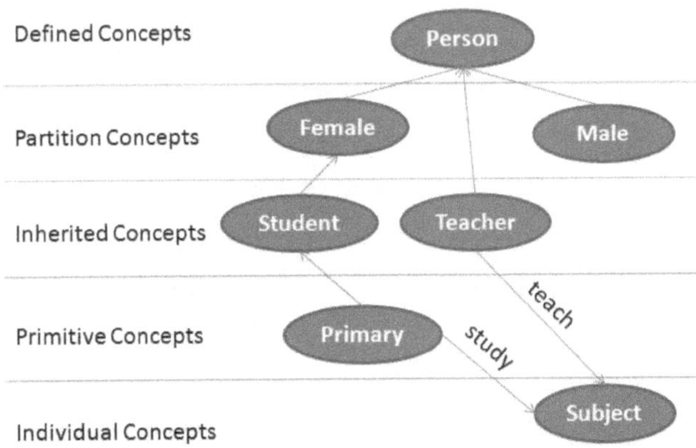

Fig. 3. A classification of concepts of Ontology E_1.

Definition 13 (Possible Similarity Set (PSS)). *For two given ontologies O_1 and O_2, a concept c belongs to the ontology O_2. If the ontologies must be integrated, the* possible similarity *set of the concept c is defined as the set of concepts belong to ontology O_1 and the concepts are the same type as concept c.*

Example 8. We consider the following ontology E_2 written in OWL:

$Individual\{hasFingerprint, name\}$
$FemalePerson\{Individual \wedge hasGender = Female\}$
$MalePerson\{Individual \wedge hasGender = Male\}$
$Teacher\{Individual, IdTeacher, TeachTo\}$
$Learner\{FemalePerson, IdLearner\}$

$PreSchool\{Learner \wedge Level = 1, LearnTo\}$
$Subject\{Name, Credit\}$

The attributes *hasFingerprint*, *IdTeacher*, *IdLearner* are owl:DatatypeProperty with three restrictions: owl:FunctionalProperty, owl:InverseFuncti-onalProperty, and owl:ca-rdinality = 1. So they are identities. Based on the above definitions and the position of the concepts in the hierarchy of the ontology, we can classify them into the following five groups:

$C_{E_2}^{DC} = \{Individual\}$
$C_{E_2}^{PC} = \{FemalePerson, MalePerson\}$
$C_{E_2}^{IC} = \{Learner, Teacher\}$
$C_{E_2}^{PvC} = \{PreSchool\}$
$C_{E_2}^{IvC} = \{Subject\}$
Similar to the ontology E_1 (see figure 3), it follows that:
$C_{E_1}^{DC} = \{Person\}$
$C_{E_1}^{PC} = \{Female, Male\}$
$C_{E_1}^{IC} = \{Student, Teacher\}$
$C_{E_1}^{PvC} = \{Primary\}$
$C_{E_1}^{IvC} = \{Subject\}$

If two ontologies must be integrated, only two concepts belonging to the same disjoint groups of ontologies can match. This is direct matching between the same types of concepts. It is shown in *Figure* 4.

We consider Inha University library (InhaLib) and Wroclaw University library (WrocLib). We are assuming that the libraries have the same concept of *Book* owning the unique identification of *BookID*. We refer to an instance of *Book* called the *Artificial Intelligence* book:

{ In the InhaLib, the *Artificial Intelligence* has the *BookID* of *IH-00012-AI*;
In the WrocLib, the *Artificial Intelligence* has the *BookID* of *WL-003334*.

Based on the above we know that the *BookID* is a local identity in each library. This means that it is possible for the *BookID* of the same instance *Artificial Intelligence* to differ from the InhaLib to the WrocLib. However, while we refer to the instance as two concepts *Information (Paperback, Publisher, Language, ISBN, Product Dimensions, Shipping Weight, Average Customer Review)* belonging to the InhaLib and *PublishedInf(NumberPage, PublisherName, Language, ISBN, Dimensions, Weight)* belonging to the WrocLib, they represent the super-concept of *Book*, thus:

{ In the InhaLib, the *Artificial Intelligence* has the *ISBN* of *978-0070522633*;
In the WrocLib, the *Artificial Intelligence* has the *ISBN* of *978-0070522633*.

Based on this we can specify that the *ISBN* is a key identity in each library. This means that the *ISBN* is identifiable in the real world.

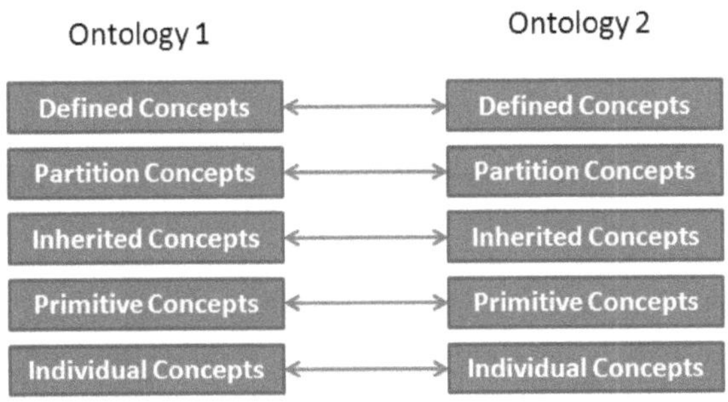

Fig. 4. Direct matching between the same type of concepts.

While calculating the similarity between the two concepts *Information* and *PublishedInf*, although we compare all associated attributes, the degree of similarity is low and the number of comparisons is high (7 x 6= 42). This is because they have many different attribute names and the number of properties belonging to each concept is high. According to the characteristics of the identities of concepts, instead of using blind or exhaustive comparisons among all properties of concepts, we simply focus on those properties that identify them. For example, we only compare the associated key identity of $ISBN$ while computing the similarity between the two concepts $Information$ and $PublishedInf$. The degree of similarity is high (one), and the number of comparisons is one and is independent of the number of properties belonging to each concept. Similarly, for the concept *Book* case, we simply compare associated identities $BookID$ and $ISBN$.

Thus we have a heuristic matching method that differs from previous the studies as follows:

Proposition 2 (Identity-based Similarity). *For two given concepts* $(z_{c_1}, A^{c_1}, V^{c_1})$ *belong to ontology* O_1*, and* $(z_{c_2}, A^{c_2}, V^{c_2})$ *belong to ontology* O_2.

1. *For any two concepts* $c_1, c_2 \in C_{DC}$*, if* c_1*'s key identity is equivalent to* c_2*'s then* c_1 *is equivalent to* c_2*,*
2. *For any two concepts* $c_1, c_2 \in C_{HC}$*, if* c_1*'s (Local & Inheritance) identities are equivalent to* c_2*'s then* c_1 *is equivalent to* c_2*,*
3. *For any two concepts* $c_1, c_2 \in C_{PvC}$ *or* $c_1, c_2 \in C_{PC}$*, if* c_1*'s inheritance identity is equivalent to* c_2*'s and* c_1*'s constant attribute is equivalent to* c_2*'s then* c_1 *is equivalent to* c_2*,*

5 Importance Concept-Based Similarity

Cognitive support for ontology integration emphasizing *importance concepts* has not been explored. However, *importance concepts* have been considered for ontology understanding in several early studies e.g. [14]. They suggest that the importance measurement of a concept must take into account the contributions from all the other concepts in the ontology via characterization of four features of potentially important concepts and relations, which drive the drifting stream of consciousness:

- A concept is more important if there are more relations originating from it.
- A concept is more important if there is a relation originating from this concept to a more important one.
- A concept is more important if it has a higher relation weight with respect to any other concept.
- A relation weight is higher if it originates from a more important concept.

Here, the term has in three senses. First, it explains what is important (or interesting). The term *importance* is used as a metric for measuring the extent to which the ontology creator suggests a concept or relation to users. Second, a concept is regarded as a source that owns a set of relations related to other concepts. Finally, concepts and relations exhibit a mutually reinforcing relationship.

Our work [12] differs from the above approach, because our aim is to design a method of reducing the computational *complexity* in ontology integration. First, we agree that the *importance measurement* of a concept must take into account the contributions from all the other concepts in the ontology based on all types of relations, including subsumption and non-subsumption. However for our purposes, the term *importance of concepts* is used to identify the concept's position in the hierarchy. It provides the importance level of the concept with respect to the same type of concepts. Therefore, the *importance concepts* can be used for priority matching between the same types of concepts. Here, we address four more features of potentially important concepts and relations as follows:

- A concept is more important if there are more relations originating from other concepts to it.
- A concept is more important if there are more important concepts which are partitions of it.
- A concept is more important if there are more important concepts which are inherited from it.
- A concept is more important if there are more important concepts which are defined from it.

Definition 14 (Imported Concepts). *The* imported concepts *of the concept c (I_c) are the set of concepts defined from the concept c.*

- *If c is a DC, the imported concepts of c include the partition concepts of c, the inherited concepts of c and the primitive concepts defined from c.*
- *If c is a PC, the imported concepts of c consist of the inherited concepts of c and the primitive concepts defined from c.*
- *If c is a IC, the imported concepts of c consist of the inherited concepts of c and the primitive concepts defined from c.*
- *If c is a PvC, the imported concepts of c comprise the primitive concepts defined from c.*

Definition 15 (Forward concepts). *The* forward concepts *of the concept c (F_c) are the set of concepts of the* Range *of properties belonging to the concept c.*

Definition 16 ((Backward concepts)). *The* backward concepts *of the concept c (B_c) are the set of concepts that have relations/properties to the concept c.*

Let r(c_i) be a function of an importance weight of concept c_i, r_i=r(c_i) be an importance weight of the concept c_i, w(c_i, c_j) be a relation weight function, and $w_{i,j}$=w(c_i, c_j) be the weight of all relations from c_i to c_j. It is possible that there exists more than one relation from concept c_i to concept c_j. For example, the concept *Student* may have two relations *study* and *learn* with the concept *Subject*. Some students may have a part time relation with an assistant professor, so instead of mere study they may also teach a given subject. Therefore, $r_j w_{i,j}$ is the total importance of all the relations from concept c_i to concept c_j.

According to the last hypothesis in [14], we present a similar recursive formula which computes the weight of the relations from concept c_i to concept c_j at the (k+1)th iteration. The weight is proportional to the importance of c_i and is the inverse ratio of the sum of all the importance values of c_j's backward concepts at the kth iteration.

$$w_{k+1}(c_i, c_j) = \frac{r_k(c_i)}{\sum_{t_i \in B_i} r_k(t_i)} \tag{12}$$

Based on our four hypotheses combined with the first three hypotheses in [14], we present recursive formulae which calculate the importance of concept c_i at the (k+1)th iteration. The importance consists of two parts; all the importance values of c_i's imported concepts with probability α, and the weight of relations from c_i to the forward concepts with probability λ. However it is only applied if the concept $c_i \in C \backslash C^{IvC}$. The importance weight of the concept c_i is as follows:

$$r_{k+1}(c_i) = \alpha \sum_{c_j \in I_i} r_k(c_j) + \lambda \sum_{c_j \in F_i} w_{k+1}(c_i, c_j) r_k(c_j), \alpha + \lambda = 1 \tag{13}$$

If the concept $c_i \in C^{IvC}$, the importance weight of the concept is computed by the weight of relations from the backward concepts to c_i:

$$r_{k+1}(c_i) = \sum_{c_j \in B_i} w_{k+1}(c_j, c_i) r_k(j) \tag{14}$$

Definition 17 (Prior Matching Set (PMS)). *The* Priority matching set *of concept c is the possible similarity set of c in ascending order of the distance between the importance weight of each concept belonging to the PSS and the importance weight of concept c.*

Proposition 3. *Assuming two ontologies O_i and O_j must be integrated, the concept c_i belonging to ontology O_i should be priority matched to the concept c_j or neighbors of c_j belonging to PMP of c_i, where the distance between c_i's importance weight and c_j's is minimal.*

In this paper, the term *importance* is a measurement used for priority matching between concepts of the same type. The measurement is based on the following consideration: First, a concept is regarded as a source owning a set of relations with other concepts. For example, the concept *Teacher* of ontology E_1 owns the relation *teach* to the concept *Subject*. Second, a concept has a set of attributes with associated restrictions. It is useful to identify the type of concepts. For example, the concept *Person* with the attribute *hasFingerPrint* has associated retrictions: owl:DatatypeProperty, owl:FunctionalProperty, owl:InverseFunctionalProperty, and owl:cardinality = 1. So concept *Person* may be a *DC*. Finally, concepts with associated attributes and relations exhibit a mutually reinforcing relationship used to identify the concept's importance weight (measurement). In our ongoing example, two ontologies E_1 and E_2 must be integrated. First, we classify all concepts belonging to each ontology into five disjoint groups, as shown in *Example* 8. Then we compute the importance weight of each concept based on formulas 3-5. We consider that if the set of $C_{E_2}^{PC} = \{FemalePerson, MalePerson\}$ and $C_{E_1}^{PC} = \{Female, Male\}$, they should be matched. According to the above measurement of the importance of concepts, it is clear that the distance between the $Female$'s importance weight and $FemalePerson$'s is less than the distance between $Male$'s and $FemalePerson$'s. This is because $FemalePerson$'s position in E_2 is more similar to $Female$'s than $Male$'s in ontology E_1. This is similar to the case of the two concepts $Male$ and $MalePerson$. So priority matching between the two sets of C_{E_1} and C_{E_2} is as follows: The two concepts $Male$ and $Female$ must be matched to those of $MalePerson$ and $MalePerson$ respectively.

6 Conflict in Ontology Integration

6.1 Conflicts on Instance-Level

On this level we assume that two ontologies differ only in the values of their instances. This means that they may have the same concepts and relations.

Definition 18. *Let* O_1 *and* O_2 *be* (**A, V**)*-based ontologies. Let concept* (z_c, A^c, V^c) *belong to both ontologies and let the same instance* i *belong to concept* c *in each ontology, that is* $(i, v_1) \in \text{Ins}(O_1, c)$ *and* $(i, v_2) \in \text{Ins}(O_2, c)$*. There is a conflict if* $v_1 \neq v_2$*.*

From Definition 20 it follows that referring to instance i the instance integration condition is not satisfied. Lets consider an example.

Example 9. As an example lets consider the ontologies of two information systems; those of a university and company. Consider the concept *Student* belonging to the university system ontology, which has the following structure, ({*St–id, Name, Age, Address, Specialization, Results*}, $V^{Student}$) and the concept Employee belonging to the company system ontology, which has the following structure, ({*Emp–id, Name, Address, Position, Salary*}, $V^{Employee}$). Assume that

Table 2. An instance of the concept Student

St–id	*Name*	*Age*	*Address*	*Specialization*
1000	Nowak	20	Wroclaw	Information Systems

Table 3. An instance of the concept Employee

Emp–id	*Name*	*Address*	*Position*	*Salary*
1000	Nowak	Warsaw	Designer	15.000

there is an instance i which belongs to both concepts (i.e., somebody is both a student of the university and an employee of the company). The value of this instance referring concept *Student* is given by $Table$ 2 and the value of this instance referencing concept Employee is given by $Table$ 3:

Thus there is an inconsistency on the instance level, because the instance in the university ontology referencing attribute Address has value *Wroclaw* and that in the company ontology referring to the same attribute has value *Warsaw*.

To solve conflicts of ontologies on the instance level, consensus methods seem to be very useful. Different criteria, data structures and algorithms have been determined [27]. In our conflict profile we have to deal with a set of versions of the same instance. A version of an instance (i, x) can be represented by the value x. The task is to determine a version (value) which best represents those given. To achieve this kind of consistency, the consensus problem can be defined as follows.

Given a set of values $X = \{v_1, \ldots, v_n\}$ where v_i is a tuple of type A^c, that is:

$$v_i : A^c \rightarrow V^c \tag{15}$$

for $i = 1, \ldots, n$; $A^c \subseteq \boldsymbol{A}$ and $V = \bigcup_{a \in A^c} V_a$ we must find the tuple v of type A, such that one or more selected postulates consensus are satisfied [27].

One popular postulate requires minimizing the following sum.

$$\sum_{i=1}^{n} d(v, v_i) = \min_{v' \in T(A^c)} \sum_{i=1}^{n} (v', v_i) \tag{16}$$

where $T(A^c)$ is the set of all tuples of type A^c.

6.2 Conflict on Concept-Level

On this level, Nguyen [27] assumes that two ontologies differ in terms of the structure of the same concept. This means that they contain the same concept but its structure differs according to the ontology. The definition of a concept on the concept-level is as follows:

Definition 19. *Let O_1 and O_2 be* (**A, V**)*-based ontologies. Let concept* $(z_{c_1}, A^{c_1}, V^{c_1})$ *belong to O_1 and concept* $(z_{c_2}, A^{c_2}, V^{c_2})$ *belong to O_2. There is a conflict on the concept level if $c_1 = c_2$ but $A^{c_1} \neq A^{c_2}$ or $V^{c_1} \neq V^{c_2}$.*

Example 10. Concept *Person* in one ontology may be defined by the following attributes: {*Name, Age, Address, Sex, Job*} whereas in the other it is defined by attributes: {*Id, Name, Address, Date–of–birth, TIN,1 Occupation*}. Another example refers to the situation in which the sets of attributes are identical, but their domains vary. We consider the attribute *Age*: for the concept *Person* one ontology assigns domain $V^{Age} = [1, 17]$ and the other assigns domain $V^{Age} = [18 - 65]$. In this case there is an inconsistency.

Here we have considered conflict on the concept level in the following cases:

1. *Multiple forms of the same concept* means ontologies define the same concept in different ways. For example, the concept *Person* in one ontology may be defined by attributes: *Name, Age, Address, Sex, Job*, while in another it is defined by attributes: *Id, Name, Address, Date_of_birth, Taxpayer identification number, Occupation.*
   The problem: *For a given set of pairs $X = \{(A^i, V^i) : (A^i, V^i)$ is the structure of concept c belonging to the ontology O_i* for $i = 1, \ldots, n\}$, *we need to determine the pair (A^*, V^*) which best represents those given.*
   Algorithm 2 is an overview of the solution.

input : $C^* = \bigcup c^i, i = 1.\ldots.n$ is the set of concepts that are recognized as the same concept, but the associated structures $\{(A^i, V^i), i = 1 \ldots n\}$ are different.
output: Pair (A^*, V^*) which is the integration of the given pairs $\{(A^i, V^i), i = 1 \ldots n\}$.

$A^* = \bigcup A^i, i = 1.\ldots.n$ where A^i is the set of attributes of the concept $c^i \in C^*$;
foreach *pair* $a_1, a_2 \in A^*$ **do**
  if $R(a_1, \Leftrightarrow, a_2)$ **then** $A^* \setminus \{a_2\}$; /* eg., *job* $\Leftrightarrow$ *occupation* */
  if $R(a_1, \sqsubseteq, a_2)$ **then** $A^* \setminus \{a_1\}$; /* eg., *age* $\sqsubseteq$ *birthday* */
  if $R(a_1, \sqsupseteq, a_2)$ **then** $A^* \setminus \{a_2\}$; /* eg., *sex* $\sqsupseteq$ *female* */
  if $R(a_1, \perp, a_2)$ **then** $A^* \setminus \{a_1\}$; /* eg., *single* $\perp$ *married* */
end
foreach *attribute a from set* A^* **do**
  if *the number of occurrences of a in pairs* (A^i, V^i) *is smaller than n/2* **then** set $A^* := A^* \setminus \{a\}$;
end
foreach *attribute a from set* A^* **do**
  determine its domain V_a as the sum of its domains in pairs (A^i, V^i);
end
Return((A^*, V^*));

Algorithm 2. Multiple forms of the same concept

2. *Overlapping but different concepts* means that ontologies define different concepts with the same name and structure. As an example, we consider two concepts with the same name and structure which both contain knowledge about *student*. The first the concept structures its knowledge of *female student*, whereas the second one structures its knowledge of *male student*.
   The problem: *For a given set of concepts for which the names and structures are equivalent* $T = \{c_1, \ldots c_n\}$ *where* c_i *in ontology* O_i, *we need to determine the set tuple* $C^* = \{(c_i\ r\ c_j)$: *the concept* c_i *is relation* r *to* c_j *where* r *is one of the following relations: more general, equivalent, and disjoint*$\}$.
   Algorithm 3 is an overview of the solution.

input : $T = \bigcup c_i, i = 1.\ldots.n$ is the set of concepts for which the names and structures are equivalent.
output: $C^* = \{(c_i r c_j)$: the concept c_i is relation r to c_j where r is one of the following relations: more *general, equivalent,* and *disjoint*$\}$

foreach *pair* $c_i, c_j \in T$ **do**
    if $\exists\ a \in A^{c_i} \cap A^{c_j}$ *and* $z_{c_i}(a) \perp z_{c_j}(a)$ **then** $C^* = \bigcup\{(c_i \perp c_j)\}$;
    else if $\exists\ a \in A^{c_i} \cap A^{c_j}$ *and* $z_{c_i}(a) \sqsupseteq z_{c_j}(a)$ **then**
        $C^* = \bigcup\{(c_i \sqsupseteq c_j)\}$;
        else if $\exists\ a \in A^{c_i} \cap A^{c_j}$ *and* $z_{c_i}(a) \sqsubseteq z_{c_j}(a)$ **then**
            $C^* = \bigcup\{(c_j \sqsupseteq c_i)\}$;
            else if $\exists\ c$ *and* $c' \in C \cap C'$ *where* c *and* c' *are a super-concept corresponding to* c_1 *and* c_2, *and* $c \perp c'$ **then**
                $C^* = \bigcup\{(c_i \perp c_j)\}$;
                else
                    $C^* = \bigcup\{(c_i \Leftrightarrow c_j)\}$;
end
return (C^*);

Algorithm 3. Overlapping but different concepts

Apart from the aforementioned conflict problems, we also point out the following three other cases of conflicts on the concept-level:

- The *same concept but different names* means that ontologies define the same concept with different names. For example, for the concept *Person*, one ontology defines the concept name of *Individual* but another defines the concept name of *Homo*.
- The *same name but different concepts* means that ontologies define different concepts with the same name. For example, for the same name of *Master Course*, one ontology considers the concept of subjects for a master student, while another considers the concept of a master student.
- The *Multiple concepts of the same form* means that ontologies define different concepts with the same structure.

```
input  : Given two ontologies O1 and O2
output: Pairs of concepts are equvalent
dJ ⟵ clfConcepts(O2);
wC ⟵ wConcepts(dJ);
foreach concept c belonging to ontology O1 do
    PSS ⟵ getPSS(c, dJ);
    PMS ⟵ getPMS(c,wC, PSS);
    for i ← 1 to size(PMS) do
        match ← Matching (c,PMS [i]);
        if match ≥ threshold then
            smatch ⟵ ∪ (c, PMS [i]);
            break;
        end
    end
end
cConflict(smatch);
Return(smatch);
```

Algorithm 4. Direct Matching Algorithm (DMA)

The first two problems can be solved by combining the text corpus and WordNet-based method, which is presented in [9]. The third problem can be solved by using the same multiple concepts for the same form method.

7 Effective Algorithm for Ontology Integration

In this section, we apply the aforementioned methods to design an effective algorithm for ontology matching. An overview of the algorithm as follows:

1. Function clfConcepts(O_2) is used to classify all concepts belonging to the ontology O_2 into disjoint groups (dJ). We apply the knowledge from section 3.1 to clfConcepts(O_2). The main problem here is how to identify the concept's identities. We assume the ontologies were written in the OWL language. Identities are distinguished via the following two steps:
   (1) collecting the necessary and sufficient properties of each concept.
   (2) considering an identity as a property of the concept and distinguishing it from other properties based on the characteristics of the one-to-one function between its domain and range. The identities can be written in OWL based on owl:Data-typeProperty with three restrictions: owl:Functional-Property, owl:InverseFuncti-onalProperty, and owl:cardinality = 1.

   Notice that we use the following heuristic to distinguish a *DC*: If a concept is a top-most taxonomy in a given ontology and it contains at least one identity, it must be a *DC*.
2. Function wConcepts(dJ) is used to compute the importance weights of each concept belonging to each group in the set dJ. It returns an importance

weight of vectors (wC) set corresponding to dJ. The algorithm is similar to [14].

3. Function getPSS(c, dJ)is used to generate a set of possible similarities of concept c from dJ.
4. Function getPMS(c, wC, PSS) is used to obtain a set of priority matching of concept c from its PSS.
5. Function Matching(c, $PMS[i]$) is used to find the degree of similarity between the concept c and concept $PMS[i]$. Here we apply the *content-based similarity* and *identity-based similarity* techniques.
6. Function cConflict is used to deal with the aforementioned problem of conflict in ontology integration and to solve the problem of conflict on the concept-level.

8 Experiments

In Table 2, we compare the techniques of similarity analysis used in existing mapping tools with our approach based on the novel *Identity-based similarity* method. Note that to find similarities between concepts, most existing mapping methods compare all properties belonging to each concept, while the *identity-based similarity* method simply focuses on those identities belonging to each concept.

Table 4. Comparative techniques of similarity analysis

Matching Methods	*Instance -based*	*Lexical -based*	*Schema -based*	*Taxonomy -based*	*Identity -based*
PROMPT	Y	Y	Y	Y	N
MAFRA	Y	Y	Y	Y	N
RiMOM	Y	Y	Y	Y	N
GLUE	Y	Y	Y	Y	N
Our	Y	Y	Y	Y	Y

According to our studies of ontology integration, methods of reducing the complexity of ontology integration have not yet been explored. Therefore we simply present the comparative complexity between our method of matching equality and the *content-based matching* approach as follows: Suppose that N_c, N_p, and $N_i$are the maximum numbers of nodes, properties (attributes), and instances. Let us assume that the complexity of comparing two attribute values between two instances is O(1). Then, the complexity of calculating the similarity between two instances is O(N_p^2). The complexity of calculating the similarity between two nodes is O(N_p^2 x N_i). Finally, matching between two ontologies costs O(N_c x N_p^2 x N_i). In order to compare the content-based method with our matching method, we substitute N for every parameter. Then the cost of the content-based method is O(N^4), using the direct matching algorithm for the content-based method, *DMAContent-based*, whereas our matching method costs

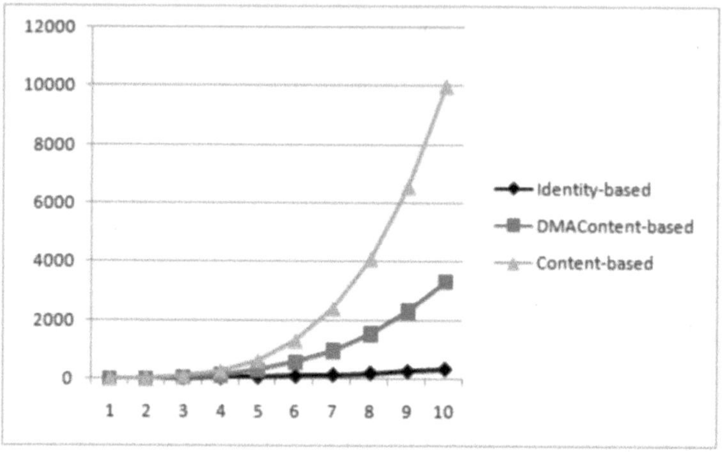

Fig. 5. Complexity comparison between Identity-based and Content-based method.

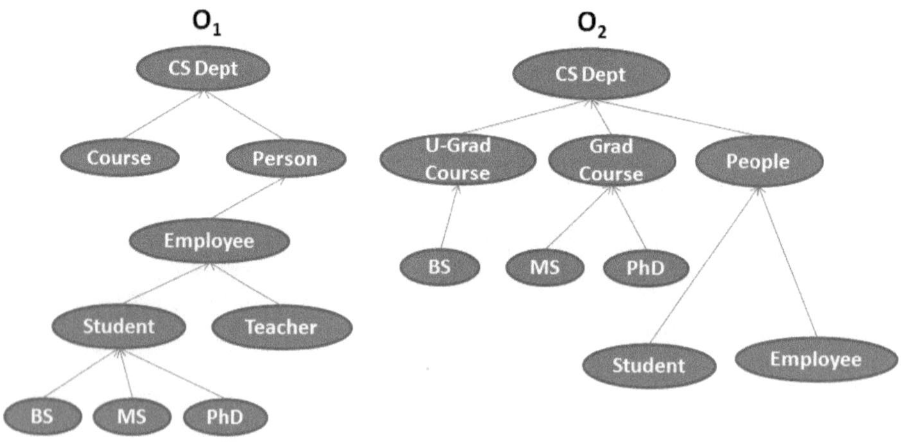

Fig. 6. Possible conflict on concept-level.

O(N^3 x $logN$), because the method involves direct matching between concepts of the same type. If we apply *identity-based similarity* for matching between concepts of the same type, the cost is O(N^2 x $logN$), because it does not require a comparison of all properties belonging to each concept. *Figure* 5 illustrates the complexity difference between our methods of matching and content-based matching in a line chart. The chart shows that the complexity differs according to the number of properties, assuming that the number of concepts and instances are equal in each case. The number of properties belonging to concepts is proportional to the complexity difference.

Moreover, our method avoids many cases of conflict between concepts. *Figure* 6 shows the top-level view of source ontologies in PROMPT [28]. Most of the above

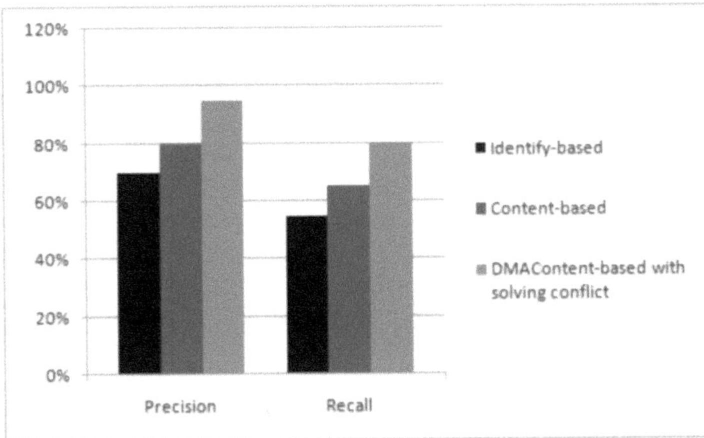

Fig. 7. Comparison of our system and content-based system.

matching methods produce incorrect matching from O_1:*BS*, O_1:*MS*, O_1:*PhD* and O_1:*Student* to O_2:*BS*, O_2:*MS*, O_2:*PhD* and O_2:*Student*, respectively. However, Our approach recognizes that these are cases of conflict between the labels and contents of concepts.

We collected many ontologies from the Internet (URL: http://www.aifb.uni-karlsruhe.de/W- BS/meh/foam/ontologies.htm) and composed their corresponding ontologies. We also modified the ontologies to create cases of conflict. Each sample included at least three ontologies. N_{total} denotes the total number of pairs used for matching concepts between the candidate ontologies by experts, $N_{correct}$ and $N_{incorrect}$ correspond to the number of correct and incorrect pairs for matching concepts sought by our system, respectively. *Precision* $(= \frac{N_{correct}}{N_{correct}+N_{incorrect}})$ is used to evaluate the ratio of incorrectly extracted relationships. *Recall* $(= \frac{N_{correct}}{N_{total}})$ is used to evaluate the ratio of correct matching sought by the system. Figure 7 illustrates the comparative experimental results between the *Identity-based* and *Content- based* and *DMSContent-based* methods.

9 Conclusion

The direct matching algorithm is a smart approach to ontology integration. It combines the *types of concepts* and *importance of concepts* in order to directly match concepts of the same type, instead of using blind or exhaustive matching among all concepts. The *Identity-based similarity* method is a novel method of heuristic matching. Its advantage is that the complexity is initially reduced by comparing only those properties which identify each concept, instead of matching all properties belonging to each concept. The proposed solutions to the problem of conflicts prevent many cases of conflicts in ontology integration. In future work, we will meticulously explore the idea of complexity in ontology integration.

References

1. Bagui, S.: Mapping XML Schema to Entity Relationship and Extended Entity Relationship Models. International Journal of Intelligent Information and Database Systems 1(3), 325–345 (2007)
2. Castano, S., Ferrara, A., Montanelli, S.: Matching ontologies in open networked systems: Techniques and applications. Journal on Data Semantics 3870, 25–63 (2006)
3. Do, H.H., Rahm, E.: COMA – a system for flexible combination of schema matching approaches. In: Proc. 28th International Conference on Very Large Data Bases (VLDB), Hong Kong, pp. 610–621 (2002)
4. Dou, D., McDermott, D., Qi, P.: Ontology translation on the semantic web. Journal on Data Semantics 3360, 35–57 (2005)
5. Doan, A.H., Domingos, P., Halevy, A.: Reconciling Schemas of Disparate Data Sources: a machine learning approach. In: Proc. of ACM SIGMOD Conference, pp. 509–520
6. Doan, A.H., Madhavan, J., Domingos, P., Halevy, A.: Learning To Map between Ontologies on the Semantic Web. In: Proc. of WWW 2002, pp. 662–673. ACM Press, New York (2002)
7. Doan, A.H., Domingos, P., Halevy, A.: Learning to match the schemas of data sources: A multistrategy approach. Machine Learning 50(3), 279–301 (2003)
8. Doan, A.H., Madhavan, J., Domingos, P., Halevy, A.: Ontollogy matching: a machine learning approach, pp. 385–404. Springer, Berlin
9. Duong, T.H., Nguyen, N.T., Jo, G.S.: A Method for Integration across Text Corpus and WordNet-based Ontologies. In: IEEE/ACM/WI/IAT 2008, Workshops Proceedings, pp. 1–4. IEEE Computer Society, Los Alamitos (2008)
10. Duong, T.H., Nguyen, N.T., Jo, G.-S.: A method for integration of wordNet-based ontologies using distance measures. In: Lovrek, I., Howlett, R.J., Jain, L.C. (eds.) KES 2008, Part I. LNCS (LNAI), vol. 5177, pp. 210–219. Springer, Heidelberg (2008)
11. Duong, T.H., Nguyen, N.T., Jo, G.S.: A Method for Integrating Multiple Ontologies. Cybernetics and Systems 40(2), 123–145 (2009)
12. Duong, T.H., Jason, J.J., Nguyen, N.T., Jo, G.S.: Complexity Analysis of Ontology Integration Methodologies: a Comparative Study. Appear in Journal of Universal Computer Science (2009)
13. Ehrig, M., Sure, Y.: Ontology mapping - an integrated approach. In: Bussler, C.J., Davies, J., Fensel, D., Studer, R. (eds.) ESWS 2004. LNCS, vol. 3053, pp. 76–91. Springer, Heidelberg (2004)
14. Gang, W., Juanzi, L., Ling, F., Kehong, W.: Identifying Potentially Important Concepts and Relations in an Ontology. In: Sheth, A.P., Staab, S., Dean, M., Paolucci, M., Maynard, D., Finin, T., Thirunarayan, K. (eds.) ISWC 2008. LNCS, vol. 5318, pp. 33–49. Springer, Heidelberg (2008)
15. Gangemi, A., Pisanelli, D.M., Steve, G.: Ontology Integration: Experiences with Medical Terminologies. In: Formal Ontology in Information Systems, pp. 163–178. IOS Press, Amsterdam
16. Guarino, N., Welty, C.: Ontological Analysis of Taxonomic Relationships. In: Laender, A.H.F., Liddle, S.W., Storey, V.C. (eds.) ER 2000. LNCS, vol. 1920, pp. 210–224. Springer, Heidelberg (2000)
17. Hearst, M.A.: Automatic Acquisition of Hyponyms from Large Text Corpora. In: Proceedings, 14th International Conference on Computational Linguistics (COLING 1992), Nantes, France, pp. 539–545 (1992)

18. Hearst, M.A.: Automated Discovery of WordNet Relations. In: Fellbaum, C. (ed.) WordNet: An Electronic Lexical Database, pp. 131–151. MIT Press, Cambridge (1998)
19. Jiang, J., Conrath, D.: Semantic similarity based on corpus statistics and lexical taxonomy. In: Proceedings on International Conference on Research in Computational Linguistics, pp. 19–33 (1997)
20. Lee, J., Chae, H., Kim, K., Kim, C.H.: An Ontology Architecture for Integration of Ontologies. In: Mizoguchi, R., Shi, Z.-Z., Giunchiglia, F. (eds.) ASWC 2006. LNCS, vol. 4185, pp. 205–211. Springer, Heidelberg (2006)
21. Leacock, C., Chodorow, M.: Combining local context and WordNet similarity for word sense identification. In: Fellbaum, C. (ed.) WordNet: An electronic lexical database, pp. 265–283. MIT Press, Cambridge (1998)
22. Lin, D.: An information-theoretic definition of similarity. In: Proceedings of the International Conference on Machine Learning (1998)
23. Madhavan, J., Bernstein, P., Rahm, E.: Generic schema matching with Cupid. In: Proc. 27th International Conference on Very Large Data Bases (VLDB), pp. 48–58. Morgan Kaufmann Publishers Inc., San Francisco (2001)
24. Madhavan, J., Bernstein, P., Doan, A., Halevy, A.: Corpus-based schema matching. In: Proc. 21st International Conference on Data Engineering (ICDE), pp. 57–68. IEEE Computer Society Press, Los Alamitos (2005)
25. Maedche, A., Motik, B., Silva, N., Volz, R.: MAFRA-An ontology MApping FRAmework in the context of the semantic web. In: Gómez-Pérez, A., Benjamins, V.R. (eds.) EKAW 2002. LNCS (LNAI), vol. 2473, pp. 235–250. Springer, Heidelberg (2002)
26. Morin, E., Jacquemin, C.: Automatic acquisition and expansion of hypernym links. Computer and the Humanities 38(4), 363–396 (2004)
27. Nguyen, N.T.: Advanced Methods for Inconsistent Knowledge Management. Springer, London (2008)
28. Noy, N.F., Musen, M.A.: The PROMPT Suite: Interactive Tools For Ontology Merging And Mapping. In International Journal of Human-Computer Studies 59, 983–1024 (2003)
29. Pedersen, T., Patwardhan, S., Michelizzi, J.: WordNet:Similarity-measuring the relatedness of concepts. In: Proceedings of NAACL (2004)
30. Pinto, H.S., Martins, J.P.: A Methodology for Ontology Integration. In: Proceedings of the First International Conference on Knowledge Capture, pp. 131–138. ACM Press, New York
31. Resnik: Using information content to evaluate semantic similarity in a Taxonomy. In: Proceedings of the 14th International Joint Conference on Artificial Intelligence, Montreal, Canada, pp. 448–453 (1995)
32. Tang, J., Li, J., Liang, B., Huang, X., Li, Y., Wang, K.: Using Bayesian decision for ontology mapping. Journal of Web Semantics 4(4), 243–262 (2006)
33. Tozicka, J., Rovatsos, M., Pechoucek, M., Urban, S.: MALEF: Framework for distributed machine learning and data mining. International Journal of Intelligent Information and Database Systems 2(1), 6–24 (2008)
34. Zhang, W.R.: Concepts, challenges, and prospects on Multiagent Data Warehousing (MADWH) and Multiagent Data Mining (MADM). International Journal of Intelligent Information and Database Systems 2(1), 106–124 (2008)
35. Wu, Z., Palmer, M.: Verb semantics and lexical selection. In: 32nd Annual Meeting of the Association for Computational Linguistics, pp. 133–138 (1994)

Cluster Analysis in Personalized E-Learning Systems

Danuta Zakrzewska

Institute of Computer Science Technical University of Lodz,
Wolczanska 215, 90-924 Lodz, Poland
dzakrz@ics.p.lodz.pl

Abstract. Personalized e-learning system should be tailored into student needs, which usually differ even among learners, who attend the same course and have similar technical skills. In the chapter, it is proposed the system architecture, in which teaching paths as well as proper layouts are adjusted to groups of students with similar preferences, created by application of clustering techniques. Learner models are based on dominant learning style dimensions, according to which students focus on different types of information and show different performances in educational process. Extension of the model by including usability preferences is investigated. There are examined different clustering techniques to obtain groups of the best quality. It is presented the algorithm that will fulfill tutor requirements especially concerning the choice of parameters. Some experimental results for real groups of students and different algorithms are described and discussed.

1 Introduction

Performance of an e-learning system, depends significantly on its adaptation possibilities. Educational software designers should take into account different student needs and preferences, even while considering learners, who study the same subject at the same university and who have similar technical skills. Application of intelligent methods allows to equip the system with personalization features, as well as to tailor it into student requirements. If preferences of each user were taken into consideration, multiple versions of teaching paths and materials should have been created. Dividing learners into groups of similar preferences will allow to limit the recommendations to the certain numbers without loosing personalization features.

The aim of the chapter is to investigate application of clustering techniques to find groups of students with similar preferences. Individual models are based on preferred learning style dimensions as well as usability preferences, according to which students focus on different types of information and show different performances in an educational process. It is proposed the system architecture, in which teaching paths as well as proper layouts are adjusted to groups of students according to their learning styles and usability preferences. Considered student models are built on the basis of Felder and Silverman learning style dimensions,

N.T. Nguyen and E. Szczerbicki (Eds.): Intel. Sys. for Know. Management, SCI 252, pp. 229–250.
springerlink.com

together with student color choices. In the chapter, there are examined different well-known clustering techniques to obtain groups of the best quality. It is presented the algorithm that will fulfill tutors' requirements especially concerning the choice of parameters. Some experimental results for real groups of students and different algorithms are described and discussed.

The chapter is organized as follows. In the next section, literature review concerning intelligent e-learning systems as well as application of clustering methods in educational software are presented. Then, student models, and attributes such as dominant learning styles or usability preferences, which decide on learner needs, are depicted. The aim of the following two sections is an investigation of application of cluster analysis for student grouping; in Section 4 well known algorithms and their main features are considered, while in Section 5, it is described the algorithm that fulfils tutor needs and may be applied for learning style attributes as well as for the characteristics connected with usability. Evaluation of considered methods is done in the next section, where results obtained for real students' data, for different algorithms as well as choice of parameters are compared and discussed. Finally some concluding remarks and future research are presented.

2 Related Work

The importance of personalization features of Web-based systems was emphasized by many researchers in recent years. The overview of technologies supporting adaptivity in educational systems was presented by Brusilovsky & Peylo [1], who analyzed their variety and provided software samples. Different authors considered student characteristics that decide on their requirements and preferences, and there exist several papers concerning students' modeling for the purpose of adaptive intelligent educational systems (see for example [2,3]).However there is still no agreement which features can and should be used. Brusilovsky [4] defined user individual traits as a group name for user features, such as cognitive factors, or simply learning styles, which are stable and usually extracted by specially designed psychological tests. Santally & Alain [2] considered student model components such as learning styles, cognitive styles and cognitive controls for the purpose of Web-based learning environments' personalization. They built the adaptation framework and observed the importance of different components in the instructional design process.

Personal learning styles were considered, by many authors, as factors that should be taken into account during adaptation of teaching process into individual needs, as they may influence learners' individual attitudes towards using the educational software (see [5,6]). Lu et al. [5] identified the impact of learning styles, learning patterns and other selected factors on students' learning performance in a Web Course Tools (WebCT) MIS graduate course. Lee [6] showed that student styles influence their academic achievement and should be taken into account to improve the quality of Web-based education. The study was based on the tests provided at 11 Korean universities. Beaudoin [7] examined, if the preferred learning styles of students influence their online behavior. He concluded that establishing a learning environment that depends on student profiles is the necessary condition for a good performance of online courses. Li et al.[8], in their intelligent tutoring system, considered tailoring into learner needs both: user environment and pedagogical environment . The authors presented a formal model of a system, which constructs the user environment based on cognitive abilities, knowledge levels, learning styles and psychology characteristics. They showed practical significance of the model. Cha et al. [9] investigated interface customization to students' preferences while building an intelligent learning environment. The authors examined efficiency of a development of an intelligent learning system by providing effective user interfaces and learning contents, which depend on the learner's preferences. They diagnosed individual learning styles by learner behavior patterns using Decision Tree and Hidden Markov Model methods.

An overview of the research concerning building adaptive systems by taking into consideration learning styles can be found in [3]. In their study, Stash et al. incorporated also different learning styles in two systems: an adaptive hypermedia and a high-level adaptive authoring tool [3]. They emphasized the importance of providing users with different teaching strategies while using applications. In many systems, adaptivity according to learning styles consisted in static assignments into groups of different needs. As the most representative systems in that area there should be mentioned: Arthur [10], 3DE [11] or AES-CS [12]. Gilbert & Han [10] developed Arthur as a Web-based system that provides adaptive instruction, where each instructor represented different teaching method and each course was divided into small units called concepts. Such approach allowed to use a many-to-one instructor/learner relationship to accommodate individual learning styles. The 3DE system, in turn, was enhanced in the tool [11], which was to choose the most suitable for learners course packages depending on their individual learning styles. Triantafillou et al. [12] based adaptivity of their AES-CS on instructional strategies, which supported students according to their cognitive styles.

In [13], it was stated, that the creation of multiple versions of an educational system, especially regarding its interface, can be extremely costly and it was proposed, to divide students into groups of similar characteristic features. For the purpose of web-based educational systems, students were grouped according to learning actions to discover their sequential patterns [14]. Talavera & Gaudioso clustered students according to their behaviors in unstructured collaboration spaces [15], defining attributes on the basis of the most often interactions with the system, such as number of messages, sessions, posts etc. They built descriptive group models and compared them with external features. Tang et al. considered student clustering according to the sequence and the contents of the pages they visited [16]. In [17], authors proposed using fuzzy clustering algorithm for both course materials' difficulty and learners' ability. The first parameter was determined by experts, while the second one included such factors as: time spent on each material, questionnaire feedback, quiz feedback, times clicking and random mouse move and click. After reevaluating learners' abilities and recalculating the course materials' difficulty levels, the system recommended appropriate materials to learners. Merceron & Yacef, in turn, clustered students according to the mistakes they made, for identifying different types of learners [18]. The authors reported the experiment, in which students were grouping using their mistakes made with the web-based tutoring tool, the Logic-ITA. Perera et al. [19] used clustering technique to identify different features of groups in the context of senior software development project. The aim of the research was to find out measures, which can help in group advising at the start and early identification of effective and poor practices. Clustering was applied to find both groups of similar teams and similar individual members. Broad review of application of clustering techniques in e-learning systems was presented in [20].

3 Student Models

In [21], it was proposed the architecture of intelligent e-learning system, where students were clustered according to their learning style preferences, what allowed to assign appropriate teaching paths for groups of students with similar preferences. Adding usability needs to the model enabled to include also interface adaptivity into the personalization features [22].

3.1 Dominant Learning Styles

The relationship of individual learning styles and distance education was investigated by different authors, who indicated that learning styles are valuable features characterized learners in a teaching process (see for example [23]). Graf & Kinshuk [24] showed that students with different learning styles have different needs and preferences. Rovai [25] stated that teaching methods should vary according to learning styles of learners, the significant impact of which, on Web based courses performances, was identified by Lu et al. [5] and Lee [6]. Authors examined different techniques for students' learning styles investigations. Alfonseca et al. [26] used descriptive statistics for student grouping in collaborative learning. García et al. [27] applied Bayesian networks for detecting students' learning styles on the basis of their behaviors. Xu et al. [28] built fuzzy models on the basis of learning activities and interaction history. Viola et al. [29], in turn, showed the effectiveness of data-driven methods for pattern extractions.

There exist different models of learning styles, as the most frequently used, there should be mentioned the ones defined by: Kolb [30], Honey & Mumford [31], Dunn & Dunn [32] and Felder & Silverman [33]. The last model, which is considered in our investigations, was very often applied for providing adaptivity regarding learning styles in e-learning environments [29]. It is based on *Index of Learning Style* (ILS) questionnaire, developed by Felder & Soloman [34].

The results of ILS questionnaire indicate preferences for 4 dimensions of the Felder & Silverman model, from among excluding pairs: *active* vs. *reflective*, *sensing* vs. *intuitive*, *visual* vs. *verbal*, and *sequential* vs. *global*. The index obtained by each student has the form of an odd integer from the interval [-11,11], assigned for all of the four dimensions.

Each student, who filled ILS questionnaire, can be modeled by a vector SL of 4 integer attributes:

$$SL = (sl_1, sl_2, sl_3, sl_4) = (l_{\mathrm{ar}}, l_{\mathrm{si}}, l_{\mathrm{vv}}, l_{\mathrm{sg}}) \ , \tag{1}$$

where l_{ar} means scoring for *active* (if it has negative value) or *reflective* (if it is positive) learning style, and respectively $l_{\mathrm{si}}, l_{\mathrm{vv}}, l_{\mathrm{sg}}$ are points for all the other dimensions, with negative values in cases of *sensing*, *visual* or *sequential* learning styles, and positive values in cases of *intuitive*, *verbal* or *global* learning styles.

Score from the interval [-3,3] means that the student is fairly well balanced on the two dimensions of that scale. Values -5,-7 or 5,7 mean that student learns more easily in a teaching environment which favors the considered dimension; values -9,-11 or 9,11 mean that learner has a very strong preference for one dimension of the scale and may have real difficulty learning in an environment which does not support that preference [34].

3.2 Usability Preferences

Experiment results, described in [35], showed that students with different dominant learning style dimensions indicated as important different usability features. Color preferences are strictly connected with usability needs and are important factors in interface designing. Suitable usage of colors can make interfaces easier to understand and use [36]. On the other hand colors used in layouts, usually do not address users' requirements and only few authors consider user oriented color models [37]. In the paper [35], there were examined student models based on both factors: dominant learning styles and preferable colors.

In our investigations, we will focus on color choices as usability preferences and it will be considered extended student model SM represented by two elements: learning styles and color preferences.

$$SM = (SL, SC) \ , \tag{2}$$

where SL is defined by (1) and

$$SC = (sc_1, sc_2) \ . \tag{3}$$

SC represents numbers of the first (sc_1) and the second (sc_2) choices of preferable colors of each student. Both of the attributes are of categorical type. To obtain the attribute values, students should be asked to choose two the most preferable colors' compositions, from among all that may be used in the system layout, and to indicate the rank for each.

4 Cluster Analysis for Student Grouping

4.1 System Architecture

In the current study, the aim of the use of unsupervised classification is finding groups of students of similar preferences for the purpose of courses' adaptation according to learner needs. The proposed system allows to adjust learning paths together with layouts into individual preferences of students, taking into account their dominant learning styles and usability requirements. In the pre-processed phase, the database of learning styles and usability preferences of the sample group of students is created. On the basis of that data set, by using unsupervised classification, students are divided into groups, each of which characterized by different preferences. Now, teaching materials as well as information content may be adjusted to needs of every group and different learning paths may be created. Each new student fills the questionnaire to determine his (her) learning style together with usability choices, and according to them, the proper group is chosen and personalized learning materials and contents are assigned. The proposed personalized approach allows to build individual paths and to make changes in teaching materials as well as to determine the ways of their presentations for students groups with different preferences. However, the personal content is static for each student during the course, it may change while starting the new one.

The system is based on three main data sets containing students individual features, teaching materials and layouts. The last two should be created for different courses separately, as students attending different classes, may be characterized by different features. The application is equipped in modules of learner profiling by determining their learning styles and usability preferences as well as the cluster analysis tool for building student groups and assigning new students into them. In the presented system, it is possible to limit considered personalized features only to dominant learning styles. Then, only teaching paths will be adjusted to student needs.

According to the main rule of the system, each new student, after filling the questionnaires should be assigned to the proper group. However there may exist students (outliers), who do not fit to any clusters, they should be considered individually. The overview of the simplified system architecture (without outliers) is presented on Fig. 1. In further considerations we will use color choices as examples of usability preferences.

The performance of the proposed system, which may be measured by the effectiveness in achievement of learning outcomes, depends on the quality of obtained clusters, which means similarity of group members.

4.2 Problem Statement

The problem statement, which consists in finding good quality groups of students with similar learning styles and usability preferences by unsupervised classification, imposes requirements for characteristic features of a clustering algorithm.

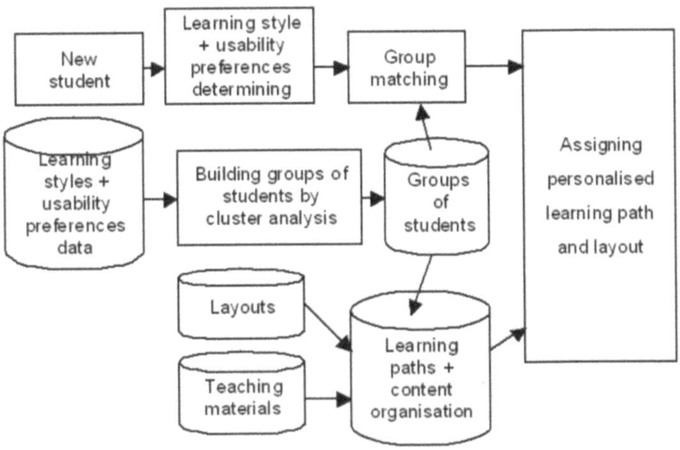

Fig. 1. System architecture [22]

The applied grouping technique should be fully self-generating and order independent with possibilities of changing similarity threshold value as well as determining the significance of each attribute value (each learning style dimension or usability preference may be of different importance for disparate courses). There should be no requirements for determining, the final number of clusters, in advance (the amount of learners' groups may change for different data), however it cannot exceed the maximal value, determined by a tutor. What is more, there should be included the possibility of clustering according to attributes represented by variables of different types (dominant learning styles are numeric, while color preferences are categorical for example), The last feature may be fulfilled by using similarity measures for mixed type data. Finally, dealing with outliers, which may signify outstanding students, will be an advantage of an applied method.

4.3 Clustering Algorithms

Once we have defined the preferable features of the clustering algorithm, we may consider finding the method that will satisfy tutors. In spite of the most often required features concerning the power and flexibility of data mining methods, in e-learning the emphasis is done on their simplicity. As the main goal of data mining, and in particular of clustering, is the improvement of courses. Educators need to understand the methods they apply. Most of the current data mining tools are too complex for teachers to use and their features do not meet their needs [38]. Consequently, in further considerations we will limit into the simplest approaches and check if they fulfill the requirements defined in the Section 4.2.

Partitioning Approach. K-means is one of the most known technique from among partitioning methods [39]. Algorithm consists in assigning data into the given number of clusters. At the beginning, clusters are randomly selected. In each iteration, observations are reassigned by moving them into the nearest cluster. New cluster centers are recalculated. The process is continued until all the observations are situated in the closest cluster. The method is simple and effective on large data sets, but its results depend significantly on initial assignments. The performance of K-means was investigated in [40] on students learning styles dimensions data sets and in [35] for all the students' attributes (including color choices).

The main parameters that influence significantly clustering effects by K-means, are initial assignments for cluster centers and the number of desired clusters. The algorithm is noisy sensitive and even a presence of one outlier may result in deformation of obtained clusters, what will take place in case of an outstanding student or if mistakes were made while filling questionnaires.

In [41], it was presented the modified method, where cluster centers are calculated after every object's allocation, what effects in outlier detection ability. Further improvement of the method by combining with agglomerative hierarchical clustering gave better results for real student learning style data [40]. However the combined method still required the input parameters: number of clusters and initial cluster centroid assignments, which influence the final results. That feature should be mentioned as the main disadvantage of K-means and its modifications, especially from the point of view of tutors who will have to determine at least the number of clusters in advance.

Statistical Approach. The goal of statistical models is to find the most likely set of clusters on the basis of training data and prior expectations. Expectation - Maximization algorithm (EM) uses the finite Gaussian mixtures model to generate probabilistic descriptions of clusters in terms of means and standard deviations [39]. The big advantage of EM algorithm is possibility of the selection of number of clusters by cross validation techniques, what allows to obtain their optimal value [42]. That feature allows not to determine the number of clusters at the beginning. Similarly to K-means method, parameters are recomputed until the desired convergence value is achieved. Experiments, described in [40], showed also noise sensitiveness of EM for student learning style data, the presence of outliers completely changed obtained effects: the number of clusters decreased, what resulted in building groups of lower quality.

Density Based Approach. Density-based methods discover clusters as dense regions of objects that are separated by regions of low density. Density-Based Spatial Clustering of Applications with Noise (DBSCAN) algorithm defines a cluster as a maximal set of so called density-connected points [43]. It allows to find clusters, for large spatial databases, of arbitrary shape and detect outliers. As the additional advantage of DBSCAN, it should be mentioned law number of required input parameters. However, on the other hand, the ones, which are necessary, are strictly connected with the density of considered objects, they should be chosen for every data set separately and they are difficult to determine. Even in the case of learning style data, where the range of the attribute values is limited, what may facilitate the choice of parameters, it is not easy to find out the parameters that guarantee clusters of good qualities. Experiments, presented in [40], have shown that DBSCAN did not give acceptable results as it performed the tendency to divide data into big clusters and to indicate many outliers.

Hierarchical Techniques. From among hierarchical algorithms, in spite of the agglomerative approach used so far (see [44]), it has been chosen Farthest First Traversal Algorithm, based on the strategy introduced in [45].This divisive approach guarantees very good performance in comparison with agglomerative methods [46]. Similarly to K-means, it is very simple and requires the desired number of clusters as an input parameter. Hierarchical algorithms are also noise sensitive even in the case of the farthest first traversal strategy, which secures the best effectiveness in the hierarchical approach.

Dealing with outliers. From among the considered algorithms, only DBSCAN has the ability to find outliers, however bad choice of input parameters may entail in indicating wrong objects as outliers. The other techniques, in the presence of noise changed their performance [40]. The one possible solution of withdrawing the difficulties consists in including preprocess phase to find out and remove outliers, by using one of the techniques for detecting outliers. The broad review of such methods was presented in [47]. Such approach was proposed in [21], where it was considered application of Farthest First Traversal Algorithm, for data previously cleaned out from outliers by COF technique [48], to find out groups of students with similar learning style dimensions. That method occurred to be effective [21], however tutors had to determine the required number of learners' groups in advance. Good effects in case of data containing noise are obtained by combining modified K-means with the agglomerative hierarchical algorithm, proposed in [41]. Such approach allows not only to indicate instances containing noise but also objects that outstand from the others [40]. However the number of clusters is still required as input data.

Comparison of Methods. Performance of the techniques considered in the previous paragraphs, as well as some of their features were concluded on the basis of the experiments described in [40] and their characteristics. Table 1 shows all the algorithm features, which are important from tutors' point of view and clustering techniques fulfilling them.

The content of the Table 1 shows that none of the algorithms fulfill all the requirements. The most effective are combinations of two algorithms, however even if they deal with outliers, the final number of clusters is required as input data. Some more experimental results concerning presented methods, which confirmed information contained in the table will be described in Section 6.

In the present study, it is proposed to apply two-phase hierarchical algorithm, which deals with outliers and allows for determining weights for different learning styles dimensions, with number of clusters depending on a required threshold. The detailed algorithm will be presented in Section 5.

Table 1. Required features and clustering techniques

Feature	Technique
Final number of clusters not required	EM DBSCAN
Limits for maximal number of clusters	Only as final number of clusters: K-means Hierarchical Approach
Dealing with outliers	DBSCAN modif. K-means combined with hierarch. agglomerative
Attributes of mixed type	all, depending on distance function difficult choice of parameters for DBSCAN
Different significance of each attribute	all except of DBSCAN, by including weights into distance function
Self generating	all except DBSCAN
Order independent	all

5 Two Phase Hierarchical Clustering

In [49], it was considered building special clustering algorithm that will fulfill all the requirements of tutors for student grouping according to learning styles and usability features, and will be not too complex for educators to use.

5.1 Algorithm and Its Parameters

The aim of the proposed two-phase hierarchical algorithm was to obtain clusters of good qualities and fulfill tutors' requirements, defined in Section 4.2, at the same time. The review of clustering algorithms done in Section 4.3 showed that combinations of two algorithms performed better than each of them independently. The idea of proposed solution consists in dividing the process into two stages: the first one aims finding groups of students of big similarity and detecting the ones suspected to be outliers; during the second stage groups are merged into bigger ones if it is necessary and outliers are indicated. The first stage consists of a single layer clustering, during the second phase clusters are formed into hierarchical structure. The algorithm may be used for both of student models: SL defined by (1) or SM determined by (2).

The following parameters are necessary for the algorithm:

- Clustering threshold T
- Minimal clustering threshold $MINT$
- Maximal number of clusters $KMAX$.

T indicates minimal similarity of two students, who belong to the same group. *MINT* should be determined by taking into account the requirements connected with the considered courses and should guarantee that students grouped into the same clusters are of sufficiently similar characteristics. *MINT* value decides, how far from any of groups student should be to be qualified as the outlier. *KMAX* is connected with number of teaching paths, which together with different versions of layouts, a tutor is able to prepare for the educational process. All the outliers require individual paths as it always takes place with outstanding students.

The steps of the algorithm may be described as follows:

Phase I: Single Layer Clustering
Input: A set of N SM data, threshold T, maximal number of clusters $KMAX$, minimal clustering threshold $MINT$.
Output: Set of clusters SCM
Steps:

```
1. Assign SM_1 as the first cluster
2. Calculate the similarity between i-th student's SM
   and the centroid of each existing cluster,
3. Assign the student into the closest cluster, that
   the similarity is greater than T,
   calculate the new centroid of the cluster
   otherwise SM_i  initiates a new cluster
4. Repeat 2 and 3 for each student SM_i, i = 2,...N
5. If there exist clusters containing one element,
   repeat 2 and 3 until all clusters stabilize.
```

Phase II: Hierarchical Agglomerative Clustering
Output: The set of clusters SCM and the set of outliers.
Steps:

```
1. Remove all 1-element clusters  from SCM,
   let they consist the set SCM1
2. If number of clusters in SCM less equal KMAX
   go to 4,
3. Repeat: merge the closest clusters
   until KMAX is achieved
4. For each cluster in SCM1, find the closest in SCM
   if similarity of the centroids, greater than MINT
   merge them otherwise
   indicate 1-element clusters as outliers.
```

Instead of using required number of clusters, the algorithm is based on the clustering thresholds, which are easy to determine in advance, in the case of data of limited range and differences between their values. The hierarchical clustering algorithm of this kind was presented in [50] for web page clustering, with divisive hierarchical approach in the second phase. Wang proved [51] that this kind of hierarchical clustering algorithm is independent of the order of the input data, if they are well normalized. Normalization of data is guaranteed by the properly defined similarity measure, which should also allow for using weights and for application to data of mixed types, as it was stated at the beginning of this section. In the algorithm, there is applied similarity function, introduced by Gower [52], that may be used for both types of data: numerical and categorical, at the same time determined as follows:

Definition 1. *Let SM_i, SM_j denote two d-dimensional objects. Then the general similarity coefficient sim_g is defined as:*

$$sim_g(SM_i, SM_j) = \frac{1}{\sum_{k=1}^{d} w_k} \sum_{k=1}^{d} w_k s_k \ , \tag{4}$$

where s_k is the similarity of the k-th attribute of SM_i, SM_j and w_k is either one or zero depending on whether or not a comparison is valid for the k-th attribute of the two objects. For numerical attributes s_k is defined as

$$s_k = 1 - \frac{| \, sl_{i_k} - sl_{j_k} \, |}{R_k} \ , \tag{5}$$

where R_k is the range of k-th attribute. For categorical values s_k has the form:

$$s_k = \begin{cases} 1 & \textit{if both attributes have the same value,} \\ 0 & \textit{otherwise.} \end{cases} \tag{6}$$

Using of w_k enables assigning weights for different attributes, the ones not valid in the classification process should be equal to 0. Such defined similarity function imposes the way of calculating attribute values for the centroid of each cluster. The required values should be calculated differently for the attributes of numerical type and for the ones, which are categorical. For the first 4 attributes of SM model, their values for a cluster centroid are just the averages of the respective attribute values of all the cluster members. The last two attributes of SM model, of the cluster centroid, should represent values of the majority of cluster objects and will be defined by using frequency vectors [53].

Definition 2. *Let SLM denote the cluster of d-dimensional objects of mixed: numerical and categorical types. Let o means the number of all possible categorical values for attributes of SM (2). Then, as cluster frequency vector for each categorical attribute, we will consider o-dimensional vector $cf = [cf_1, cf_2, ..., cf_o]$, containing the number of occurrences of each categorical value $c_k, k = 1, 2, ..o$ of the considered categorical attribute in the cluster SCM. Then the cluster centroid $SCMC$ will be the d-dimensional object, which elements of numerical types are the average value of respective elements of all the cluster members. Centroid attribute values for variables of categorical type are equal to the categorical values of the maximal frequency for the respective attribute.*

Let the considered cluster contains M objects, then $i-th$ element of the centroid SCMC may be defined as:

$$SCMC_i = \frac{1}{M}\sum_{k=1}^{M} sl_{i_k} \ , \quad (7)$$

for $i = 1, ..., 4$ *and*

$$SCMC_i = c_{i_K} \ where\ K is\ such\ that\ cf_{i_K} = max\{cf_{i_k}, 1 \leq k \leq o\} \ , \quad (8)$$

where

$$\sum_{k=1}^{o} cf_{i_k} = M, \quad (9)$$

for $i = 5, 6$.

In case of the application of the proposed technique into the model limited to SL attributes, the similarity function will take the form of (4), and centroid attribute values will be calculated as the averages for all the cluster objects.

5.2 Algorithm Modification

As the algorithm, presented in the previous section, consists of two stages and the vector SM is built of two models: SL and SC, it may be considered different approaches for application of the technique. The first one consists in using all of the components of SM model in both of the phases, while in the second approach elements SL and SC are separated, to use independently in each phase of the algorithm. In that case single layer clustering is applied for components of SL, and then output clusters are merged according to SC components. In that case the parameters, may change their form and have the following meanings:

- Clustering threshold T depends on considered courses and is connected with sufficiently similar learning styles,
- Minimal clustering threshold $MINT$ concerns color preferences,
- Maximal number of clusters $KMAX$ is still connected with number of teaching materials combined with different layouts, which tutors are able to prepare together for the educational process.

That version of proposed algorithm takes the following form:

Phase I: Single Layer Clustering
Input: A set of N students' SL and SC data, threshold T, maximal number of clusters $KMAX$, minimal clustering threshold $MINT$.
Output: Set of clusters SCM
Steps:

1. Assign SL_1 as the first cluster and its centroid,
2. Calculate the similarity to SL_i(4),(5), and the centroid of each existing cluster,

```
3. Assign the student into the closest cluster, for which
   the similarity is greater than T
   otherwise SL_i initiates a new cluster
4. Repeat 2 and 3 for each student SL_i, i = 2,...N
5. If there exist 1-element clusters,
   repeat 2 and 3 until all clusters stabilize.
```

Phase II: Hierarchical Agglomerative Clustering
Output: The set of clusters *SCM* and the set of outliers.
Steps:

```
1. Remove all 1-element clusters from SCM,
   let they consist the set SCM1
2. If number of clusters in SCM less equal KMAX
   go to 4,
3. Repeat: merge the closest clusters
   until KMAX is achieved
4. For each cluster from SCM1, find the closest one in SCM
   if similarity of the centroids, is greater than MINT
   merge them otherwise
   indicate 1-element clusters as outliers.
```

The first phase and the proper choice of T should guarantee sufficient similarity of students' learning styles in each cluster. The second stage together with suitable *MINT* value should fulfill the requirements of the same color choices among the most possible number of students with similar dominant learning style preferences. Similarly to the previous version the measure defined by (4) ensures good data normalization and independence of the algorithm of input data order [51]. Using w_k enables to assign weights for attributes, the ones not valid in the classification process should be equal to 0. Weights may be also used while taking into account color ranks in student choices.

6 Experimental Results and Discussion

The aim of the experiments was to examine the performance of the proposed clustering technique for different student data sets depending on the choice of parameters. The investigations were done in two stages, in each of which, considering different models. The first one, in which students are represented by their learning style dimensions and the second one with attributes based on both: learning style preferences and color choices. The aim of the first stage was to check the effectiveness of the technique and to compare the results with the ones obtained by the other algorithms, for the simple model, represented by numerical attributes. The goal of the second stage was to examine the usage of the technique for the data of mixed type, taking also into account algorithm modification.

To obtain the most representative the possible test results, the special emphasis was done on the differentiation of student groups that took part in the experiments. Students were chosen from among those, who took part in collaboration online and used educational environment based on Moodle. Data were collected during three academic years from students attending different courses of engineering and master levels, different years of studies, including also part-time and evening courses. However all of the test participants studied Computer Science, but the ones from the second cycle of studies especially those who studied during weekends were graduated also in the other programmes.

6.1 Learning Style Model

During the experiments, student model SL, described by (1) was considered. There were taken into account two data sets of 125 and 71 instances, from among all of the real students' data. The second set consisted of volunteers who partcipated also in the tests concerning color preferences. The performance of the algorithm was checked, depending on the choice of parameters. The results were compared with the ones obtained by using well known K-means and hierarchical Farthest First Traversal algorithm, both from the Open Source Weka software [42].

Test results showed, for both of data sets that the threshold value has the big influence on the quality of results and the number of clusters obtained; for $T = 0.9$, the algorithm assigned all the students, except outliers, into one cluster after the first phase; if $T = 0.92$ the number of clusters created was equal to two, and all the assignments were also done during the first phase. In the case of the high value of T, parameter $KMAX$ may decide of the final number of clusters. Table 2 presents the final number of clusters, depending on the assumed threshold value, for both considered data sets: A of 71 instances and B of 123 instances. As it may be easily noticed for both of the cases, obtained effects are similar, however greater number of instances entails in greater number of clusters.

Threshold value depends on the required similarity of the attributes and, in fact, in the case of learning style dimensions may be easily calculated. If we assume that the average difference between student learning style attributes, should be less than 4 , after summing up, it will give 16 for all the attributes, then taking into account equations (4) and (5) we will obtain that the similarity

Table 2. Number of clusters depending on the threshold value

Threshold	Data set	Number of clusters
0.92	Set A (71)	1
	Set B (123)	2
0.94	Set A	2
	Set B	3
0.96	Set A	5
	Set B	5

Table 3. Number of instances for different cluster numbers(data set A)

Number of clusters	Algorithm	Instances
2	Two-phase	48,23
	FFT	67,4
	K-means	36,35
5	Two-phase	31,15,19,3,3
	FFT	49,3,8,7,4
	K-means	20,13,12,21,5

(threshold) cannot be less then 0.96. If our similarity requirements are not so strict, the assumed threshold may be less, for example threshold of 0.9 is connected with the attribute differences not greater than 8. Tutors can easily determine that value in advance, depending on their requirements. The number of obtained clusters cannot be easily determined as they are connected not only with the threshold, but also with the data structure.

Examination of the technique, using different values of weights, showed the big influence of each dimension on the final effects. Setting any of weights equal to zero, completely changed the structure of obtained groups. It means that in ane e-learning system, there should be taken into account only these dimensions that are expected to have an influence on students' perception and activity during the course.

The algorithm expressed also the big ability for finding outliers, comparing with the performance of Connectivity Outlier Factor algorithm considered in [21], especially taking into account difficulties with the choice of parameters, that will guarantee the success. In the presented approach $MINT$ value, which decides on the final indications of outliers, may be easily calculated similarly to the threshold T .

The structure of the groups received (clusters of different sizes), was similar to the ones obtained by hierarchical Farthest First Traversal, while K-means divided students into almost equal clusters, what is presented in Table 3 and Table 4.

6.2 Extended Student Model

The experiments were done for two data sets: the one of Computer Science students' data and the other artificially generated; their goal was to compare the performance of two versions of the two-phase approach. In the first version all the attributes were used in both phases, while in the second one, they were separated, for applying two-phase algorithm modification.

As the first data set, the sample group of 71 students was considered. Students examined color compositions of the portal layout, where their activities took place. They were asked to choose 2 from 9 colors as preferable and indicate ranks for them. They filled also ILS questionnaire. The second data set, of 73 instances, was generated artificially.

Table 4. Number of instances for different cluster numbers(data set B)

Number of clusters	Algorithm	Instances
2	Two-phase	116,5
	FFT	108,15
	K-means	68,55
3	Two-phase	67,23,33
	FFT	88,12,23
	K-means	43,40,40
5	Two-phase	63,33,19,5,3
	FFT	83,9,22,5,4
	K-means	27,21,30,18,27

The performance of the proposed method was examined by evaluation of the quality of obtained clusters, taking into account two points of view: color choices and learning styles. In the first case, it is measured by number of instances with different color preferences assigned into the same clusters, while in the second one the quality is measured by differences among dominant learning styles dimensions inside clusters. Research was done for different number of clusters as well as depending on inclusion of color rank into weight coefficients w_k. Thresholds T and $MINT$ were experimentally chosen (in case of the model including categorical attributes calculating the threshold value is not so obvious as it was for numerical attributes) and constant during all experiments: T equal to 0.92 in the first version and 0.96 in the second one, for the first data set, and respectively 0.78 and 0.93 for artificially generated data; $MINT$ was equal to 0.75 in all the cases.

Table 5 presents percentage of students, whose color preferences differ from the majority of the cluster, for two versions of the algorithm, various number of clusters, taking or not taking into account color ranks. Result analysis shows that the best quality of clusters, from the point of view of color preferences, was obtained by the first version of the algorithm, what means using all the attributes during the whole clustering process; and for different number of clusters, in case of usage of color ranks. Though on the other hand distinguishing the first color choice may have bad influence on cluster quality connected with the other attributes. Comparison of percentage values for both considered versions, and both data sets, shows that using of color preferences attributes in the first phase of the algorithm definitely improves the quality of obtained clusters.

Quality of clusters, from the point of view of learning style dimensions, presents completely disparate image. Detailed analysis of obtained groups, for both of the data sets, shows that, in some of them, assigned objects are characterized by learning style dimensions of significantly different score values. For example, in case of the first version of the algorithm with number of obtained clusters equal to 5, and including ranks, in one of the cluster of 13 students, the strongly active ones together with strongly reflective as well as sequential together with global were placed. It may mean that the second version gives better results from the point

Table 5. Percentage of cluster objects with different color preferences

Algorithm	Rank	Number of clusters	Real data	Artificial data
VER.1	No	3	25.35%	30.14%
		4	12.68%	30.14%
		5	14.08%	24.66%
	Yes	3	28.17%	31.51%
		4	19.72%	30.14%
		5	11.27%	26.03%
VER.2	No	3	28.17%	36.99%
		4	26.76%	34.25%
		5	40.84%	34.25%
	Yes	3	28.17%	32.88%
		4	28.17%	32.88%
		5	40.84%	32.88%

of view of learning styles, while the first one, if focus is done on color preferences, however, to obtain general conclusions further investigations may be necessary.

7 Conclusion and Future Research

In the chapter, it was considered application of cluster analysis for student grouping according to their dominant learning styles and usability preferences. In the examined student model, as usability preferences, there were considered color choices. It was proposed intelligent e-learning system, in which personalized teaching paths as well as information content organization are adjusted to student groups of different requirements. Using of unsupervised classification enables to create student groups dynamically, depending on the courses they attend. Presented investigations focused on finding clustering technique, which guarantee clusters of good quality and fulfils tutor requirements at the same time. The review of well known algorithms enabled to indicate their main disadvantages, regarding an application in e-learning system as well as tutors requirements. It may be concluded that the new approach is necessary.

To fulfill the presented conditions, it was proposed to apply the two-phase hierarchical algorithm for students' grouping. That specially constructed technique enables to find student groups according to their individual learning style and usability preferences, what is more, all the students who differ significantly from their colleagues (outliers) are indicated. Tutors do not need to determine the exact number of students' groups in advance. The technique is based on the clustering thresholds, which decide on cluster qualities. The maximal number of required clusters, is only the upper limit depending on the number of possible teaching paths or personalized layouts, which may be created. For the purpose of the tests there were taken rather small values, but in reality the parameter may be not limited, if there are such possibilities. The proposed similarity function allows not only to differentiate the importance of every attribute, by using weights or even not taking into account certain dimensions, but it is also adapted for the usage of the data of mixed type.

In the chapter, it was also considered usage of the modified version of the two-phase hierarchical clustering algorithm, in which, in each of the phases different attributes are taken into account. In the first stage clusters are built according to learning style preferences, while in the second one similarity on the basis of usability preferences is considered. Investigations showed that different versions of the algorithm should be chosen depending on which part of students' model the emphasis is done, however further research concerning the role of weights may be also useful.

Experiments, showed the good performance of the algorithm, for student grouping according to their preferences, taking into account also the data sets containing outliers. Experiments done for real and artificial data allowed to indicate parameters, which guarantee the best results, but some modifications of the algorithm and further research concerning choice of parameters may be necessary.

Proposed student model, depends on color preferences as usability needs. In the next step development of models by taking into account other preferences, like navigational paths may be considered. The future research should also consist of further investigations of the examined algorithm and its possible modifications, with the emphasis on the choice of parameters, application of different similarity functions, for different data sets, as well as of comparing its effectiveness with other algorithms. Obtaining students' clusters of good quality is the first stage. Then there should be investigated defining representative features for each group, what will enable tutors to personalize the system. However, many research needs to be done also in that area.

References

1. Brusilovsky, P., Peylo, C.: Adaptive and intelligent web-based educational systems. International Journal of Artificial Intelligence in Education 13, 156–169 (2003)
2. Santally, M.I., Alain, S.: Personalisation in web-based learning environments. International Journal of Distance Education Technologies 4, 15–35 (2006)
3. Stash, N., Cristea, A., De Bra, P.: Authoring of learning styles in adaptive hypermedia: Problems and solutions. In: Proc. WWW Conf., N.Y., pp. 114–123 (2004)
4. Brusilovsky, P.: Adaptive hypermedia. Use Model. User-Adap. 11, 87–110 (2001)
5. Lu, J., Yu, C.S., Liu, C.: Learning style, learning patterns and learning performance in a WebCT-based MIS course. Inform. Manage. 40, 497–507 (2003)
6. Lee, M.: Profiling students adaptation styles in web-based learning. Comput. Educ. 36, 121–132 (2001)
7. Beaudoin, M.F.: Learning or Lurking? Tracking the "Invisible" Online Student. Internet & Higher Educ. 5, 147–155 (2002)
8. Li, Z., Sun, Y., Liu, M.: A web-based intelligent tutoring system. In: Artificiall Intelligence and Innovations AIAI 2005. IFIP International Federation for Information Processing, vol. 187, pp. 583–591. Springer, Boston (2005)
9. Cha, H.J., Kim, Y.S., Park, S.H., Yoon, T.B., Jung, Y.M., Lee, J.-H.: Learning Styles Diagnosis Based on User Interface Behaviors for Customization of Learning Interfaces in an Intelligent Tutoring System. In: Ikeda, M., Ashley, K.D., Chan, T.-W. (eds.) ITS 2006. LNCS, vol. 4053, pp. 513–524. Springer, Heidelberg (2006)

10. Gilbert, J.E., Han, C.Y.: Adapting Instruction in Search of a Significant Difference. J. Netw. Comput. Appl. 22, 149–160 (1999)
11. Saarikoski, L., Salojärvi, S., Del Corso, D., Ovein, E.: The 3DE: an environment for the development of learner-oriented customized educational packages. In: Proc. of ITHET 2001, Kumamoto (2001)
12. Triantafillou, E., Pomportsis, A., Georgiadou, E.: AES-CS: Adaptive educational system base on cognitive styles. In: AH Workshop, Malaga, pp. 10–20 (2002)
13. Gonzalez-Rodriguez, M., Manrubia, J., Vidau, A., Gonzalez-Gallego, M.: Improving accessibility with user-tailored interfaces. Appl. Intell. 30, 65–71 (2009)
14. Shen, R., Han, P., Yang, F., Yang, Q., Huang, J.: Data mining and case-based reasoning for distance learning. Journal of Distance Education Technologies 1, 46–58 (2003)
15. Talavera, L., Gaudioso, E.: Mining student data to characterize similar behavior groups in unstructured collaboration spaces. In: Workshop on Artificial Intelligence in CSCL. 16th European Conference on Artificial Intelligence, pp. 17–23 (2004)
16. Tang, T., McCalla, G.: Student modeling for a web-based learning environment: A data mining approach. In: Eighteenth National Conference on Artificial Intelligence, Menlo Park, CA, USA, pp. 967–968 (2002)
17. Lu, F., Li, X., Liu, Q., Yang, Z., Tan, G., He, T.: Research on personalized e-learning system using fuzzy set based clustering algorithm. In: Shi, Y., van Albada, G.D., Dongarra, J., Sloot, P.M.A. (eds.) ICCS 2007, Part III. LNCS, vol. 4489, pp. 587–590. Springer, Heidelberg (2007)
18. Merceron, A., Yacef, K.: Clustering students to help evaluate learning. In: Courtiat, J.-P., Davarakis, C., Villemur, T. (eds.) Technology Enchanced Learning, vol. 171, pp. 31–42. Springer, Heidelberg (2005)
19. Perera, D., Kay, J., Koprinska, I., Yacef, K., Zaïane, O.R.: Clustering and sequential pattern mining of online collaborative learning data. IEEE T. Knowl. Data En. 21, 759–772 (2009)
20. Romero, C., Ventura, S.: Educational data mining: a survey from 1995 to 2005. Expert Syst. Appl. 33, 135–146 (2007)
21. Zakrzewska, D.: Cluster Analysis for Building Personalized E-learning System. Pol. J. Environ. Stud. 16, 330–334 (2007)
22. Zakrzewska, D.: Using Clustering Technique for Students' Grouping in Intelligent E-Learning Systems. In: Holzinger, A. (ed.) USAB 2008. LNCS, vol. 5298, pp. 403–410. Springer, Heidelberg (2008)
23. Felder, R., Brent, R.: Understanding student differences. J. Eng. Educ. 94, 57–72 (2005)
24. Graf, S., Kinshuk: Considering learning styles in learning managements systems: investigating the behavior of students in an online course. In: Proc. of the 1st IEEE Int. Workshop on Semantic Media Adaptation and Personalization, Athens (2006)
25. Rovai, A.P.: The relationships of communicator style, personality-based learning style and classroom community among online graduate students. Internet & Higher Education 6, 347–363 (2003)
26. Alfonseca, E., Carro, R.M., Martin, E., Ortigosa, A., Paredes, P.: The impact of learning styles on student grouping for collaborative learning: a case study. Use Model. User-Adap. 16, 377–401 (2006)
27. García, P., Amandi, A., Schiaffino, S., Campo, M.: Evaluating Bayesian networks' precision for detecting students' learning styles. Comput. Educ. 49, 794–808 (2007)

28. Xu, D., Wang, H., Su, K.: Intelligent student profiling with fuzzy models. In: Proc. of HICSS 2002, Hawaii (2002)
29. Viola, S.R., Graf, S., Kinshuk, L.T.: Investigating Relationships within the Index of Learning Styles: a Data Driven Approach. Interactive Technology & Smart Education 4, 7–18 (2007)
30. Kolb, D.A.: Experiential Learning: Experience as a Source of Learning and Development. Prentice-Hall, Englewood Cliffs (1984)
31. Honey, P., Mumford, A.: The Manual of Learning Styles. Peter Honey, Maidenhead (1986)
32. Dunn, R., Dunn, K.: Teaching Students Through Their Individual Learning Styles: A Practical Approach. Reston Publishing, Reston (1978)
33. Felder, R.M., Silverman, L.K.: Learning and Teaching Styles in Engineering Education. Eng. Educ. 78, 674–681 (1988)
34. ILS Questionnaire, `http://www.engr.ncsu.edu/learningstyles/ilsweb.html`
35. Zakrzewska, D., Wojciechowski, A.: Identifying Students Usability Needs in Collaborative Learning Environments. In: 2008 Conference on Human System Interaction, Krakow, pp. 862–867 (2008)
36. Marcus, A.: Designing Graphical Interfaces. Unix World (October 1990)
37. Bauersfeld, P.F., Slater, J.L.: User-Oriented Color Interface Design: Direct Manipulation of Color in Context. In: SIGCHI Conference on Human Factors in Computing Systems: Reaching through Technology, New Orleans, pp. 417–418 (2001)
38. Romero, C., Ventura, S., Garcia, E.: Data mining in course management systems: Moodle case study and tutorial. Comput. Educ. 51, 368–384 (2008)
39. Han, J., Kamber, M.: Data Mining. Concepts and Techniques, 2nd edn. Morgan Kaufmann Publishers, San Francisco (2006)
40. Zakrzewska, D., Ruiz-Esteban, C.: Cluster analysis for students profiling. In: Proc. 11th Int. Conf. SMC, pp. 333–338. EXIT, Warsaw (2005)
41. Jiang, M.F., Tseng, S.S., Su, M.M.: Two-phase clustering process in outliers detection. Pattern Recogn. Lett. 22, 691–700 (2001)
42. Witten, I.H., Frank, E.: Data Mining: Practical machine learning tools and techniques, 2nd edn. Morgan Kaufmann Publishers, San Francisco (2005)
43. Ester, M., Kriegel, H.-P., Sander, J., Xu, X.: A density- based algorithm for discovering clusters in large spatial databases with noise. In: Proc. of the 2nd Int. Conf. on Knowledge Discovery and Data Mining, Portland, pp. 226–231 (1996)
44. Heffler, B.: Individual learning style and the learning style inventory. Educ. Stud. 27, 307–316 (2001)
45. Hochbaum, S.D., Shmoys, B.D.: A best possible heuristic for the k-center problem. Math. Oper. Res. 10, 180–184 (1985)
46. Dasgupta, S.: Performance guarantees for hierarchical clustering. In: Kivinen, J., Sloan, R.H. (eds.) COLT 2002. LNCS (LNAI), vol. 2375, pp. 235–254. Springer, Heidelberg (2002)
47. Hodge, V.J., Austin, J.: A survey of outlier detection methodologies. Artif. Intell. Rev. 22, 85–126 (2004)
48. Tang, J., Chen, Z., Fu, A.W., Cheung, D.W.: Enhancing effectiveness of outlier detections for low density patterns. In: Chen, M.-S., Yu, P.S., Liu, B. (eds.) PAKDD 2002. LNCS (LNAI), vol. 2336, pp. 535–548. Springer, Heidelberg (2002)
49. Zakrzewska, D.: Cluster Analysis for Users' Modeling in Intelligent E-learning Systems. In: Nguyen, N.T., Borzemski, L., Grzech, A., Ali, M. (eds.) IEA/AIE 2008. LNCS (LNAI), vol. 5027, pp. 209–214. Springer, Heidelberg (2008)

50. Zha, Y., Yu, J.X., Hou, J.: Web Communities. Analysis and Construction. Springer, Heidelberg (2006)
51. Wang, L.: On competitive learning. IEEE T. Neural Networ. 8, 1214–1217 (1997)
52. Gower, J.: A general coefficient of similarity and some of its properties. Biometrics 27, 857–874 (1971)
53. Gan, G., Ma, Ch., Wu, J.: Data Clustering: Theory, Algorithms, and Applications. ASA-SIAM. SIAM, Philadelphia (2007)

Agent-Based Modelling and Analysis of Air Traffic Organisations

Alexei Sharpanskykh

Vrije Universiteit Amsterdam, Department of Artificial Intelligence,
De Boelelaan 1081a, 1081 HV Amsterdam, The Netherlands
sharp@few.vu.nl

Abstract. A modern Air Traffic Organisation (ATO) represents a complex organisation that involves many parties with diverse goals performing for a wide range of tasks. Due to the structural and behavioural complexity of ATOs, mistakes and performance problems are not rare in such organisations. Some of these faults may seriously affect safety, causing incidents. Therefore, the possibility to perform detailed and reliable analysis is of primary importance for ATOs. To this end, this chapter introduces an automated approach for modelling and analysis of complex ATOs. The developed model incorporates all important structural and behavioural aspects of an ATO. The approach is illustrated by a case study in which ATO's tasks related to movement of aircraft on the ground and to safety occurrence reporting are considered.

1 Introduction

An Air Traffic Organisation (ATO) ensures a safe and efficient flow of aircraft both at airports and in the air. An ATO represents a complex organisation that involves many parties with diverse goals performing a wide range of tasks. Among the ATO's participants are airports, air navigation service providers (ANSP), airlines, regulators, and the government. Due to the high complexity, inconsistencies and performance bottlenecks often occur in ATOs. Some of these faults may result into performance issues, whereas others can seriously affect safety, causing incidents. Therefore, the possibility to perform detailed and reliable automated analysis aiming at detecting safety hazards in ATOs is of primary importance in the air traffic domain. Currently, formal risk assessment approaches [6, 11] are based predominantly on fault/event trees used for sequential cause-effect reasoning for accident causation. However, such trees do not encounter for complex, non-linear dependencies and dynamics inherent in ATOs. Advantages of agent-based organisation modelling that allows investigating complex emergent dynamics of a system are increasingly recognized in the domain. In particular, agent-based modelling has been proposed as a means to assess safety risk of complex emergent dynamics of air traffic operations [15, 16]. These studies focus on the risk of air traffic operations and use a plain society of agents, without considering the organisational layer. Several approaches [8, 11] consider an influence of different organisational aspects on safety, however, without providing precise details. This chapter presents the first attempt to create a formal agent-based

N.T. Nguyen and E. Szczerbicki (Eds.): Intel. Sys. for Know. Management, SCI 252, pp. 251–274.
springerlink.com

organisational model of an ATO using the developed previously organisation modelling and analysis framework [13] and the methodology from [16]. On the one hand, the framework allows specifying the prescriptive structural and behavioural dependencies of an organisation. On the other hand, it provides means to describe autonomous behaviour of the organisational actors. In contrast to many existing enterprise modelling approaches (CIMOSA [4]; ARIS [12]) this framework has a formal basis, which enables reliable analysis of models. More specifically, to express structural relations, sorted predicate logic-based languages are used, whereas the Temporal Trace Language (TTL) [2] is used for specifying dynamic aspects of organisations.

Also, more limited languages dedicated for automated modelling and analysis of particular aspects of organisations have been developed: process-oriented modelling techniques [18], organisational performance evaluation techniques [17]. However, modelling of particular organisational aspects does not allow defining interdependencies between different perspectives on organisations and to investigate a combined influence of factors from different perspectives on the organisational behavior. By considering multiple perspectives of organisations the designer is equipped with more rigorous and manifold analysis possibilities than by using analysis techniques dedicated to a particular view only.

In [5] an integrated framework for process and performance modelling is described that incorporates accounting/business parameters into a formal process modelling approach based on Petri-nets. However, key aspects as authority and power relations, organisational and individual goals, individual behavior are not considered. Another formal framework for business process modelling is described in [7] focusing on the formal goal-oriented modelling using situation calculus. Modelling and analysis of processes and other organisational concepts are not properly addressed in this framework. A formal framework for verifying models specified in Unified Enterprise Modelling Language (UEML) is proposed in [3]. It identifies a general idea to use conceptual graphs for verifying enterprise models; however, neither technical nor experimental results are provided to support this idea.

In the framework used in this chapter four interrelated views are distinguished: the *performance-oriented view* describes organisational goal structures, performance indicators structures, and relations between them; the *process-oriented view* describes organisational functions, processes, resources and relations between them; the *organisation-oriented view* describes organisational roles, their authority, responsibility and interraction relations; the *agent-oriented view* describes agents' types with their capabilities, and principles of allocating agents to roles.

The framework proposes a number of analysis techniques, the application of some of which is demonstrated in this chapter. Specifically, the constructed specification of the ATO is verified for correctness by applying the general and specific for particular views consistency verification techniques from [2, 13]. Further, the consequences of different types of agent behaviour that diverges from the prescriptive (formal) organisational specification are simulated.

The chapter is organized as follows. Section 2 introduces the description of the ATO under consideration. In Section 3 the developed model for the formal organisation is presented. A specification of the organisational agents is described in Section 4. The results of the correctness verification of the designed ATO specification are presented in Section 5. The results of the analysis by simulation are considered in Section 6. Section 7 concludes the chapter.

2 Case Study

The ATO performs a variety of tasks: development and evaluation of new air traffic operations (e.g., introduction of a new runway on an aerodrome), movement of aircraft on the ground, safety occurrence reporting and investigation, etc. In this paper we focus particularly on the ATO tasks related to movement of aircraft on the ground and safety occurrence reporting. More specifically, the taxiing of an aircraft to a designated runway and the subsequent take off from this runway are considered using input from [14, 15]. Furthermore, reporting of safety occurrences during taxiing operations near an active runway are investigated.

During the taxiing, an aircraft moves from one sector of the airport to another, until it reaches the runway designated for take off. The crew of an aircraft consists of the pilot-in-command and the second pilot. The monitoring and control over the traffic in a sector is performed by a dedicated ground controller. Also, the control over aircraft on a runway and in its surroundings is performed by a dedicated runway controller. During the taxiing control over an aircraft is handed over from one controller to another, depending on the physical position of the aircraft. Before crossing a runway on its way, the crew of a taxiing aircraft should request the controller responsible for the runway for clearance. Only when the clearance is provided, the aircraft is allowed to cross. The same holds for the take off operation. Controllers may be situated in the same or in different towers at the airdrome, each of which is guided by a Tower Controllers Supervisor.

In this operational context, safety-relevant events may occur, e.g. taxiing aircraft initiates to cross due to misunderstanding in communication. To support safety management, such events should be reported by the involved pilots and controllers. In this case we consider reporting that occurs either via formal organisational lines or via informal coordination. The formal organisation considers safety occurrence reporting at the air navigation service provider (ANSP) and at airlines, the informal path considers coordination between air traffic controllers.

The formal occurrence reporting at the ANSP starts by the creation of a notification report by the involved controller(s). This notification report is examined and possibly improved by the supervisor. The notification report is processed by the safety investigation unit (SIU) of the ANSP. The severity of the occurrence is assessed and a description of the event is stored in a safety occurrences database. In the case of single severe occurrences or in the case of a consistent series of less severe occurrences, the SIU may initiate an investigation for possible causes. The investigation results are reported to the operation management team at the ANSP.

The organisation of the safety occurrences processing at the airline starts with a notification report created by the pilots. This notification report may be provided to the airline's safety management unit or it may be directly provided to the regulator (a governmental organisation). The airline's safety management unit examines and potentially improves the report and it informs the regulator about safety occurrences at the airline. The regulator may decide on further investigation of safety occurrences by itself or by a facilitated external party.

The informal safety occurrence reporting path at the ANSP considers that controllers discuss during breaks the occurrences that happened in their shifts. If they identify potential important safety issues they inform the head of controllers, who is a member of the operation management team. This team may decide on further investigation of the issue.

3 Modelling the Formal Air Traffic Organisation

To perform analysis of the ATO's structures and processes both the specification of the formal organisation and the specifications of agents and the principles of their allocation to the roles should be developed. To design such specifications a sequence of design steps is identified in [16]. The formal organisational specification is built by executing the steps 1-9 described below. Agents that cause performance variability in the ATO are specified in the next section.

In general, organisation modelling is a challenging task that requires a close investigation of organisational documents (e.g., policies, job prescriptions), interaction with organisational actors (e.g., by organising interviews and formulating questionnaires).

Step 1. The identification of the organisational roles
In this step organisational roles are identified, both simple and composite ones and subrole-relations are established. A *role* represents a set of functionalities of (part of) an organisation abstracted from specific actors who fulfil them. Each role can be composed by several other roles. A role composed of subroles is called a composite role. In the considered organisation roles can be represented at three aggregation levels (see Fig. 1-3). For example, at the aggregation level 1 the Air Navigation Service Provider is considered as one composite role interacting with other roles. The subroles of the Air Navigation Service Provider (e.g., the Control Unit) are described at the aggregation level 2, and so forth. Note that based on the introduced generic roles role instances may be defined for particular applications (e.g., in simulations). Each role has an input and an output interface facilitating in interaction with other roles.

A special role type is the environment (env). The environment for the case study consists of two sectors of an airdrome, each of which is controlled by the corresponding ground controller role. The sectors adjoin a runway that is in control of the runway controller role.

Step 2. The specification of the interactions between the roles
In this step, interaction relations between roles, roles and the environment are identified. To specify interaction relations, the interfaces of the roles and the environment are formalised by interaction ontologies. For a role, input and output ontologies are specified referred to as interface ontologies, which are used to describe interactions with other roles. Generally speaking, an input ontology determines what types of information are allowed to be transferred to the input of a role (or of the environment), and an output ontology predefines what kinds of information can be

generated at the output of a role (or of the environment). For specifying communications the interface ontologies for all roles include the following predicate:

communication_from_to: ROLE x ROLE x MSG_TYPE x CONTENT

Here the first argument denoted the role-source of information, the second – the role-recipient of information, the third argument denoted the types of the communication (which may be one of the following {observe, inform, request, decision, readback}) and the fourth – the content of the communication. The sort ROLE is a composite sort that comprises all subsorts of the roles of particular types (e.g., CONTROLLER). The sort CONTENT is also the composite sort that comprises all names of terms that are used as the communication content. Such terms are constructed from sorted constants, variables and functions in the standard predicate logic way.

Relations between roles are represented by interaction and interlevel links. An interaction link is an information channel between two roles at the same aggregation level. An interlevel link connects a composite role with one of its subroles. It represents information transition between two adjacent aggregation levels. The interaction relations for the ATO have been identified and formalised at each aggregation level (see Fig. 1-3). An ANSP interacts with Crews of aircraft being guided by air traffic controllers. A regulator performs safety assessment of ANSP's operations and procedures regularly. In some ANSPs Ministry of Justice may be involved in the safety investigation of severe occurrences.

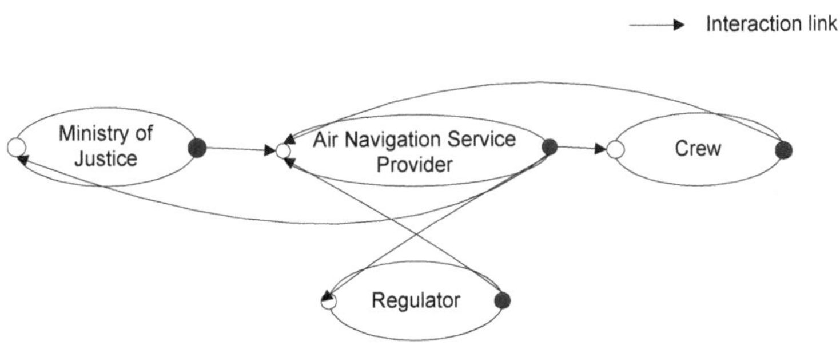

Fig. 1. The interaction relations between the generic roles at the aggregation level 1

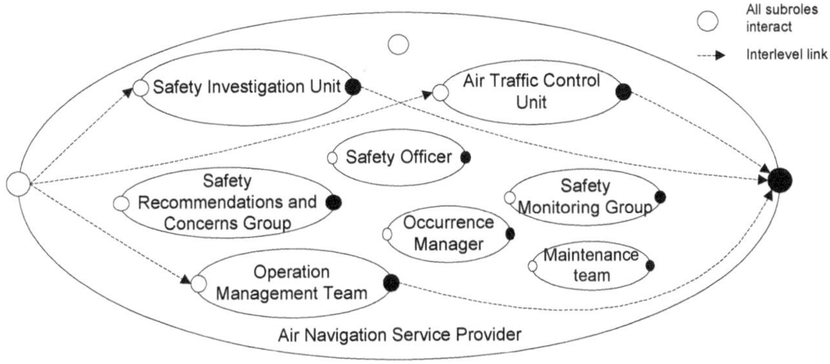

Fig. 2. The role Air NavigationService Provider considered at the aggregation level 2

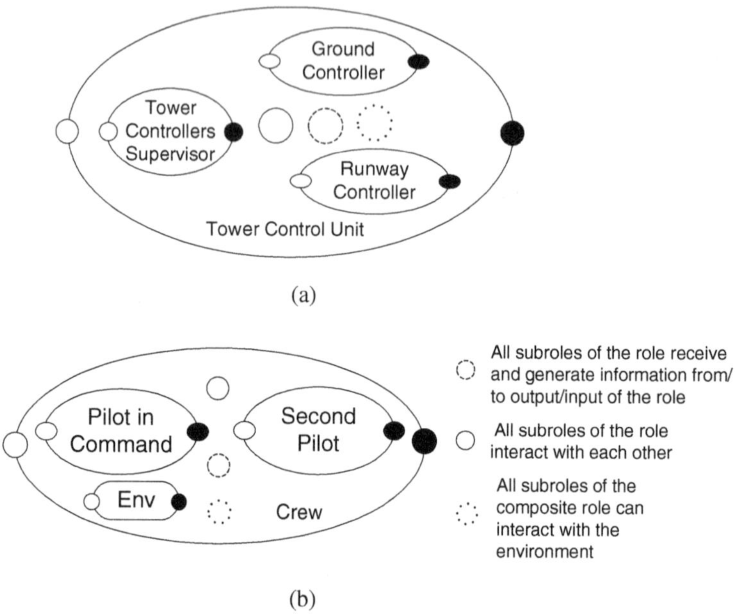

Fig. 3. Interaction relations at the aggregation level 3: (a) within the Tower Control Unit (subrole of the ANSP) and (b) within Crew role (subrole of the Airline).

Step 3. The identification of the requirements for the roles
In this step the requirements on knowledge, skills and personal traits of the agent implementing a role at the lowest aggregation level are identified. Knowledge-related requirements define facts and procedures that must be well understood by an agent. Skills describe developed abilities of agents to use their knowledge for tasks performance. For example the following requirements for the air traffic controller role are defined: (1) passed a medical examination; (2) 2 or 4 year college degree before initiation of training; (3) thorough knowledge of the air traffic management system and the flight regulations; (4) computer training; (5) air traffic control training; (6) excellent listening and communication skills; (7) quick decision-making skills.

Step 4. The identification of the organisational performance indicators and goals
In this step, organisational goals, performance indicators (PIs) and relations between them and organisational roles are identified. A *PI* is a quantitative or qualitative indicator that reflects the state/progress of the company, unit or individual. PIs can be hard (e.g., taxiing time) or soft, i.e., not directly measurable, qualitative (e.g., level of collaboration between controllers). PIs can be related through various relationships. The following relations are considered in the framework: (strongly) positive/negative causal influence of one PI on another, positive/negative correlation between two PIs, aggregation – two PIs express the same measure at different aggregation levels.

Goals are objectives that describe a desired state or development and are defined as expressions over PIs. The characteristics of a goal include, among others: *priority*; *horizon* – for which time point/interval should the goal be satisfied; *hardness* – hard or soft (for which instead of satisfaction, degrees of *satisficing* are defined). A goal can be refined into subgoals forming a hierarchy. Some examples of the goals and PIs of the ATO are given in Table 1. Here goals 25, 26, 27, 28, 16 and 21 are subgoals of goal 10, and the PIs, on which these subgoals are based, are related to the PI of goal 10 by the aggregation relation. Goals are related to roles. For example, goal 10 is associated with Airline, Tower Control Unit and Safety Investigation Unit roles.

Step 5. The specification of the resources

In this step organisational resource types and resources are identified, and characteristics for them are provided, such as: *name*, *category*: discrete or continuous, *measurement unit*, *expiration duration*: the time interval during which a resource type

Table 1. Examples of goals and PIs of the ATO

#	Goal	Based on the PI
10	It is required to maintain a high level of safety of execution of tasks related to the air traffic management	The level of safety of execution of tasks related to the air traffic management
16	It is required to maintain a high level of robustness and unambiguousness of the control (coordination) structure for the execution of tasks	The level of robustness and unambiguousness of the control (coordination) structure for the execution of tasks
20	It is required to maintain a sufficient level of autonomy of decision making and the operation execution for the roles involved into the air traffic management	The level of autonomy of decision making and the operation execution for the roles involved into the air traffic management
21	It is required to maintain unambiguousness, consistency, correctness and timeliness of information exchanged between agents	The unambiguousness, consistency, correctness and timeliness of information exchanged between agents
23	It is desired to increase the volume of passengers, departing/arriving from/to an airport	The volume of passengers, departing/arriving from/to an airport
25	It is required to maintain a high level of conformance of all roles involved into the air traffic management to the formal norms and regulations defined for their tasks	The level of conformance of all roles involved into the air traffic management to the formal norms and regulations defined for their tasks.
26	It is required to maintain a high (sufficient) level of proficiency of pilots	The level of proficiency of pilots
27	It is required to maintain a high (sufficient) level of proficiency of controllers	The level of proficiency of controllers
28	It is required to maintain the high quality and reliability of the hardware used in the air traffic control management	The quality and reliability of the hardware used in the air traffic control management

Table 2. Examples of the tasks of the ATO in relation to goals and resources

#	Task name	Uses	Produces	Durations
1	Taxiing the aircraft to the designated runway	All resources of the subtasks	All resources of the subtasks	Depends on the durations of subtasks
1.1	Taxiing the aircraft on a taxiway	Airport's diagram, the taxi instructions, compass, radar, aircraft	-	Depends on a particular taxiway
Goal: 29, 30, 31, 32, 33, 34, 20				
1.2	Switching to the frequency of another controller	Data about the new frequency	-	Min: 1 sec Max: 5 sec
Goal: 30, 31, 33				
1.3	Inquiry for the clearance for crossing an active runway	Observations, the taxi instructions,communication R/T system	A request for clearance	Min: 2 sec Max: 5 sec
Goal: 30, 31, 33				
1.4	Making and communicating the decision on a request for crossing a runway	Data about the current state of the runway, a request for for clearance to cross, communication R/T system	'Position and hold' or 'clearance is provided'	Min: 3 sec Max: 11 sec
Goal: 35, 36, 32, 34, 20				
1.5	Crossing a runway	Clearance to cross, airport's diagram, taxiing instructions, radar	'Clear of the runway'	Min: 30 sec Max: 60 sec
Goal: 30, 31, 34, 33, 20				
2	Safety occurrence reporting and the report handling	All resources of the subtasks	All resources of the subtasks	Depends on the duration of the subtasks
2.1	Create a notification report	The observation from the environment of an occurrence that may be classified as an incident/accident, the incident classification database	A notification report	Min:3 min Max: 24 hours
2.2	Preliminary assessment of an occurrence	A processed notification report	A preliminary safety occurrence assessment report	For severe occurrences: Max: 48 hours For less severe occurrences: Max: 72 hours
2.3	Investigation of an occurrence	A preliminary safety occurrence assessment report, additional data about the occurrence (optional)	An interrim safety occurrence assessment report	Min: 3 days Max: 90 days

can be used; *location*; *sharing*: some processes may share resources. Examples of resource types of the ATO are: airport's diagram, aircraft, incident classification database, clearance to cross a runway, an incident investigation report.

Step 6. The identification of the organisational tasks, the relations between the tasks, and relations between the tasks, the resources and the goals
A *task* represents a function performed in the organisation and is characterized by *name*, *maximal* and *minimal duration*. Tasks can be decomposed into more specific ones using AND- and OR-relations forming hierarchies. Each task performed in an organisation should contribute to the satisfaction of one or more organisational goals. Examples of the ATO's tasks in relation to goals and resources are given in Table 2.

Step 7. The specification of the authority relations
In this step authority (i.e., informal power) relations of an organisation are identified: superior-subordinate relations on roles with respect to tasks, responsibility relations, control for resources, authorization relations. Organisational roles may have different rights and responsibilities with respect to different aspects of task execution, such as execution, passive monitoring, consulting, making technological decisions (i.e., decisions that concern technical questions related to the task content) and making managerial decisions (i.e., decisions that concern general organisational issues related to the task). Examples of responsibility relations in the air traffic organisational model are presented in Table 3.

Step 8. The specification of the flows of control
In this step dynamic structures (called workflows or flows of control) are defined that represent temporal execution sequences of processes of an organisation in particular scenarios. The framework allows representing all commonly used workflow templates

Table 3. The responsibility relations of the roles on different aspects of some identified tasks

Task	Execution	Monitoring	Consulting	Technological decisions	Managerial decisions
1.1	Crew	Ground Controller, Tower Controllers Supervisor	Ground Controller	Crew	Ground Controller, Tower Controllers Supervisor
1.2	Crew		Ground Controller	Crew	
1.3	Crew				
2.1	Safety Investigator	Safety Manager; Safety Investigator	Safety Manager	Safety Investigator	
2.3	Safety Investigator	Safety Manager, Occurrence Manager	Safety Manager	Safety Investigator	

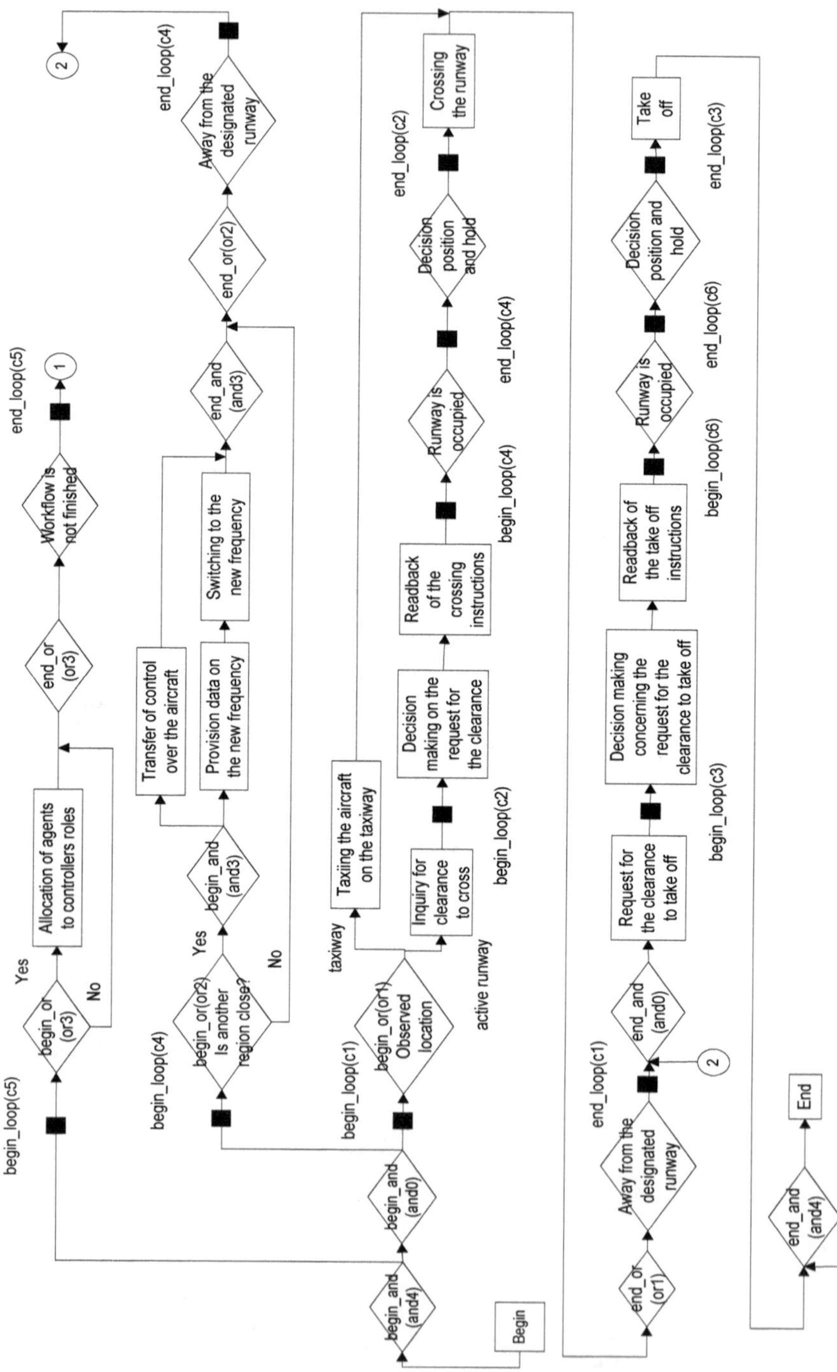

Fig. 4. Workflow for an aircraft taxiing to and taking-off from a designated runway

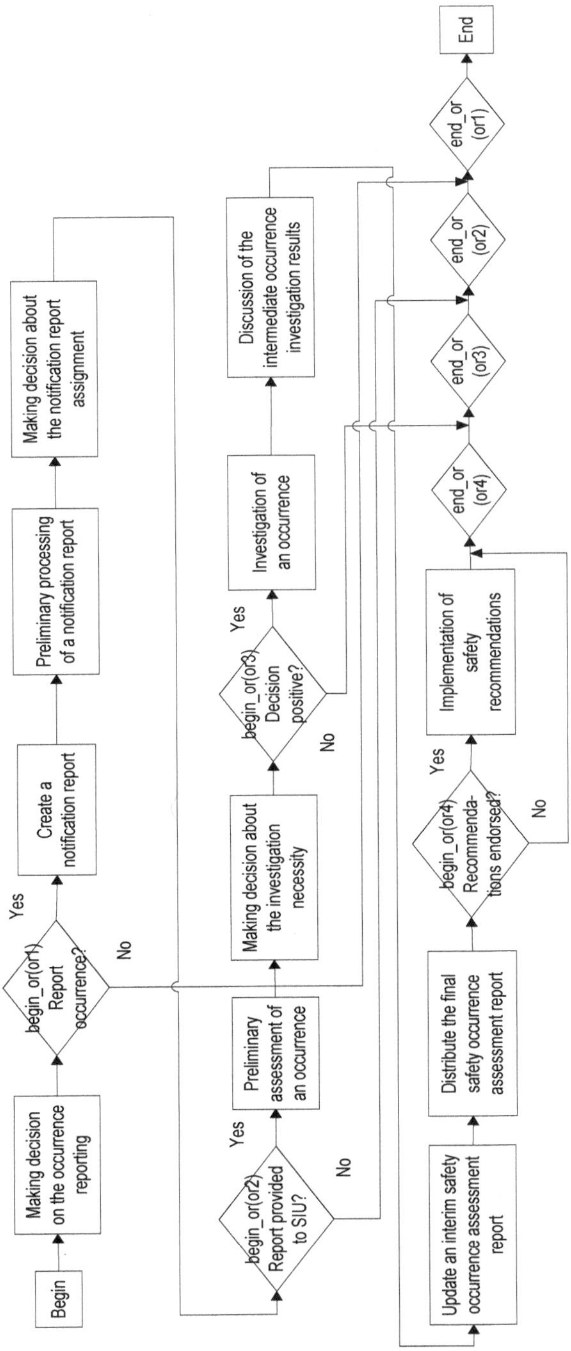

Fig. 5. The workflow that defines the execution of the safety occurrence reporting and the report handling task initiated by a controller

described in [18]. In Fig.4 a worklow is given that describes the process of taxiing of an aircraft to and taking-off from a designated runway. Fig. 5 describes the execution of the formal occurrence reporting initiated by a controller.

Step 9. The identification of the generic and domain-specific constraints
In this step generic and domain-specific constraints imposed on a formal organisational specification are identified. Generic constraints need to be satisfied by any organisational specification. Domain-specific constraints are dictated by the application domain and may be changed by the designer. A set of constraints imposed on an organisational specification is expressed using the formal languages of the framework. An organisational specification is *correct* if every constraint in the corresponding set is a logical consequence of this specification. The framework provides means for automated checking of the correctness of a specification. Consider examples of the domain-specific constraints of the ATO obtained from the formal regulations of a controller and of a pilot:

C1: When an aircraft is approaching to an active runway, the pilots should cease all processes not related to the taxiing.

C2: The pilots of a crew should verbally share information about the instructions of controllers.

C3: A controller may guide maximum two aircrafts at the same time.

C4: A controller is not allowed to issue any new clearances for some runway until this runway is vacated by the aircraft that had received the last clearance from the controller.

C5: Perform the allocation of agents-controllers to the aircraft monitoring processes in such a way that the number of processes executed at the same time by each controller is less than four.

Furthermore, the ANSP's reprimand policies related to reporting were formalized as constraints.

4 Modelling of Agents

The specification of a formal organisation forms a part of an overall organisational specification. Another part describes characteristics and behavior of agents and their allocation to the organisational roles. Thus, an overall organisational specification combines prescriptive aspects of a formal organisation with the specification of the autonomous behavior of agents. Using such specifications, investigations of different scenarios of organisational behavior can be performed by simulation.

Agent models are formally grounded in order-sorted predicate logic with finite sorts. More specifically, the static properties of a model are expressed using the traditional sorted first-order predicate logic, whereas dynamic aspects are specified using the Temporal Trace Language (TTL) [2, 13], a variant of the order-sorted predicate logic. In TTL, the dynamics of a system are represented by a temporally ordered sequence of states. Each state is characterized by a unique time point and a set of state properties that hold, specified using the predicate at: STATE_PROPERTY x TIME. Dynamic properties are defined in TTL as transition relations between state properties. For example, the property that for all time points if an agent ag believes that action a is rewarded with r, then ag will eventually perform a, is formalized in TTL as:

∀t:TIME [at(internal(ag, belief(reward_for_action(r, a))), t) →
∃t1 > t at(output(i, performed_action(a)), t1)]

The behavior of an agent can be considered from external and internal perspectives. From the external perspective the behavior can be specified by temporal correlations between agent's input and output states, corresponding to interaction with other agents and with the environment. An agent perceives information by observation and generates output in the form of communication or actions.

From the internal perspective the behavior is characterized by a specification of direct causal relations between internal states of the agent, based on which an externally observable behavioral pattern is generated. It is assumed that agents create time-labeled internal representations (*beliefs*) about their input and output states, which may persist over time:

∀ag:AGENT ∀p:STATE_PROPERTY ∀t:TIME at(input(ag, p), t)→ at(internal(ag, belief(p, t), t+1))

Information about observed safety occurrences is stored by agents as beliefs: e.g., belief(observed_occurrence_with(ot:OCCURRENCE_TYPE, ag:AGENT)), t:TIME). Besides beliefs about single states, an agent forms beliefs about dependencies between its own states, observed states of the environment, and observed states of other agents (such as expectancies and instrumentalities from the following section):
belief(occurs_after(p1:STATE_PROPERTY, p2:STATE_PROPERTY, t1:TIME, t2:TIME), t:TIME), which expresses that state property p2 holds t' (t1 < t' < t2) time points after p1 holds.

In social science behavior of individuals is considered as *goal-driven*. It is also recognised that individual goals are based on *needs*. Different types of needs are distinguished: (1) *extrinsic needs* associated with biological comfort and material rewards; (2) *social interaction* needs that refer to the desire for social approval and affiliation; in particular own group approval and management approval; (3) *intrinsic needs* that concern the desires for self-development and self-actualization; in particular contribution to organisational safety-related goals and self-esteem, self-confidence and self-actualization needs. Different needs have different priorities and minimal acceptable satisfaction levels for individuals.

Furthermore, agents are characterised by sets of skills and personal traits that influence their behavior and performance in the organisation. For the ATO a number of agent types have been identified, among which: Controller, Pilot, and Manager.

Table 4. Skills and influence of the agents-controllers.

Agent	Skill(development level)	Influence
ag_controllerA	atc (2)	0.3
ag_controllerB	atc (3)	0.6
ag_controllerC	atc (2)	0.3
ag_controllerD	atc (4)	1
ag_controllerE	atc (3)	0.6
ag_controllerF	atc (4)	1
ag_controllerG	employee management(4) atc (4)	1

Based on agent type Controller, 7 instances have been defined with varying development levels of the skills given in Table 5. All the agents-controllers possess the aggregated air traffic control skill (atc), which allows them to be assigned either to Runway or Ground Controller roles. The agent ag_controllerG also possesses the skill employee management, which allows allocating this agent to role Tower Controllers Supervisor. Based on observations in the air traffic control domain, it is assumed that the development level of the atc skill forms the basis for influence (informal power) of controllers: the higher the development level of the controller's atc, the more influence s/he has in the organisation (the assumed influence levels for this study are given in Table 4). For plausible modelling of the organisation authority structure both informal power (influence) and formal power (authority) of the organisation (identified in Step 7 in Section 3) should be considered explicitly, as they both influence the execution of tasks.

In particular, the level of influence of an agent-controller plays an important role in the propagation of information about potential safety problems to the management level of the ANSP.

A prerequisite for the allocation of an agent to a role is the existence of a mapping between the capabilities and traits of the agent and the role requirements. In the considered case study at the beginning of each day, three agents controllers from Table 4 are chosen randomly to be allocated to two ground controllers and the runway controller roles. The traffic flow in the surroundings of the runway is assumed to be 30 aircraft per hour, 12 hours per day. For each aircraft a crew role is introduced, to which properly qualified agents pilots are assigned.

Controller and crew agents are able to react to 6 types of safety-related occurrences that may happen during the execution of taxiing operations. Table 5 shows the events with the probability values of their occurrence assumed in the simulation studies considered in Section 6.

Some event types can be observed by the agents allocated to particular roles only (see Table 6). Moreover, agents may not always recognize and report observed events correctly (see Table 7). A sufficient number of observed occurrences of a particular type results into the initiation of a formal reporting process, more specifically: 1 event of type (a); 3 of type (b), 6 of type (c), 55 of type (d), 55 of type (e), and 6 of type (f).

Table 5. Safety-relevant events and their probability values per taxiing operation.

Event	Probability
(a) Runway incursion	5e-6
(b) Taxiing aircraft stops progressing on the runway crossing only after the stopbar and due to a call by the runway controller	2e-5
(c) Taxiing aircraft makes wrong turn and progresses towards the runway crossing	1e-4
(d) Taxiing aircraft makes wrong turn and progresses on a wrong taxiing route	2e-4
(e) Taxiing aircraft has switched to a wrong frequency	1e-3
(f) Taxiing aircraft initiates to cross due to misunderstanding in communication	1e-4

Table 6. The observation possibilities of safety-relevant events by controllers and crews.

Event	Identification by			
	Runway Controller	Ground Controller	Crew takingoff	Crew taxiing
a	Yes	No	Yes	Yes
b	Yes	Maybe	Yes	Maybe
c	Yes	Maybe	Maybe	No
d	No	Maybe	Maybe	No
e	Maybe	Maybe	Maybe	No
f	Yes	No	Maybe	Maybe

Table 7. The probability values for recognition and registration of events by controllers and crews.

Event	Probability of correct event recognition		Probability of the event registration	
	Controller	Crew	Controller	Crew
a	$1 - 10^{-5}$	$1 - 10^{-5}$	$1 - 10^{-5}$	$1 - 10^{-5}$
b	$1 - 10^{-5}$	$1 - 10^{-5}$	0.99	0.99
c	0.99	0.98	0.9	0.9
d	0.95	0.8	0.5	0.5
e	0.7	0.9	0.5	0.5
f	0.99	0.9	0.99	0.99

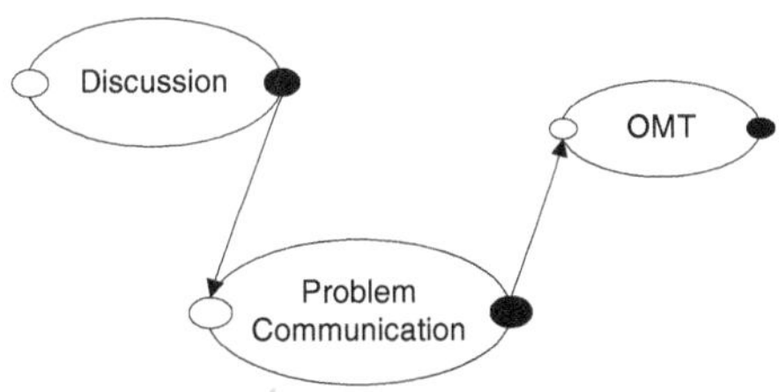

Fig. 6. Interaction relations in the ATO considered at the aggregation level 1

According to the formal organisation a shift of a controller consists of three sessions. The duration of each session is 1 hour. After each session, a break with the duration 1 hour follows. During breaks controllers discuss occurrences observed during their shifts. Since the results of such discussions may initiate the path of informal occurrence reporting, interaction between controllers during such discussions should be modeled explicitly. To this end, the role Discussion is introduced that contains subroles Participant 1…N. The agent controller with the highest influence level in Discussion role has also the joint allocation to subrole Problem Informant in Problem Communication role (see Fig. 6).

Thus, this agent represents Discussion role in the interactions with the management (more specifically, Operational Management (OMT) role). OMT has the formally defined authority to make decisions based on information provided by controllers (e.g., the decision about the beginning of an occurrence investigation). The provision of relevant and reliable information about safety-related occurrences to OMT depends greatly on the informal influence relations that exist among controllers. More specifically, the relevant information is propagated if the controllers involved in the discussion are sufficiently influential and possess sufficient knowledge about occurrences. To create a quantitative model for informal incident reporting, the motivation model by Vroom [9] is used. The motivation model defines the motivational force of an agent to perform some action as:

$$F_i = f\left(\sum_{j=1}^{n} E_{ij} \times V_j\right), \quad V_j = \sum_{k=1}^{m} V_{jk} \times I_{jk}$$

Here, E_{ij} is the strength of the expectancy (belief) that act i will be followed by outcome j; V_j is the valence (i.e., perceived importance/the desire level) of first-level outcome j; V_{jk} is the valence of second-level outcome k that follows first-level outcome j; I_{jk} is perceived instrumentality (belief about the likelihood) of outcome j for the attainment of outcome k.

This model is used to represent the motivation of the agent allocated to a participant role (within Discussion role) with the highest influence level to propagate information about a safety-related issue (see Fig. 7). In the model two second-level outcomes are identified related to the needs of a controller: positive impact on the organisational goals (intrinsic needs) and group acceptance (social interaction needs). The parameters of the motivation are defined as follows: instrumentalities I11 and I12 are assigned high values (0.9), as the controllers involved in the discussion believe that the identified safety related issue will contribute to the satisfaction of the organisational safety-related goals. Both second-level outcomes have a high level of priority for the controllers (valence value = 1).

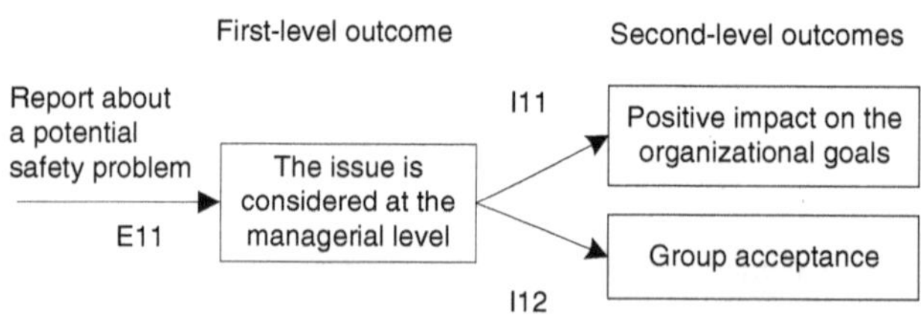

Fig. 7. The motivation model of a controller for reporting about a potential safety problem

Expectancy E11 is proportional to the influence levels of the controllers in the discussion as well as to the quotient of the number of occurrences required for the investigation and the number of occurrences observed by the controllers involved in the discussion so far:

$$E11(occur\_type, CD) = ac(occur\_type) \times \sum_{C_i \in CD} influence\_level(C_i)$$

where CD is the set of the controllers involved in the discussion and ac(occur_type) is defined as:

$$ac(occur\_type) = \begin{cases} 1, & N(occur\_type) \le N(occur\_type)_{curr} \\ \dfrac{N(occur\_type)_{curr}}{N(occur\_type)}, & N(occur\_type) > N(occur\_type)_{curr} \end{cases}$$

with $N(occur\_type)$ the number of occurrences of the type occur_type required for the investigation (the same as for the formal incident reporting) and $N(occur\_type)_{curr}$ the number of occurrences of the type occur_type observed by the controllers involved in the discussion so far.

Thus, the motivation force to report about a possible problem based on the observations of events of type occur_type is

$$F(occur\_type, CD) = 1.8 * E11(occur\_type, CD)$$

If $F(occur\_type, CD) > 1.8$, the problem will be reported to the management by the controller representative of Discussion role. After that, the problem will be discussed at the nearest OMT meeting and the occurrence investigation will be initiated.

5 Correctness Verification

In this Section the results of the correctness verification of the designed ATO specification are presented. As a result of the automated analysis a number of inconsistencies have been identified. In particular, identification of (potential) conflicts in the goal structure has been performed using the corresponding PI structure based on the following principle. If goals are related by a refinement relation, then the PIs corresponding to these goals are related by a certain (positive or negative) causality relation. To determine the exact type of causality, goal expressions should be analyzed. Typically a goal is defined based on a PI expression in the form of (a conjunction of) clauses in the form [pi op value], where op is one of the relations =, <, or >, and in the form of individual PIs involved in the optimized goal pattern. By analyzing a goal expression target/unsatisfactory values of the PI that contribute to the satisfaction/failure of the goal can be determined on the PI's scale. For example the target value of the PI 'the level of safety of execution of tasks related to the air traffic management' from goal 10 'It is required to maintain a high level of safety of execution of tasks related to the air traffic management' is 'high', whereas the unsatisfactory values of the PI are 'average' and 'low' (given a three-valued qualitative scale).

When the target and unsatisfactory values of a PI are identified, a label ('↑' indicating growth of a PI; '↓' indicating decrease of a PI) may be determined for the PI that indicates the dynamics of change of the PI, when the goal's satisfaction degree changes from denied (failed) to satisfied. For example, for the PI 'the level of safety of execution of tasks related to the air traffic management' of the goal 10 label ↑ is identified. A PI used for a goal pattern of the type minimize (maximize) can be labelled by ↓ (↑) even without considering its values. Note that in some cases additional information is required to identify the label. For example, for the goal 'it is required to maintain the density of traffic = 100' the target value of the PI is 100, whereas the unsatisfactory values are in principle all values on the PI's scale ≠ 100 (i.e., on the scale used both PI values > 100 and < 100 are possible). To determine the label in this case, information is needed, for example, on the unsatisfactory PI's values that can actually be observed in reality.

Algorithm to verify the consistency of goal and PI structures

For each subgoal relation from the goal hierarchy perform the following steps:

1. If the goal pattern of the subgoal is of type minimize (maximize), then the PI of the subgoal is labelled by ↓ (↑); proceed with Step 4.

2. Identify the target and unsatisfactory values of the PI of the subgoal.

3. Identify the label for the PI of the subgoal. If the label is unknown, return UNKNOWN.

4. If the goal pattern of the parent goal is of type minimize (maximize), then the PI of the parent goal is labelled by ↓ (↑); proceed with Step 7.

5. Identify the target and unsatisfactory values of the PI of the parent goal.

6. Identify the label for the PI of the parent goal. If the type is unknown, return UNKNOWN.

7a. If the labels of the PIs of the subgoal and of its parent goal are the same,
and the subgoal has a high degree of influence on its parent goal (i.e., the goals
are related by satisfices or is_subgoal_of relations),
and the PIs of the subgoal and the parent goal are related by a negative causality
relation in the PI structure,
then return INCONSISTENT_HIGH_CONFIDENCE (i.e., inconsistency is identified that concerns the considered subgoal relation and the relation from the PI structure between the PIs on which the goals are based)
else

7b. If the labels of the PIs of the subgoal and of its parent goal are the same,
and the subgoal has a low degree of influence on its parent goal (i.e., the goals are related by contributes_to relation),
and the PIs of the subgoal and the parent goal are related by a negative causality
relation in the PI structure,
then return INCONSISTENT_LOW_CONFIDENCE
else

7c. If the labels of the PIs of the subgoal and of its parent goal are opposite,
and the subgoal has a high degree of influence on its parent goal,
and the PIs of the subgoal and the parent goal are related by a positive causality
relation in the PI structure
then return INCONSISTENT_HIGH_CONFIDENCE
else

7d. If the labels of the PIs of the subgoal and of its parent goal are different,
and the subgoal has a low degree of influence on its parent goal,
and the PIs of the subgoal and the parent goal are related by a positive causality
relation in the PI structure
then return INCONSISTENT_LOW_CONFIDENCE
else

7e. return CONSISTENT

If the same labels are identified for the PIs of a goal and of its subgoal, then one may assume that the PIs are related by a positive causality relation, in case of different labels the PIs are related by a negative causality relation. The higher the degree of influence of a subgoal on its parent goal, the higher the confidence that the identified relation indeed holds.

Note that since the designer has a lot of freedom in specifying goal expressions, there is no guarantee that inconsistencies identified in a PI structure are valid. Therefore, all automatically identified inconsistencies in goal and PI structures still need to be confirmed by the designer. Below the algorithm is given to verify the consistency of goal and PI structures by processing individual subgoal relations in the goal hierarchy.

Using this algorithm several (potential) conflicts have been identified in the goal structure from the case study: between goals 27 and 37; between goals 38 and 37; between goals 20 and 25; and between goals 23 and 25. The goals that are in conflict cannot be satisfied at the same time. For example, goal 25 ensures adherence of the roles to the safety-related norms, which may not always be optimal from the performance point of view (goal 23). Besides execution of tasks, the formal authority relations influence the satisfaction of organisational goals. For example, to achieve the satisfaction of the goal 20, the crews and the controllers should be provided sufficient decision making power with respect to their tasks.

To check the constraints introduced in step 9 in Section 3 a number of dedicated algorthims have been developed. In particular, consider the algorithm for checking constraint C5 for a controller role r.

The presented algorithm proceeds under the assumption that any organisational process p may be executed at any time point during the interval $[est_p, let_p]$. Thus, the satisfaction of constraint C5 is checked for all possible executions (combinations of intervals) of the processes allocated to an agent. To this end, instead of the calculation

Algorithm to verify the constraint C5

1. Identify the set of processes for which the role is responsible: PROC_REL={p:PROCESS| ∃a∈ ASPECT is_responsible_for(r,a, p)}
2. For each process p∈ PROC_REL identify the execution interval $[est_p, let_p]$ and write the values est_p and let_p in a new row of the allocation matrix M of role r. If a role has an allocation for a part of the execution interval of a process, then the time points of the beginning and end of this allocation is written in M.
3. Process the obtained allocation matrix M row by row. For each row identify the existence of non-empty intersections with intervals represented by other rows of M. An intersection of the intervals represented by rows i and j is nonempty if $\neg((m_{i2} < m_{j1}) \vee (m_{j2} < m_{i1}))$ is true. When a row is processed, it is not taken into account in any other evaluations. If for a row the amount of non-empty intersections is greater than 4, then **C5 is not satisfied, exit**.
4. When all rows are processed, **C5 is satisfied.**

of combinations of processes at each time step (as in state-based methods), the algorithm establishes the existence of non-empty intersections of the complete execution intervals of processes in a more efficient way. As shown in [10] the time complexity for the calculation of the execution bounds for all processes of a workflow for the worst case is not greater than $O(|P|^2 Cw)$, where P is the set of processes of the workflow, and Cw is the set of constraints on the workflow.

The automated verification of this constraint identified many situations in which the same controller role was allocated to more than four aircraft's monitoring processes, thus violating constraints C3 and C5, and sacrificing the satisfaction of goals 10 and 24. Sometimes the management to keep the satisfaction of C5, allocates not (completely) qualified agents to the controller roles, thus, causing the dissatisfaction of goal 10. Obviously, the satisfaction of the important safety-related goal 10 is sacrificed in both solutions. The lack of the consideration for the safety-related goals may cause incidents or even accidents.

6 Analysis by Simulation

Using the developed model of the ATO, simulation of different scenarios can be performed. In section 6.1 the simulation results of the runway incursion scenario considered in the case study are presented. Then, the simulation results of the safety occurrence reporting scenario are considered in section 6.2.

6.1 Simulation Results for the Runway Incursion Scenario

In this scenario based on the joint decision of the Airport's Management, the ANSP and the largest airlines, the new runway runway1 has been introduced. Due to its physical position, most of the aircraft taxiing to other runways need to cross runway1 on its way (runway1 can be crossed at one place only, whereas may be approached using two taxiways situated in two different sectors of the airport) (see Fig. 8 below).

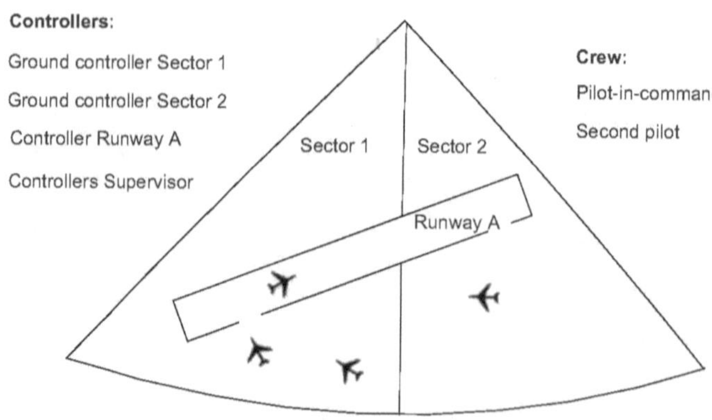

Fig. 8. A graphical layout for the airport situation considered in the scenario

The purpose of this study was to investigate the safety issues that may be caused by the introduction of runway1. Both the normal and the critical configurations have been investigated. In the normal configuration the number of aircraft guided by each controller is less than 3, whereas in the critical - the number ≥ 6 and constraint C5 cannot be satisfied for some controllers. The number of agents-controllers in both configurations is limited to 4, 1 of which is always allocated to Tower Controllers Supervisor role. This agent sees to the satisfaction of constraint C5. It is assumed that the agents are properly qualified for their roles. The behaviour of the allocated agents is defined by the ATO formal specification extended with the behavioural deviations, some of which are identified in [15] (e.g., incorrect situation awareness, mistakes) associated with the probability values. Some of these values are dependant on the agents' workload: the higher the agent's workload, the more chance of its error. In the study only serious occurrences were considered, such as an incursion of aircraft on a runway. One hundred simulation trials with the simulation time 3 years and simulation step 1 hour have been performed in the LeadsTo simulation environment [1]. The obtained simulation traces have been analyzed using the the TTL Checker software [2]. By the analysis the following results have been obtained:

(1) The agent allocated to Controller Runway1 role in all traces most of the time was monitoring at least 4 aircraft (i.e., was overloaded).
(2) The ground controller of the sector 1 was also overloaded, guiding in average three aircrafts at the same time.
(3) The incursion event on the runway1 occurred in 36 traces from 100.
(4) The number of incursions caused by the combination of the events a (the crew mistakenly recognized the runway as a taxiway) and c (the responsible ground controller forgot to inform the crew about the frequency change because of the high workload) is 30.
(5) The number of incursions caused by mistakes of the runway controller in the calculation of the separation distance between aircrafts is 5.
(6) There is only one incursion caused by a crew mistakenly reacting to the clearance for some other crew.

In the future a precise validation of the obtained results will be performed.

6.2 Simulation Results for the Safety Occurrence Reporting Scenario

For the safety occurrence reporting both the formal and informal reporting paths were simulated and compared. Based on the developed model of the ATO 100 stochastic simulations with a simulation time of maximum 3 years (12 operational hours per day) each have been performed using the simulation tool LeadsTo. When the formal or informal safety occurrence reporting has lead to the identification of a safety problem and a further investigation thereof, the simulation was halted. As a result of each simulation trial, a trace is generated by the LeadsTo. Then, such traces can be automatically analyzed using the TTL Checker software. Besides the logical analysis, the tool allows statistical post-processing of traces. For this the following functions are used:

case(logical_formula, value1, value2): if logical_formula is true, then the case function is mapped to value1, otherwise – to value2.

sum([summation_variables], case(logical_formula, value1, 0)): logical_formula is evaluated for every combination of values from the domains of each from the summation_variables; and for every evaluation when the logical formula is evaluated to true, value1 is added to the resulting value of the sum function.

In this case study a number of properties has been checked automatically on 100 generated traces, two of which are described in the following. The first property calculates the number of traces, in which the safety problem has been found based on the reported occurrences of some type. Formally, for the occurrence type a:

sum([γ:TRACE], case(∃t:TIME holds(state(γ, t, environ), problem_found_based_on(a)), true), 1, 0)) > 0

Another property calculates the mean time of the problem recognition on all traces in which the problem of a particular type has been found. Formally, for the occurrence type a:

sum([γ:TRACE], case(∃t:TIME holds(state(γ, t, environ), problem_found_based_on(a)), true), t, 0)) / sum([γ:TRACE], case(∃t:TIME holds(state(γ, t, environ), problem_found_based_on(a)), true), 1, 0))> 0

The simulation results for both formal and informal reporting cases are presented in Table 8.

Table 8. Results of the simulation experiments.

Event	Percentage of traces, in which the investigation began		Mean time of the problem recognition (days)	
	Formal	Informal	Formal	Informal
a	22%	21%	155.1	134.9
b	5%	15%	168.1	123.9
c	28%	50%	194.6	149.6
d	0%	0%	-	-
e	0%	3%	-	278.9
f	45%	11%	185.9	184.7
total	100%	100%	180.8	150.4

The mean time value of the identification of a safety related problem with respect to some event type is calculated over all traces, in which the occurrences of events of this type caused the incident investigation.

Table 8 shows that for both the formal and informal handling of safety occurrences in all simulation traces a safety investigation is initiated, however, the mean time until start of the investigation is 181 days in the formal case, whereas it is 150 days in the informal case. Considering the simulation results for the particular events, the mean time of recognition is smaller for all event types in the informal reporting path.

A main reason underlying the difference in the time until recognition of the safety problem is that situations like event *b* ("Taxiing aircraft stops progressing on the runway crossing only after the stopbar and due to a call by the runway controller") and event *c* ("Taxiing aircraft makes wrong turn and progresses towards the runway crossing") are often recognized by both ground and runway controllers and thus feed common situation awareness on safety-critical aspects in informal discussions, whereas such events are just single occurrence reports in the formal incident reporting case.

7 Conclusions

The paper presents the first attempt to create a formal agent-based organisational model of an ATO using the framework of [13]. A typical ATO has high structural and behavioural complexities that present a challenge both for modelling and analysis of such an organisation. All important aspects of the considered organisation have been identified at four complementary levels, i.e. performance-oriented, process-oriented, organisation-oriented and agent-oriented. The modelling framework used allows scalability by distinguishing four interrelated organisational views and by specifying an organisational model at different aggregation levels and identifying relations between these levels. However, for complex organisations (such as ATOs) specifications of lower aggregation levels still may be very elaborated.

Using the automated analysis techniques from [2, 13] missing and conflicting parts of the ATO model can be identified. Some examples of application of these techniques are provided in the paper. The scalability of analysis is achieved by applying dedicated techniques for verification of specifications of particular organisational views and by distinguishing aggregation levels. Another analysis type demonstrated in the paper is by simulation. It allows to evaluate how different types of divergent agent behaviour in simulation models may result into delays in executions of processes up to the level of incidents. In comparison with the simulation approach of [14, 15], the novel approach considers the organisational layer of the ATO explicitly, however at the expense of not simulating beyond the incidents. Another example of such analysis, in which the formal and informal occurrence reporting paths of the ATO were investigated, is provided in this chapter. The analysis results show that the informal safety-occurrence reporting path results in faster identification of safety-related problems than the formal reporting path. Next research steps will focus on assessing whether this important feedback on safety occurrence reporting processes is recognized in actual air traffic organisations and may be a basis for organisational change.

Acknowledgement

The author would like to acknowledge Sybert Stroeve and Henk Blom (the Dutch National Aerospace Laboratory, NLR, Amsterdam) and Barry Kirwan (Eurocontrol, Bretigny) for their cooperation. This research has partly been funded through an Eurocontrol Innovative Research grant.

References

1. Bosse, T., Jonker, C.M., Meij, L., van der Treur, J.: A Language and Environment for Analysis of Dynamics by Simulation. International Journal of Artificial Intelligence Tools 16, 435–464 (2007)
2. Bosse, T., Jonker, C.M., Meij, L., van der Sharpanskykh, A., Treur, J.: Specification and Verification of Dynamics in Agent Models. International Journal of Cooperative Information Systems 18(1), 167–193 (2009)
3. Chapurlat, V., Kamsu-Foguem, B., Prunet, F.: A formal verification framework and associated tools for enterprise modelling: Application to UEML. Computers in industry 57, 153–166 (2006)
4. CIMOSA – Open System Architecture for CIM. ESPRIT Consortium AMICE. Springer, Berlin (1993)
5. Dalal, N., Kamath, M., Kolarik, W., Sivaraman, E.: Toward an integrated framework for modelling enterprise processes. Communications of the ACM 47(3), 83–87 (2004)
6. Eurocontrol: Air navigation system safety assessment methodology. SAF.ET1.ST03.1000-MAN-01, 2.0 edn. (2004)
7. Koubarakis, M., Plexousakis, D.: A formal framework for business process modelling and design. Information Systems 27(5), 299–319 (2002)
8. Le Coze, J.: Are organisations too complex to be integrated in technical risk assessment and current safety auditing? Safety Science 43, 613–638 (2005)
9. Pinder, C.C.: Work motivation in organisational behavior. Prentice-Hall, Englewood Cliffs (1998)
10. Popova, V., Sharpanskykh, A.: Process-oriented organisation modelling and analysis. Enterprise Information Systems Journal 2(2), 157–176 (2008)
11. Reason, J., Hollnagel, E., Paries, J.: Revisiting the Swiss cheese model of accidents. Eurocontrol, EEC Note no. 13/06 (2006)
12. Scheer, A.-W., Nuettgens, M.: ARIS Architecture and Reference Models for Business Process Management. In: van der Aalst, W.M.P., Desel, J., Oberweis, A. (eds.) Business Process Management. LNCS, vol. 1806, pp. 366–389. Springer, Heidelberg (2000)
13. Sharpanskykh, A.: On Computer-Aided Methods for Modelling and Analysis of Organisations. PhD Dissertation. Vrije Universiteit Amsterdam (2008)
14. Stroeve, S.H., Blom, H.A.P., Bakker, G.J.: Safety risk impact analysis of an ATC runway incursion alert system. Eurocontrol Safety R&D Seminar, Barcelona (2006)
15. Stroeve, S.H., Blom, H.A.P., Van der Park, M.N.J.: Multi-agent situation awareness error evolution in accident risk modelling. In: Proceedings of 5th ATM R&D Seminar (2003)
16. Stroeve, S.H., Sharpanskykh, A., Blom, H.A.P.: Organisational safety modelling and analysis of an air traffic application: Eurocontrol CARE Innovative research III. National Aerospace Laboratory NLR, report CR-2007-457 (2007)
17. Tham, K.D.: Representation and Reasoning About Costs Using Enterprise Models and ABC, PhD Dissertation, University of Toronto (1999)
18. Van der Aalst, W.M.P., Van Hee, K.M.: Workflow Management: Models, Methods, and Systems. MIT press, Cambridge (2002)

Identification of Contact Conditions by Active Force Sensing

Takayoshi Yamada[1], Tetsuya Mouri[1], Nobuharu Mimura[2], Yasuyuki Funahashi[3], and Hidehiko Yamamoto[1]

[1] Dept. of Human and Information Systems, Faculty of Engineering, Gifu University, 1-1, Yanagido, Gifu 501-1193, Japan
{yamat,mouri,yam-h}@gifu-u.ac.jp

[2] Dept. of Biocybernetics, Faculty of Engineering., Niigata University, 2-8050, Ikarashi, Nishi-ku, Niigata 950-2181, Japan
mimura@eng.niigata-u.ac.jp

[3] Dept. of System Engineering of Human Body, Faculty of Life System Science and Technology, Chukyo University, 101 Tokodachi, Kaizu-cho, Toyota, Aichi 470-0393, Japan
funa-@sist.chukyo-u.ac.jp

Abstract. When a grasped object is in contact with an external environment, it is required to identify contact conditions prior to performing assembly tasks. The contact conditions include contact position, contact force, contact type, direction of contact normal, and direction of contact line between a grasped object and its environment. This article distinguishes soft-finger, line, and planar contact types in addition to point contact type. It is assured that these four contact types are distinguishable when the number of active force sensing samples is at least 6. Efficient algorithms for parameter identification and transition detection of contact conditions are explained. Effectiveness of our proposed method is verified through numerical examples.

Keywords: Robot, active force sensing, contact conditions, parameter identification, transition detection, contact force and moment.

1 Introduction

When a grasped object is in contact with an external environment, it is required to identify contact conditions prior to performing assembly tasks. In this article, we explain parameter identification of contact conditions by active force sensing.

For example, let us imagine the simple task such that a box is set on a table. Human beings can easily achieve the task changing the contact type from a vertex (point contact) to an edge (line contact), and from an edge (line contact) to a bottom (planar contact) of the box. They carry out fine positioning assembly tasks with their manual hand dexterity even if they are blindfolded. In order to achieve the assembly tasks by robots in place of humans, the contact conditions must be identified and controlled.

N.T. Nguyen and E. Szczerbicki (Eds.): Intel. Sys. for Know. Management, SCI 252, pp. 275–305.
springerlink.com

In assembly tasks, contact parameters to be identified are contact point, contact force and moment, contact type, direction of contact normal, and direction of contact line. Transition of contact conditions must be also detected. This parameter identification and transition detection is important technology for enhancement of robot dexterity.

Shimokura et al. [1], Takano et al. [2], Nakai et al. [3], Skubic et al. [4], and Takamatsu et al. [5] explored transition detection of contact type by using force and moment information. However, these methods treated the case where the transition had been planned beforehand and the object and environment shapes are known. Shutter et al. [6], Lefebvre et al. [7], and Meeussen et al. [8] identified contact conditions and detected transition of the conditions by using 12 information dimensions which include force, moment, and joint angle (velocity).

Other methods identified contact conditions by only using a 6-axes force sensor on robot wrists or fingers. These methods are free from position error caused from joint angle backlash. These methods use groping motion (active force sensing) [9] [10]. Salisbury [11] and Tsujimura et al. [12] identified contact position between a probe and an external object for point contact, and roughly acquired the object shape. Zhou et al. [13], Nagata et al. [14], Nagase et al. [15] acquired contact position on the finger using several local surfaces. Bicchi [16] obtained contact position and direction of contact normal for soft-finger contact. Murakami et al. [17, 18] obtained direction of contact line on a soft-finger surface, and assumed that the probe (finger) shape is known. Kitagaki [19] obtained contact position on a grasped object with an unknown shape for point contact and verified that at least 2 samples are required to identify position.

Mimura et al. [20] identified contact conditions between a grasped object and its environment, in which the object and environment shapes are unknown. They proved that four types of contact could be distinguished by active force sensing, and verified that the number of active force sensing samples to identify contact conditions must be at least 6. They also provided identification algorithms. However, this method is troublesome, because contact moment restrictions are implicitly formulated and measured data are noiseless.

This article performs parameter identification and transition detection of contact conditions by only using force and moment information from a 6-axes force sensor mounted on robot hand or fingers [21-24].

In Section 2, we explain problem formulation of parameter identification of contact conditions. Contact moment constraint is explicitly formulated.

In Section 3, we describe formulation including the contact conditions in order to identify not only point contact type but also the other contact types. In this formulation, noiseless data are utilized for simplicity of discussions. An efficient identification method is proposed by considering the characteristics of the four contact types.

In Section 4, we treat the case that the measurement moment is contaminated with noise. The four contact types are characterized by standard deviation of contact moment. We formulate least-squares functions including contact position, direction of contact normal, and direction of contact line. The contact position and moment are estimated by minimizing the functions. The contact type is distinguished by using three eigenvalues of a covariance matrix of the estimated moment. It is proved that consistent estimates are obtained when the number of measurement data is infinite. Moreover, a two-phase method is proposed to improve the estimated parameters in practical cases that the number of data is limited.

In Section 5, we treat transition detection of contact type. Because in practical case such as assembly tasks, the contact type changes from point to line, and line to planar contact type. In order to detect the transition quickly, the limited number of measurement data is utilized, and the weight of latest data are enlarged. To correct and reduce an error of judged contact type, continuity of time series of the judged type is checked and revised.

In Section 6, we conclude this article.

2 Problem Formulation

We assume that an object of unknown shape is grasped by a robot hand and is in contact with the environment (Fig. 1). The contact conditions are identified by active force sensing.

2.1 Symbols

We define the following symbols (Fig. 1).
c: contact point between the grasped object and its environment
o: origin of a force sensor on the hand
Σ_o : sensor coordinate frame fixed at o
Σ_c : contact coordinate frame fixed at c, where the orientation of Σ_c is the same as that of Σ_o
$\boldsymbol{f}_c$, $\boldsymbol{n}_c$: contact force and moment in Σ_c
$\boldsymbol{f}_o$, $\boldsymbol{n}_o$: measured force and moment in Σ_o
$\boldsymbol{r}_c$: position vector of c in Σ_c

2.2 Assumptions

We discuss identification problem of contact conditions under the following assumptions.

(A1) The object is firmly grasped by a robot hand and arbitrarily manipulated by its robot arm.

(A2) Force $\boldsymbol{f}_o$ and moment $\boldsymbol{n}_o$ are measurable but force $\boldsymbol{f}_c$, moment $\boldsymbol{n}_c$, and contact position $\boldsymbol{r}_c$ are unknown.

(A3) The object is in contact with the environment through a point, soft-finger, line, or planar contact type with friction. But the contact type is unknown beforehand.

2.3 Contact Type

The four contact types in (A3) are defined by Mason [25]. These are classified by degree-of-freedom (DOF) contact denoted by m, which means the number of axes at which no moment occurs.

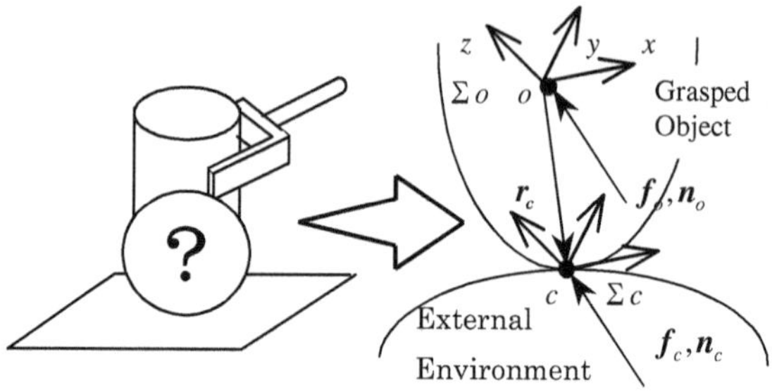

Fig. 1. Interaction between a grasped object and its environment

$$m = \begin{cases} 0: & \text{Planar contact with friction} \\ 1: & \text{Line contact with friction} \\ 2: & \text{Soft - finger contact with friction} \\ 3: & \text{Point contact with friction} \end{cases}$$

In actual practice, soft-finger contact is rare, because it theoretically has no contact area and exerts contact moment only in the direction of contact normal. Planar contact type is equivalent that the object is bound to the environment. Two-points contact is equivalent to line contact type. Three-points contact is equivalent to planar contact type.

2.4 Contact Moment Constraint

We define $\Sigma_{c'}$ as a coordinate frame concisely expressing four contact types (Fig. 2). The origin of $\Sigma_{c'}$ is fixed at c, the z' axis direction is perpendicular to the contact plane, and the x' axis direction is on the contact line. Contact moment in $\Sigma_{c'}$ is denoted by

$$\boldsymbol{n}'_c = [n'_{cx}, n'_{cy}, n'_{cz}]^T \, . \tag{1}$$

Four contact types are characterized by

$$\left.\begin{array}{lll} \text{Point contact type} & : & n'_{cx} = n'_{cy} = n'_{cz} = 0 \\ \text{Soft - finger contact type} & : & n'_{cx} = n'_{cy} = 0 \\ \text{Line contact type} & : & n'_{cx} = 0 \\ \text{Planar contact type} & : & \text{no constraints} \end{array}\right\} \tag{2}$$

2.5 Force and Moment Equilibrium

Force and moment equilibrium equations (Fig. 1) are represented by

$$\left.\begin{array}{r} \boldsymbol{f}_c = \boldsymbol{f}_o \\ \boldsymbol{r}_c \times \boldsymbol{f}_c + \boldsymbol{n}_c = \boldsymbol{n}_o \end{array}\right\}, \tag{3}$$

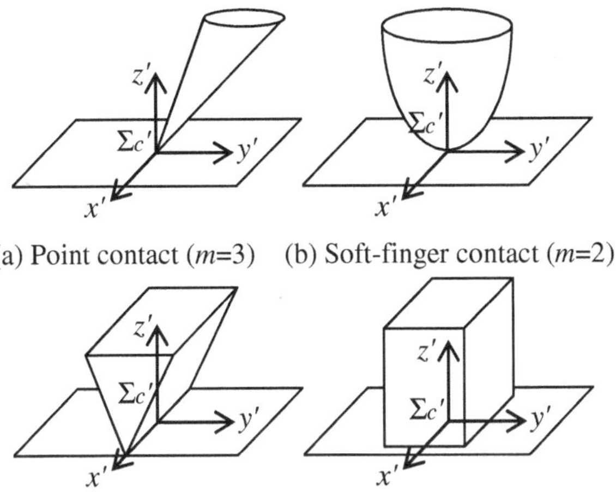

(a) Point contact (m=3) (b) Soft-finger contact (m=2)

(c) Line contact (m=1) (d) Planar contact (m=0)

Fig. 2. Contact types

Moment $\boldsymbol{n}_c$ is expressed as

$$\boldsymbol{n}_c = {}^cR_{c'}\boldsymbol{n}'_c \ , \ {}^cR_{c'} = [\boldsymbol{r}_x, \boldsymbol{r}_y, \boldsymbol{r}_z] = \begin{bmatrix} r_{x1} & r_{y1} & r_{z1} \\ r_{x2} & r_{y2} & r_{z2} \\ r_{x3} & r_{y3} & r_{z3} \end{bmatrix} \in \Re^{3\times 3}, \tag{4}$$

where matrix ${}^cR_{c'}$ is relative orientation of $\Sigma_{c'}$ for Σ_c, vector $\boldsymbol{r}_z$ is unit direction of contact normal, and vector $\boldsymbol{r}_x$ is unit direction of contact line.

2.6 Active Force Sensing

In (3), the number of unknown parameters ($\boldsymbol{f}_c, \boldsymbol{n}_c, \boldsymbol{r}_c$) outnumbers the number of measured data ($\boldsymbol{f}_o, \boldsymbol{n}_o$). To exceed the number of unknown parameters, we must increase the number of equations. Hence, we command k times of active force sensing and obtain the following equation:

$$A_i \boldsymbol{x} + \boldsymbol{n}_{ci} = \boldsymbol{b}_i, \ (i = 1, 2, \cdots, k), \tag{5}$$

where the subscript i is the sampling number, and the following notations are used for clearly dividing measured and unknown parameters.

$$A_i := [-\boldsymbol{a}_i \times], \ \boldsymbol{a}_i = \begin{bmatrix} a_{i1} \\ a_{i2} \\ a_{i3} \end{bmatrix} := \boldsymbol{f}_{oi}, \ \boldsymbol{b}_i = \begin{bmatrix} b_{i1} \\ b_{i2} \\ b_{i3} \end{bmatrix} := \boldsymbol{n}_{oi}, \ \boldsymbol{x} = \begin{bmatrix} x_1 \\ x_2 \\ x_3 \end{bmatrix} := \boldsymbol{r}_c \tag{6}$$

3 Parameter Identification from Noiseless Data

We identify contact conditions as contact type, contact point, and contact direction. To simplify and clarify our formulation, the following assumptions are taken into consideration in addition to (A1)-(A3).

(A4) Contact type remains unchanged during active force sensing.
(A5) Measured data $\boldsymbol{f}_o$ and $\boldsymbol{n}_o$ are pure (noiseless)

In this section, we will derive

(1) Identification method for contact parameters,
(2) The number of active force sensing to judge contact type, and
(3) An algorithm for identification of contact conditions.

3.1 Subdivision of Contact Type

Equation (4) is nonlinear because matrix ${}^{c}R_{c'}$ and vector $\boldsymbol{n}'_c$ are unknown and ${}^{c}R_{c'}$ is a rotation matrix which satisfies $[{}^{c}R_{c'}]^T[{}^{c}R_{c'}] = I_3$ and $\det({}^{c}R_{c'}) = 1$. To reduce unknown parameters and remove equation redundancy, four contact types are subdivided into 8 types in Table 1.

Table 1. Subdivision of contact type.

Contact type	Constraints	$\overline{\boldsymbol{n}}_{ci}$	Case
Point	-	0	PC
Soft-finger	$r_{z2}=0,\ r_{z3}=0$	$[1,0,0]^T z_i$	SC1
	$r_{z2}\neq 0,\ r_{z3}=0$	$[y,1,0]^T z_i$	SC2
	$r_{z3}\neq 0$	$[y_1, y_2, 1]^T z_i$	SC3
Line	$r_{x1}=0,\ r_{x2}=0$	$\begin{bmatrix}1&0&0\\0&1&0\end{bmatrix}^T\begin{bmatrix}z_{i1}\\z_{i2}\end{bmatrix}$	LC1
	$r_{x1}=0,\ r_{x2}\neq 0$	$\begin{bmatrix}1&0&0\\0&y&1\end{bmatrix}^T\begin{bmatrix}z_{i1}\\z_{i2}\end{bmatrix}$	LC2
	$r_{x1}\neq 0$	$\begin{bmatrix}y_1&1&0\\y_2&0&1\end{bmatrix}^T\begin{bmatrix}z_{i1}\\z_{i2}\end{bmatrix}$	LC3
Planar	-	$[z_{i1}, z_{i2}, z_{i3}]^T$	PLC

3.2 Judgment of Point Contact Type (PC)

From PC in Table 1, we have $\overline{\boldsymbol{n}}'_{ci} = 0$, (5) is written as

$$\begin{bmatrix} A_1 \\ A_2 \\ \vdots \\ A_k \end{bmatrix} \boldsymbol{x} = \begin{bmatrix} \boldsymbol{b}_1 \\ \boldsymbol{b}_2 \\ \vdots \\ \boldsymbol{b}_k \end{bmatrix} \in \Re^{3k} \tag{7}$$

Sweeping out (7) yields

$$\begin{bmatrix} 1 & 0 & 0 \\ 0 & 1 & 0 \\ 0 & 0 & 1 \\ 0 & 0 & 0 \\ \vdots & \vdots & \vdots \\ 0 & 0 & 0 \end{bmatrix} \boldsymbol{x} = \begin{bmatrix} \xi_1 \\ \xi_2 \\ \xi_3 \\ \xi_4 \\ \vdots \\ \xi_{3k} \end{bmatrix} \tag{8}$$

We check whether measured data is consistent with PC by using

$$\xi_4 = \cdots = \xi_{3k} = 0 , \tag{9}$$

Contact position is estimated as $\hat{\boldsymbol{x}} = [\xi_1, \xi_2, \xi_3]^T$.

Therefore, the number of active force sensing samples to judge point contact type and identify its contact parameters must be $k \geq 2$.

3.3 Judgment of Soft-Finger Contact Type

3.3.1 Case of Soft-Finger Contact Type 1 (SC1)

From (5) and SC1 in Table 1, we have

$$A_i \boldsymbol{x} + [1,0,0]^T z_i = \boldsymbol{b}_i , \ (i = 1,2,\cdots,k) , \tag{10}$$

From the second and third rows of (10), we have

$$\begin{bmatrix} \tilde{A}_1 \\ \tilde{A}_2 \\ \vdots \\ \tilde{A}_k \end{bmatrix} \boldsymbol{x} = \begin{bmatrix} \tilde{\boldsymbol{b}}_2 \\ \tilde{\boldsymbol{b}}_2 \\ \vdots \\ \tilde{\boldsymbol{b}}_k \end{bmatrix} \in \Re^{2k} \tag{11}$$

where

$$\tilde{A}_i := \begin{bmatrix} -a_{i3} & 0 & a_{i1} \\ a_{i2} & -a_{i1} & 0 \end{bmatrix},\ \tilde{\boldsymbol{b}}_i := \begin{bmatrix} b_{i2} \\ b_{i3} \end{bmatrix} \tag{12}$$

In a similar way of PC, sweeping out (11), the right side vector is denoted by $[\xi_1, \xi_2, \cdots, \xi_{2k}]$. If $\xi_i = 0$, $(i \geq 4)$, measured data is consistent with SC1, contact position is estimated as $\hat{\boldsymbol{x}} = [\xi_1, \xi_2, \xi_3]^T$. Using the first row of (10), $\bar{z}_i$ is estimated as

$$\hat{z}_i = b_{i1} - A_{i1}\hat{\boldsymbol{x}},\ (i = 1,2,\cdots,k), \tag{13}$$

where

$$A_{i1} := [0, a_{i3}, -a_{i2}].$$

Therefore, the number of active force sensing samples to judge SC1 and identify its contact parameters must be $k \geq 2$.

3.3.2 Case of Soft-Finger Contact Type 2 (SC2)

From (5) and SC2 in Table 1, we have

$$A_i\boldsymbol{x} + [y,1,0]^T z_i = \boldsymbol{b}_i,\ (i = 1,2,\cdots,k). \tag{14}$$

From the third row of (14), we have

$$\begin{bmatrix} a_{12} & -a_{11} \\ a_{22} & -a_{21} \\ \vdots & \vdots \\ a_{k2} & -a_{k1} \end{bmatrix} \begin{bmatrix} x_1 \\ x_2 \end{bmatrix} = \begin{bmatrix} b_{12} \\ b_{22} \\ \vdots \\ b_{k2} \end{bmatrix} \in \Re^k. \tag{15}$$

Sweeping out (15), the right side vector is denoted by $[\xi_1, \xi_2, \cdots, \xi_k]^T$. If $\xi_i = 0$, $(i \geq 3)$ is satisfied, measured data is consistent with SC2, and we obtain $\hat{x}_1 = \xi_1$ and $\hat{x}_2 = \xi_2$. Eliminating z_i in the first and second rows of (14) yields

$$-a_{i2}x_3 + (b_{i2} + a_{i3}\hat{x}_1)y - a_{i1}x_3 y = b_{i1} - a_{i3}\hat{x}_2,\ (i = 1,2,\cdots,k), \tag{16}$$

By using matrix form, (16) is rewritten as

$$\begin{bmatrix} -a_{12} & b_{12} + a_{13}\hat{x}_1 & -a_{11} \\ -a_{22} & b_{22} + a_{23}\hat{x}_1 & -a_{21} \\ \vdots & \vdots & \vdots \\ -a_{k2} & b_{k2} + a_{k3}\hat{x}_1 & -a_{k1} \end{bmatrix} \begin{bmatrix} x_3 \\ y \\ x_3 y \end{bmatrix} = \begin{bmatrix} b_{11} - a_{13}\hat{x}_2 \\ b_{21} - a_{23}\hat{x}_2 \\ \vdots \\ b_{k1} - a_{k3}\hat{x}_2 \end{bmatrix} \tag{17}$$

Sweeping out (17) yields

$$\begin{bmatrix} 1 & 0 & 0 \\ 0 & 1 & 0 \\ 0 & 0 & 1 \\ 0 & 0 & 0 \\ \vdots & \vdots & \vdots \\ 0 & 0 & 0 \end{bmatrix} \begin{bmatrix} x_3 \\ y \\ x_3 y \end{bmatrix} \boldsymbol{x} = \begin{bmatrix} \xi'_1 \\ \xi'_2 \\ \xi'_3 \\ \xi'_4 \\ \vdots \\ \xi'_k \end{bmatrix} \tag{18}$$

We check whether measured data is consistent with PC2 by using

$$\xi'_1 \xi'_2 - \xi'_3 = 0 \,, \; \xi'_i = 0 \,, \; (i \geq 4) \,. \tag{19}$$

Hence, parameters x_3, y, and z_i are estimated as

$$\hat{x}_3 = \xi'_1 \,, \; \hat{y} = \xi'_2 \,, \; \hat{z}_i = b_{i2} + a_{i3}\hat{x}_1 - a_{i1}\hat{x}_3 \,, \; (i = 1,2,\cdots,k) \tag{20}$$

Therefore, the number of active force sensing samples to judge SC2 and identify its contact parameters must be $k \geq 4$.

3.3.3 Case of Soft-Finger Contact Type 3 (SC3)

From (5) and SC3 in Table 1, we have

$$A_i \boldsymbol{x} + [y_1, y_2, 1]^T z_i = \boldsymbol{b}_i \,, \; (i = 1,2,\cdots,k) \,, \tag{21}$$

Eliminating z_i in the first and second rows of (21) yields

$$a_{i1}x_2y_1 - a_{i2}(x_3 + x_1y_1) + a_{i3}x_2 + b_{i3}y_1 = b_{i1} \,, \; (i = 1,2,\cdots,k) \tag{22}$$

$$a_{i1}(x_3 + x_2y_2) - a_{i2}x_1y_2 - a_{i3}x_1 + b_{i3}y_2 = b_{i2} \,, \; (i = 1,2,\cdots,k) \tag{23}$$

By using matrix form, (22) and (23) are respectively written as

$$\begin{bmatrix} a_{11} & -a_{12} & a_{13} & b_{13} \\ a_{21} & -a_{22} & a_{23} & b_{23} \\ \vdots & \vdots & \vdots & \vdots \\ a_{k1} & -a_{k2} & a_{k3} & b_{k3} \end{bmatrix} \begin{bmatrix} x_2 y_1 \\ x_3 + x_1 y_1 \\ x_2 \\ y_1 \end{bmatrix} = \begin{bmatrix} b_{11} \\ b_{21} \\ \vdots \\ b_{k1} \end{bmatrix} \tag{24}$$

$$\begin{bmatrix} a_{11} & -a_{12} & -a_{13} & b_{13} \\ a_{21} & -a_{22} & -a_{23} & b_{23} \\ \vdots & \vdots & \vdots & \vdots \\ a_{k1} & -a_{k2} & -a_{k3} & b_{k3} \end{bmatrix} \begin{bmatrix} x_3 + x_2 y_2 \\ x_1 y_2 \\ x_1 \\ y_2 \end{bmatrix} = \begin{bmatrix} b_{12} \\ b_{22} \\ \vdots \\ b_{k2} \end{bmatrix} \tag{25}$$

Equations (24) and (25) are linear estimation problem to determine

$$\boldsymbol{x}_{s31} := [x_2 y_1, x_3 + x_1 y_1, x_2, y_1]^T,\ \boldsymbol{x}_{s32} := [x_3 + x_2 y_2, x_1 y_2, x_1, y_2]^T \quad (26)$$

By sweeping out (24) and (25), these right side vectors are denoted by $[\xi_1, \xi_2, \cdots, \xi_k]$ and $[\xi_1', \xi_2', \cdots, \xi_k']$, respectively. We check whether measured data is consistent with SC3 by using the following constraints:

$$\xi_1 - \xi_3 \xi_4 = 0,\ \xi_i = 0,\ i \geq 5 \quad (27)$$

$$\xi_2' - \xi_3' \xi_4' = 0,\ \xi_i' = 0,\ i \geq 5 \quad (28)$$

$$(\xi_1' - \xi_3 \xi_4') - (\xi_2 - \xi_4 \xi_3') = 0 \quad (29)$$

Contact parameters are estimated as

$$\hat{x}_1 = \xi_3',\ \hat{x}_2 = \xi_3,\ \hat{x}_3 = \xi_1' - \xi_3 \xi_4',\ \hat{y}_1 = \xi_4,\ \hat{y}_2 = \xi_4' \quad (30)$$

Therefore, the number of active force sensing samples to judge SC3 and identify its contact parameters must be $k \geq 5$.

3.4 Judgment of Line Contact Type

3.4.1 Case of Line Contact Type 1 (LC1)

From (5) and LC1 in Table 1, we have

$$A_i \boldsymbol{x} + \begin{bmatrix} 1 & 0 & 0 \\ 0 & 1 & 0 \end{bmatrix} \begin{bmatrix} z_{i1} \\ z_{i2} \end{bmatrix} = \boldsymbol{b}_i,\ (i = 1,2,\cdots,k). \quad (31)$$

The number of unknown parameters $(x_1, x_2, x_3, z_{i1}, z_{i2})$ is $3 + 2k$ but the number of equations is 3 if $k = 1$ and $2 + 2k$ if $k \geq 2$. Hence, the number of unknowns always outnumbers the number of equations even if k is increased, so contact parameters are not uniquely determined. This corresponds to contact point arbitrarily locating on the contact line.

Because the third row of (31) is the same form as (15), we check whether measured data is consistent with LC1 and obtain $\hat{x}_1$ and $\hat{x}_2$ if $k \geq 3$. From the first and second rows of (31), we have

$$\boldsymbol{x} = \begin{bmatrix} \hat{x}_1 \\ \hat{x}_2 \\ 0 \end{bmatrix} + \begin{bmatrix} 0 \\ 0 \\ 1 \end{bmatrix} x_3,\ \begin{bmatrix} z_{i1} \\ z_{i1} \end{bmatrix} = \begin{bmatrix} b_{i1} - a_{i3}\hat{x}_2 \\ b_{i2} + a_{i3}\hat{x}_1 \end{bmatrix} + \begin{bmatrix} a_{i2} \\ -a_{i1} \end{bmatrix} x_3 \text{ for } \forall x_3 \quad (32)$$

The second term of $\boldsymbol{x}$ is the estimated direction of contact line. Parameter x_3 is arbitrary value on the contact line.

3.4.2 Case of Line Contact Type 2 (LC2)

From (5) and LC2 in Table 1, we have

$$A_i \boldsymbol{x} + \begin{bmatrix} 1 & 0 & 0 \\ 0 & y & 1 \end{bmatrix} \begin{bmatrix} z_{i1} \\ z_{i2} \end{bmatrix} = \boldsymbol{b}_i \,,\ (i = 1,2,\cdots,k)\,. \tag{33}$$

Eliminating z_{i2} in the second and third rows of (33) yields

$$a_{i1}(x_3 + x_2 y) - a_{i2}x_1 y - a_{i3}x_1 + b_{i3}y = b_{i3} \,,\ (i = 1,2,\cdots,k) \tag{34}$$

Because (34) is a similar form as (25), we check whether measured data is consistent with LC2, and obtain $\hat{x}_1 = \xi'_3$ and $\hat{y} = \xi'_4$ if $k \geq 5$. Parameter x_3 is represented as $x_3 = \xi'_1 - \xi'_4 x_2$, in which x_2 is arbitrary value. Hence, we have

$$\boldsymbol{x} = \begin{bmatrix} \hat{x}_1 \\ 0 \\ \xi'_1 \end{bmatrix} + \begin{bmatrix} 0 \\ 1 \\ -\hat{y} \end{bmatrix} x_2 \,,\ \begin{bmatrix} z_{i1} \\ z_{i2} \end{bmatrix} = \begin{bmatrix} b_{i1} + a_{i2}\xi'_1 \\ b_{i3} + a_{i2}\xi'_3 \end{bmatrix} - \begin{bmatrix} a_{i3} + a_{i2}\xi'_4 \\ -a_{i1} \end{bmatrix} x_2 \text{ for } \forall x_2 \tag{35}$$

3.4.3 Case of Line Contact Type 3 (LC3)

From (5) and LC3 in Table 1, we have

$$A_i \boldsymbol{x} + \begin{bmatrix} y_1 & 1 & 0 \\ y_2 & 0 & 1 \end{bmatrix} \begin{bmatrix} z_{i1} \\ z_{i2} \end{bmatrix} = \boldsymbol{b}_i \,,\ (i = 1,2,\cdots,k)\,. \tag{36}$$

Eliminating z_{i1} and z_{i2} in (36) yields

$$\begin{bmatrix} a_{11} & -a_{12} & a_{13} & b_{12} & b_{13} \\ a_{21} & -a_{22} & a_{23} & b_{22} & b_{23} \\ \vdots & \vdots & \vdots & \vdots & \vdots \\ a_{k1} & -a_{k2} & a_{k3} & b_{k2} & b_{k3} \end{bmatrix} \begin{bmatrix} -x_3 y_1 + x_2 y_2 \\ x_3 + x_1 y_2 \\ x_2 + x_1 y_1 \\ y_1 \\ y_2 \end{bmatrix} = \begin{bmatrix} b_{11} \\ b_{21} \\ \vdots \\ b_{k1} \end{bmatrix} \tag{37}$$

Sweeping out (37), the right side vector is denoted by $[\xi_1, \xi_2, \cdots, \xi_k]$, we check whether measured data is consistent with LC3 by using

$$\xi_1 + \xi_2\xi_4 - \xi_3\xi_5 = 0 \,,\ \xi_i = 0 \,,\ i \geq 6 \tag{38}$$

Hence, contact parameters are estimated as

$$\boldsymbol{x} = \begin{bmatrix} 0 \\ \xi_3 \\ \xi_2 \end{bmatrix} + \begin{bmatrix} 1 \\ -\hat{y}_1 \\ -\hat{y}_2 \end{bmatrix} x_1 \,,\ \begin{bmatrix} \hat{y}_1 \\ \hat{y}_2 \end{bmatrix} = \begin{bmatrix} \xi_4 \\ \xi_5 \end{bmatrix},$$

$$\begin{bmatrix} z_{i1} \\ z_{i2} \end{bmatrix} = \begin{bmatrix} b_{i2} + a_{i1}\xi_2 \\ b_{i3} + a_{i1}\xi_3 \end{bmatrix} - \begin{bmatrix} -a_{i3} - a_{i1}\xi_3 \\ a_{i2} + a_{i1}\xi_4 \end{bmatrix} x_1 \text{ for } {}^{\forall}x_1 \tag{39}$$

3.5 Judgment for Planar Contact Type (PLC)

From (5) and PLC in Table 1, we have

$$A_i \boldsymbol{x} + [z_{i1}, z_{i2}, z_{i3}]^T = \boldsymbol{b}_i \,, \ (i = 1,2,\cdots,k) \,. \tag{40}$$

Because the number of unknown parameters always outnumbers the number of equations even if k is increased, the parameters are indeterminate.

3.6 Algorithm for Identification of Contact Conditions

From the above discussions, we have Table 2 summarizing the number of active force sensing samples to judge contact type and identify its contact parameters. The symbol '-' means identifiable but the symbol '*' means indeterminate. Hence we propose an identification algorithm as follows.

Table 2. Necessary number of active force sensing samples.

k	Point	Soft-finger	Line	Planar
2	PC	SC1	*	*
3	-	*	LC1	*
4	-	SC2	*	*
5	-	SC3	LC2	*
6	-	-	LC3	*

Step 1: Command 2 times of active force sensing.
Step 2: If (9) is satisfied, then contact type is judged as PC.
Step 3: If (11) is satisfied, then contact type is judged as SC1.
Step 4: Command 2 times of active force sensing.
Step 5: If (15) is satisfied, then go to **Step 6**, else **Step 7**.
Step 6: If (19) is satisfied, then contact type is judged as SC2, else LC1.
Step 7: Command 1 time of active force sensing.
Step 8: If (28) is satisfied, then go to **Step 9**, else **Step 10**.
Step 9: If (27) and (29) are satisfied, then contact type is judged as SC3, else LC2.
Step 10: Command 1 time of active force sensing.
Step 11: If (38) is satisfied, then contact type is judged as LC3, else PLC.

This algorithm has the following advantages:

(1) Distinction between SC2 and LC1 is possible.

(2) For planar contact, its contact parameters are indeterminate but the contact type is distinguishable.

(3) Subdivided 8 contact types are distinguishable when $k \geq 6$.

Numerical examples are omitted because of lack of pages.

4 Parameter Identification from Contaminated Data

In actual practice, measured data are contaminated with noise, so sweeping-out method is unusable. In this section, we will derive

(1) Identification method from contaminated data of contact moment.
(2) An algorithm for identification of contact conditions.

4.1 Assumptions

The following assumptions are taken into consideration in addition to (A1)-(A4).

(A5) Measured moment $\boldsymbol{n}_o$ is contaminated with noise, and the variance of its noise is smaller than that of its pure moment.

(A6) The number of active force sensing samples is sufficient large, and our analysis is based on Ergodic hypothesis.

The noise in (A5) is denoted by ε, and its covariance matrix is denoted by

$$\mathrm{Cov}[\varepsilon] = \mathrm{diag}[\gamma_x^2, \gamma_y^2, \gamma_z^2], \tag{41}$$

where diag[•] is a diagonal matrix, $\gamma_\bullet$ standard deviation of noise. The maximum and minimum values of noise are denoted by

$$\gamma_{\max} := \max\{\gamma_x, \gamma_y, \gamma_z\},\ \gamma_{\min} := \min\{\gamma_x, \gamma_y, \gamma_z\} \tag{42}$$

4.2 Contact Moment Representation

The constraint of $\boldsymbol{n}'_c$ is defined by its covariance matrix as

$$\mathrm{Cov}[\boldsymbol{n}'_c] = \mathrm{diag}[\sigma_x^2, \sigma_y^2, \sigma_z^2]. \tag{43}$$

where $\sigma_\bullet$ is standard deviation of contact moment $\boldsymbol{n}'_c$. Four contact types are characterized by

$$\left.\begin{array}{lll} \text{Point contact} & : & \sigma_x = \sigma_y = \sigma_z = 0 \\ \text{Soft - finger contact} & : & \sigma_x = \sigma_y = 0 \\ \text{Line contact} & : & \sigma_x = 0 \\ \text{Planar contact} & : & \text{no constraints} \end{array}\right\} \tag{44}$$

4.3 Calculation of Force and Moment Deviation

By considering the noise, (5) is replaced with

$$A_i \boldsymbol{x} + \boldsymbol{n}_{ci} = \boldsymbol{b}_i - \varepsilon_i,\ (i = 1, 2, \cdots, k), \tag{45}$$

Force and moment deviation from data average are used for eliminating measured data offset. Data average $\boldsymbol{b}_0$ and deviated data $\overline{\boldsymbol{b}}_i$ are calculated by

$$\boldsymbol{b}_0 := \frac{1}{k}\sum_{i=1}^{k}\boldsymbol{b}_i\,, \tag{46}$$

$$\overline{\boldsymbol{b}}_i := \boldsymbol{b}_i - \boldsymbol{b}_0\,,\ (i = 1,2,\cdots,k)\,. \tag{47}$$

Note that the average of deviated data $\overline{\boldsymbol{b}}_i$ is zero. Similarly, $\overline{A}_i$, $\overline{\boldsymbol{n}}_{ci}$, and $\overline{\varepsilon}_i$ are obtained and (45) is transformed as

$$\overline{A}_i\boldsymbol{x} + \overline{\boldsymbol{n}}_{ci} = \overline{\boldsymbol{b}}_i - \overline{\varepsilon}_i\,,\ (i = 1,2,\cdots,k)\,. \tag{48}$$

We define contaminated contact moment $\overline{\nu}_i := \overline{\boldsymbol{n}}_{ci} + \overline{\varepsilon}_i$, where the average of $\overline{\nu}_i$ is zero. Therefore, the problem of identifying contact conditions is transformed to that of finding $\boldsymbol{x}$ and $\{\overline{\nu}_i\}_{i=1}^{k}$ for given $\{\overline{A}_i, \overline{\boldsymbol{b}}_i\}_{i=1}^{k}$.

4.4 Estimation of Contact Point and Contact Moment

To minimize the distribution of $\overline{\nu}_i$, we employ the following performance index:

$$J(\boldsymbol{x}) := \frac{1}{k}\sum_{i=1}^{k}\left\|\overline{\boldsymbol{b}}_i - \overline{A}_i\boldsymbol{x}\right\|^2 \tag{49}$$

Minimizing the index (49), contact position $\boldsymbol{x}$ is estimated as

$$\hat{\boldsymbol{x}}(k) := \left(\frac{1}{k}\sum_{i=1}^{k}\overline{A}_i^T\,\overline{A}_i\right)^{-1}\left(\frac{1}{k}\sum_{i=1}^{k}\overline{A}_i^T\overline{\boldsymbol{b}}_i\right). \tag{50}$$

This contact point is a contact centroid, because the index corresponds to point contact type ($\overline{\boldsymbol{n}}_{ci} = 0$) and the contact point is concentrated from contact area for the other contact types. Hence, the remainder of (49) corresponds to contact moment. Contaminated contact moment $\overline{\nu}_i$ is estimated as

$$\hat{\overline{\nu}}_i(k) := \overline{\boldsymbol{b}}_i - \overline{A}_i\hat{\boldsymbol{x}}(k)\,. \tag{51}$$

From Assumption (A6), estimates $\hat{\boldsymbol{x}}(k)$ and $\hat{\overline{\nu}}_i(k)$ are asymptotically unbiased.

$$\lim_{k\to\infty}\hat{\boldsymbol{x}}(k) = \boldsymbol{x}^*\,,\ \lim_{k\to\infty}\hat{\overline{\nu}}_i(k) = \overline{\nu}_i^* \tag{52}$$

where $\bullet^*$ is true value. Therefore, it is assured that consistent estimator is obtained when the number of active force sensing samples is infinite.

4.5 Contact Type Judgment

4.5.1 Space Average of Contaminated Contact Moment

We define matrix M as the covariance matrix of $\overline{\nu}$.

$$M := \mathrm{Cov}[\overline{\nu}] = \mathrm{E}[\overline{\nu}\overline{\nu}^T] = \mathrm{E}[\overline{\boldsymbol{n}}_c\overline{\boldsymbol{n}}_c^T] + \mathrm{E}[\overline{\boldsymbol{n}}_c\overline{\varepsilon}^T] + \mathrm{E}[\overline{\varepsilon}\overline{\boldsymbol{n}}_c^T] + \mathrm{E}[\overline{\varepsilon}\overline{\varepsilon}^T] \tag{53}$$

where $\mathrm{E}[\bullet]$ is an average of $\bullet$. From (4), (41), (43), and (A6), we have

$$M = [{}^cR_{c'}]\mathrm{diag}[\sigma_x^2, \sigma_y^2, \sigma_z^2][{}^cR_{c'}]^T + \mathrm{diag}[\gamma_x^2, \gamma_y^2, \gamma_z^2] \tag{54}$$

If measured data is noiseless, we have $\gamma_x^2 = \gamma_y^2 = \gamma_z^2 = 0$, so the relation between contact type and the rank of matrix M is described by

$$\mathrm{rank}\, M = \begin{cases} 0 & \text{for point contact type} \\ 1 & \text{for soft - finger contact type} \\ 2 & \text{for line contact type} \\ 3 & \text{for planar contact type} \end{cases} \tag{55}$$

From Weyl theorem [26], eigenvalues μ_1, μ_2, μ_3 $(\mu_1 \ge \mu_2 \ge \mu_3 \ge 0)$ of matrix M have the following condition.

$$\sigma_j^2 + \gamma_{\max}^2 \ge \mu_j \ge \sigma_j^2 + \gamma_{\min}^2, \ (j = 1,2,3) \tag{56}$$

where σ_j is the j-th largest value of σ_x, σ_y, and σ_z. From (56), eigenvalues μ_1, μ_2, and μ_3 are illustrated as Fig. 3. When a threshold θ is preset, four contact types are distinguishable by the number of eigenvalues exceeding θ.

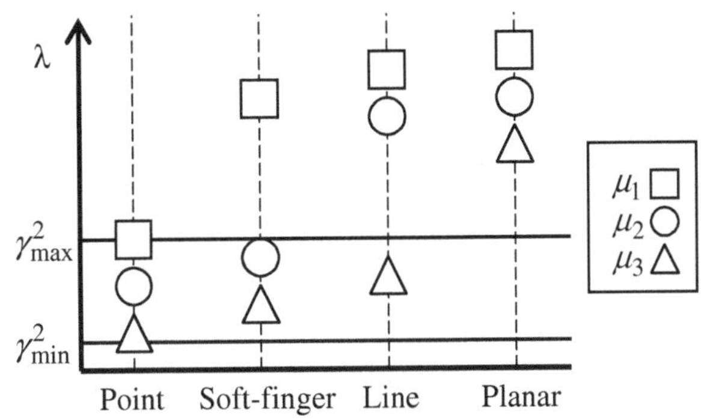

Fig. 3. Relation between contact type and eigenvalues

$$m = \begin{cases} 3 & \text{if} \quad \theta > \mu_1 \\ 2 & \text{if} \quad \mu_1 > \theta > \mu_2 \\ 1 & \text{if} \quad \mu_2 > \theta > \mu_3 \\ 0 & \text{if} \quad \mu_3 > \theta \end{cases} \tag{57}$$

The threshold θ is set exceeding $\gamma_{\max}^2$ theoretically, but can be selected from force sensor calibration beforehand in actual practice.

4.5.2 Time Series Average of Contaminated Contact Moment

We define the following symmetric matrix.

$$N(k) := \frac{1}{k}\sum_{i=1}^{k} \hat{\bar{v}}_i(k)\hat{\bar{v}}_i(k)^T \in \Re^{3\times 3} \tag{58}$$

Performing singular value decomposition, matrix $N(k)$ is rewritten as

$$N(k) = R(k)\,\mathrm{diag}[\lambda_1(k), \lambda_2(k), \lambda_3(k)]R(k)^T\,,\ (\lambda_1 \geq \lambda_2 \geq \lambda_3 \geq 0) \tag{59}$$

where λ_j is eigenvalue, $R(k) = [\boldsymbol{r}_1, \boldsymbol{r}_2, \boldsymbol{r}_3] \in \Re^{3\times 3}$ is a corresponding orthogonal matrix. From (A6), the number of active sensing samples is large enough, we have

$$N(\infty) = M \tag{60}$$

Thus, contact type is distinguishable by using the eigenvalue λ_j and (57) in which μ_j is replaced with λ_j. Eigenvalue $\lambda_j(k)$ and matrix $R(k)$ are estimates of $\sigma_\bullet$ in (44) and contact orientation ${}^{C}R_{C'}$ in Eq. (4). Hence, the contact direction is estimated as

$$\left.\begin{array}{lll} \hat{\boldsymbol{r}}_z = \boldsymbol{r}_1 & \text{for} & \text{Soft - finger contact type} \\ \hat{\boldsymbol{r}}_x = \boldsymbol{r}_3 & \text{for} & \text{Line contact type} \end{array}\right\} \tag{61}$$

This proposed method is used for the case that the number of active force sensing samples is limited in actual practice.

Contact position and contact direction estimated by (50) and (61) are called "linear estimates", because the nonlinear term (4) is omitted in the index (49).

4.6 Typed Estimates of Contact Conditions

When the number of active force sensing samples is limited, the estimation error occurs because contaminated data is used for the estimation and (4) is omitted in the index (49). To obtain more accurate estimates, we must include contact moment constraints (4) and (44). Hence, we employ the following performance index and re-identify contact parameters.

$$J_{type} := \frac{1}{k}\sum_{i=1}^{k}\left\| \overline{A}_i \boldsymbol{x} + {}^{c}R_{c'}\overline{\boldsymbol{n}}'_{ci} - \overline{\boldsymbol{b}}_i \right\|^2 . \tag{62}$$

This re-identification is executed if contact type is judged as soft-finger or line contact type, because for point contact type, (62) is reduce to (49), and for planar contact type, its contact parameters are indeterminate. Because matrix ${}^{c}R_{c'}$ and vector $\overline{\boldsymbol{n}}'_{ci}$ are unknown beforehand, (62) is a nonlinear estimation problem. Hence, linear estimates $\hat{\boldsymbol{x}}$, $\hat{\boldsymbol{r}}_x$, $\hat{\boldsymbol{r}}_z$ in (50) and (61) are assigned to the initial value for minimizing (62). Contact position and contact direction obtained by minimizing (62) are called "typed estimates". Unknown parameters in (62) are redundant, so we must subdivide the index to reduce the parameters.

4.6.1 Parameter Re-estimation for Soft-Finger Contact Type

For soft-finger contact type, we have $\overline{n}'_{cxi} = \overline{n}'_{cyi} = 0$ and $\overline{\boldsymbol{n}}_{ci} = \boldsymbol{r}_z \overline{n}'_{czi}$. To reduce unknown parameters, $\overline{\boldsymbol{n}}_{ci}$ is divided into three cases.

$$\overline{\boldsymbol{n}}_{ci} = \begin{cases} [1, y_1, y_2]^T \overline{z}_i & \text{for} \quad r_{z\max} = |r_{z1}| \\ [y_1, 1, y_2]^T \overline{z}_i & \text{for} \quad r_{z\max} = |r_{z2}| \\ [y_1, y_2, 1]^T \overline{z}_i & \text{for} \quad r_{z\max} = |r_{z3}| \end{cases} \tag{63}$$

Taking the example of $r_{z\max} = |r_{z1}|$, we have

$$J_s := \frac{1}{k}\sum_{i=1}^{k}\left\| \overline{A}_i \boldsymbol{x} + \begin{bmatrix} 1 \\ y_1 \\ y_2 \end{bmatrix} \overline{z}_i - \overline{\boldsymbol{b}}_i \right\|^2 \tag{64}$$

When the index (64) is minimized for $\boldsymbol{x}$, y_1, y_2, and $\overline{z}_s := \{\overline{z}_i\}_{i=1}^{k}$, the following conditions are satisfied.

$$\frac{\partial J_s}{\partial \boldsymbol{x}} = 0, \ \frac{\partial J_s}{\partial y_1} = 0, \ \frac{\partial J_s}{\partial y_2} = 0, \ \frac{\partial J_s}{\partial \overline{z}_i} = 0, \ (i = 1, 2, \cdots, k) \tag{65}$$

Contact position $\boldsymbol{x}$ and contact direction parameters (y_1, y_2) are estimated by

$$\hat{\boldsymbol{x}}_s := \left(\frac{1}{k}\sum_{i=1}^{k} \overline{A}_{si}^T \overline{A}_{si} \right)^{-1} \left(\frac{1}{k}\sum_{i=1}^{k} \overline{A}_{si}^T \overline{\boldsymbol{b}}_{si} \right). \tag{66}$$

Contact moment parameter $\bar{z}_i$ is estimated by

$$\hat{\bar{z}}_i := (\boldsymbol{y}_s^T \boldsymbol{y}_s)^{-1} \boldsymbol{y}_s^T (\bar{\boldsymbol{b}}_i - \bar{A}_i \boldsymbol{x}), \ (i = 1,2,\cdots,k). \tag{67}$$

In (66) and (67), the following notations are used.

$$\boldsymbol{x}_s := \begin{bmatrix} \boldsymbol{x} \\ y_1 \\ y_2 \end{bmatrix}, \ \bar{A}_{si} := \begin{bmatrix} \bar{A}_{i1} & 0 & 0 \\ \bar{A}_{i2} & \bar{z}_i & 0 \\ \bar{A}_{i3} & 0 & \bar{z}_i \end{bmatrix}, \ \bar{\boldsymbol{b}}_{si} := \begin{bmatrix} \bar{b}_{i1} - \bar{z}_i \\ \bar{b}_{i2} \\ \bar{b}_{i3} \end{bmatrix}, \ \boldsymbol{y}_s := \begin{bmatrix} 1 \\ y_1 \\ y_2 \end{bmatrix} \tag{68}$$

Unit direction vector $\boldsymbol{r}_z$ is calculated by

$$\boldsymbol{r}_z = \frac{1}{\sqrt{1 + y_1^2 + y_2^2}} \begin{bmatrix} 1 \\ y_1 \\ y_2 \end{bmatrix} \tag{69}$$

Summarizing the above discussions, we propose an algorithm for parameter estimation of soft-finger contact type.

Step 1: Initial value of $\boldsymbol{x}_s$ is set as the linear estimates $\hat{\boldsymbol{x}}$ and $\hat{\boldsymbol{r}}_z$ in (50) and (61).

Step 2: $\boldsymbol{x}_s$ is fixed and $\{\bar{z}_i\}_{i=1}^k$ is calculated by (67) for minimizing J_s.

Step 3: $\{\bar{z}_i\}_{i=1}^k$ is fixed and $\boldsymbol{x}_s$ is calculated by (66) for minimizing J_s.

Step 2 and **3** are repeated until some convergence condition is satisfied.

4.6.2 Parameter Re-estimation for Line Contact Type

For line contact type, we have $\bar{n}'_{cxi} = 0$ and $\bar{\boldsymbol{n}}_{ci} = \boldsymbol{r}_y \bar{n}'_{cyi} + \boldsymbol{r}_z \bar{n}'_{czi}$. To reduce unknown parameters, $\bar{\boldsymbol{n}}_{ci}$ is divided into three cases.

$$\bar{\boldsymbol{n}}_{ci} = \begin{cases} \begin{bmatrix} 1 & 0 & y_1 \\ 0 & 1 & y_2 \end{bmatrix}^T \begin{bmatrix} \bar{z}_{i1} \\ \bar{z}_{i2} \end{bmatrix} & \text{for} \quad r_{x\max} = |r_{x3}| \\ \begin{bmatrix} 1 & y_1 & 0 \\ 0 & y_2 & 1 \end{bmatrix}^T \begin{bmatrix} \bar{z}_{i1} \\ \bar{z}_{i2} \end{bmatrix} & \text{for} \quad r_{x\max} = |r_{x2}| \\ \begin{bmatrix} y_1 & 1 & 0 \\ y_2 & 0 & 1 \end{bmatrix}^T \begin{bmatrix} \bar{z}_{i1} \\ \bar{z}_{i2} \end{bmatrix} & \text{for} \quad r_{x\max} = |r_{x1}| \end{cases} \tag{70}$$

Taking the example of $r_{x\max} = |r_{x3}|$, we have

$$J_l := \frac{1}{k} \sum_{i=1}^{k} \left\| \bar{A}_i \boldsymbol{x} + \begin{bmatrix} 1 & 0 \\ 0 & 1 \\ y_1 & y_2 \end{bmatrix} \begin{bmatrix} \bar{z}_{i1} \\ \bar{z}_{i2} \end{bmatrix} - \bar{\boldsymbol{b}}_i \right\|^2 \tag{71}$$

When (71) is minimized, the following conditions are satisfied.

$$\frac{\partial J_l}{\partial \boldsymbol{x}} = 0\,,\ \frac{\partial J_l}{\partial y_1} = 0\,,\ \frac{\partial J_l}{\partial y_2} = 0\,,\ \frac{\partial J_l}{\partial \bar{z}_{i1}} = 0\,,\ \frac{\partial J_l}{\partial \bar{z}_{i2}} = 0\,,\ (i = 1,2,\cdots,k) \tag{72}$$

Hence, we have

$$\hat{\boldsymbol{x}}_l := \left(\frac{1}{k}\sum_{i=1}^{k} \overline{A}_{li}^T \overline{A}_{li}\right)^{-1} \left(\frac{1}{k}\sum_{i=1}^{k} \overline{A}_{li}^T \overline{\boldsymbol{b}}_{li}\right). \tag{73}$$

$$\begin{bmatrix} \hat{\bar{z}}_{i1} \\ \hat{\bar{z}}_{i2} \end{bmatrix} := (Y_l^T Y_l)^{-1} Y_l^T (\overline{\boldsymbol{b}}_i - \overline{A}_i \boldsymbol{x})\,,\ (i = 1,2,\cdots,k) \tag{74}$$

where

$$\boldsymbol{x}_l := \begin{bmatrix} \boldsymbol{x} \\ y_1 \\ y_2 \end{bmatrix},\ \overline{A}_{li} := \begin{bmatrix} \overline{A}_{i1} & \bar{z}_{li1} & \bar{z}_{li2} \\ \overline{A}_{i2} & 0 & 0 \\ \overline{A}_{i3} & 0 & 0 \end{bmatrix},\ \overline{\boldsymbol{b}}_{li} := \begin{bmatrix} \bar{b}_{i1} \\ \bar{b}_{i2} - \bar{z}_{li1} \\ \bar{b}_{i3} - \bar{z}_{li2} \end{bmatrix},\ Y_l := \begin{bmatrix} y_1 & y_2 \\ 1 & 0 \\ 0 & 1 \end{bmatrix} \tag{75}$$

Unit direction vector $\boldsymbol{r}_x$ is calculated as

$$\boldsymbol{r}_x = \frac{1}{\sqrt{1 + y_1^2 + y_2^2}} \begin{bmatrix} 1 \\ -y_1 \\ -y_2 \end{bmatrix} \tag{76}$$

In a similar way of soft-finger contact type, (71) is minimized by using (73) and (74).

4.7 Algorithm for Identification of Contact Conditions from Contaminated Data

Summarizing the above discussions, we propose an algorithm for identification of contact conditions as follows:

Step 1: Command active force sensing.
Step 2: Calculate of measured force and moment deviation by using (47).
Step 3: Estimate contact position $\hat{\boldsymbol{x}}$ by using (50).
Step 4: Estimate contact moment $\hat{\nu}_i$ by using (51).
Step 5: Calculate matrix N by using (58) and derive eigenvalues λ_1, λ_2, and λ_3.
Step 6: Judge contact type by using (57).
Step 7: If it is judged as soft-finger contact type, unknown parameters is re-estimated by using Section 4.6.1.
Step 8: If it is judged as line contact type, unknown parameters is re-estimated by using Section 4.6.2.
Step 9: Go to **Step 1**.

This algorithm has the following advantages:

(1) Contaminated measured moment is usable.
(2) The contact conditions are identified continuously by active force sensing motion.
(3) For planar contact type, unknown parameters are indeterminate but the contact type is distinguishable.

From Section 3, this algorithm is usable when $k \geq 6$, but for point contact type, contact parameters are identifiable when $k \geq 2$.

4.8 Numerical Example

We demonstrate effectiveness of our proposed method through numerical examples. That is, we verify that

(1) Contact position and contact moment are identifiable by using linear estimates,
(2) Four contact types are distinguishable by using the eigenvalues of covariance matrix of the estimated contact moment,
(3) More accurate contact parameters are obtained by reidentification.

Four contact types shown in Fig. 4 are considered. These simulation conditions are assigned as follows:

[Simulation conditions]

Contact position: $\boldsymbol{x} = \boldsymbol{r}_c = [0.1, 0.1, 0.1]^T$ [m], (77)

Contact orientation: ${}^{c}R_{c'} = [\boldsymbol{r}_x, \boldsymbol{r}_y, \boldsymbol{r}_z] = \mathrm{diag}[1, -1, -1]$, (78)

Contact force and moment in $\Sigma_{c'}$:

$$\mathrm{Cov}[\boldsymbol{f}_c'] = \mathrm{diag}[0.500^2, 0.500^2, 2.000^2]^T\,[\mathrm{N}^2],$$

$$\mathrm{Cov}[\boldsymbol{n}_c'] = \mathrm{diag}[0.100^2, 0.100^2, 0.050^2]^T\,[(\mathrm{Nm})^2] \quad (79)$$

Measurement noise in Σ_o:

$$\mathrm{Cov}[\boldsymbol{\varepsilon}] = \mathrm{diag}[0.002^2, 0.002^2, 0.003^2]^T\,[(\mathrm{Nm})^2]. \quad (80)$$

Measured force $\boldsymbol{f}_o$ and moment $\boldsymbol{n}_o$ are generated from the following steps:

Step 1: For each contact type shown in Fig. 4, contact force $\boldsymbol{f}_c'$ and moment $\boldsymbol{n}_c'$ are randomly generated by using moment constraint (44) and covariance matrix (79).
Step 2: Contact force $\boldsymbol{f}_c$ and moment $\boldsymbol{n}_c$ are calculated using (78).

$$\boldsymbol{f}_c = {}^{c}R_{c'}\boldsymbol{f}_c',\ \boldsymbol{n}_c = {}^{c}R_{c'}\boldsymbol{n}_c' \quad (81)$$

Step 3: Force $\boldsymbol{f}_o$ and moment $\boldsymbol{n}_o$ are calculated using (3) and (77).

$$\boldsymbol{f}_o = \boldsymbol{f}_c,\ \boldsymbol{n}_o = \boldsymbol{r}_c \times \boldsymbol{f}_c + \boldsymbol{n}_c \quad (82)$$

Step 4: To obtain contaminated measured moment, (80) is added to $\boldsymbol{n}_o$.

$$\boldsymbol{n}_o \leftarrow \boldsymbol{n}_o + \boldsymbol{\varepsilon} \quad (83)$$

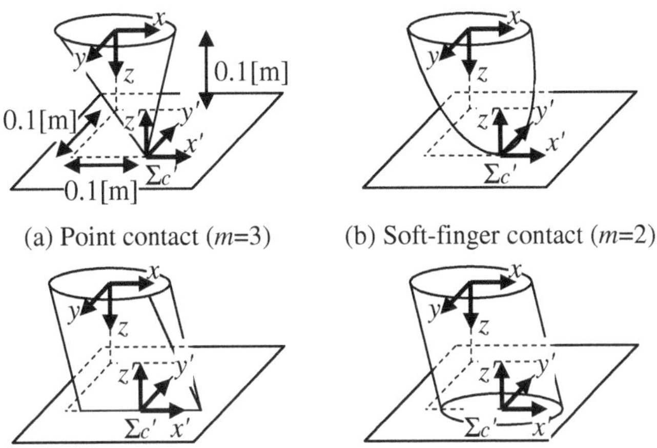

(a) Point contact (m=3) (b) Soft-finger contact (m=2)

(c) Line contact (m=1) (d) Planar contact (m=0)

Fig. 4. Contact conditions in this numerical example

The proposed algorithm (Section 4.7) is executed. Fig. 5 shows eigenvalues of (a) point contact type data, (b) soft-finger contact type data, (c) line contact type data, and (d) planar contact type data. For each contact type, 20 trials are exemplified. For each trial, 50 times of active force sensing are commanded.

In case of point contact type data, three eigenvalues converged small value corresponding to noise. In case of soft-finger contact type data, the largest eigenvalue converged contact moment corresponding to σ_z^2, and the other eigenvalues converged small value corresponding to noise. In case of line contact type data, the largest two eigenvalues converged contact moment corresponding to σ_y^2 and σ_z^2. In case of planar contact type data, three eigenvalues converged large value corresponding to σ_x^2, σ_y^2 and σ_z^2. These four contact types are distinguishable when the number of samples k is set larger than about 20 and the threshold θ is set around 0.001. Theoretically, this threshold can be set exceeding $\gamma_{\max}^2$ (Fig. 3).

Fig.6 shows measured force and moment of one trial of line contact type data (Fig. 5(c)). Figures 7 and 8 show contact parameters estimated from line contact data, and Fig. (a) and (b) show linear and typed estimates, respectively. From Fig. 7, contact point $\hat{\boldsymbol{x}}$ converged true value (0.1,0.1,0.1). From Fig. 8, direction of contact line $\hat{\boldsymbol{r}}_x$ converged true value (1,0,0). Typed estimates are more accurate than linear ones. Hence, effectiveness of our proposed method has been verified.

From these numerical examples, it has been verified that four contact types and contact parameters are identifiable from contaminated data by using our proposed method. In this numerical example, contact type is distinguishable when $k \geq 20$.

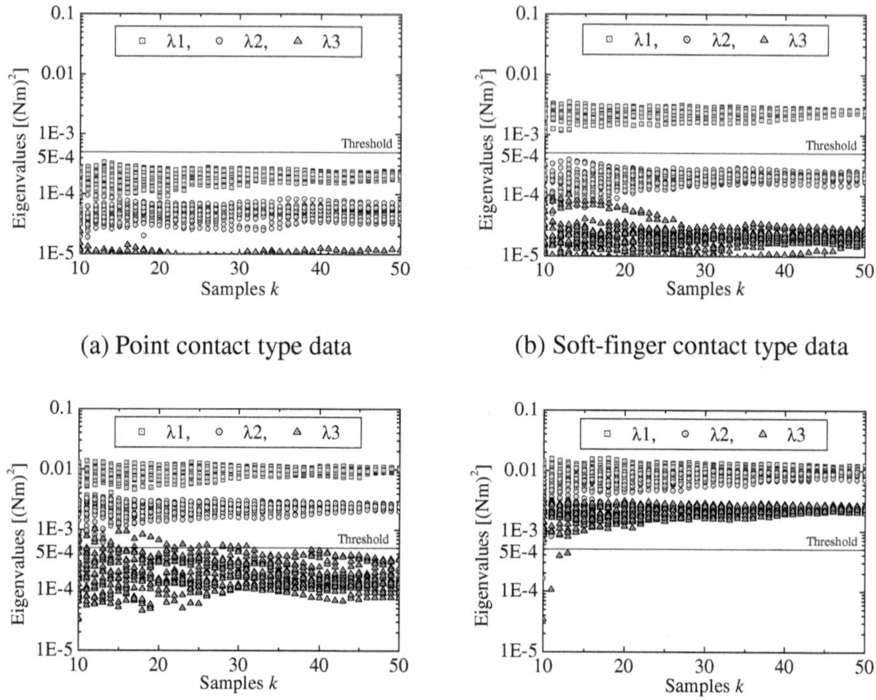

(a) Point contact type data (b) Soft-finger contact type data

(c) Line contact type data (d) Planar contact type data

Fig. 5. Eigenvalues corresponding contact type

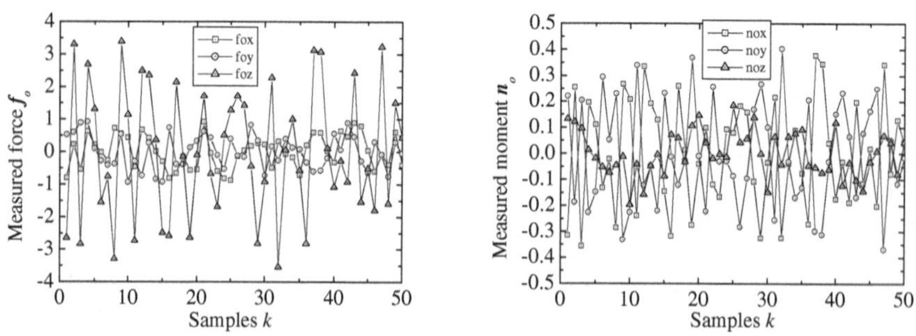

(a) Measured force f_o (b) Measured moment n_o

Fig. 6. Measured force and moment of one trial in Fig 5(c)

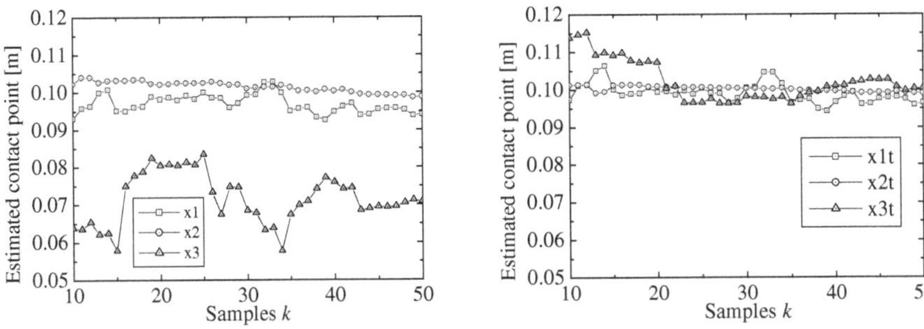

(a) Linear estimate (b) Typed estimate

Fig. 7. Estimated contact point

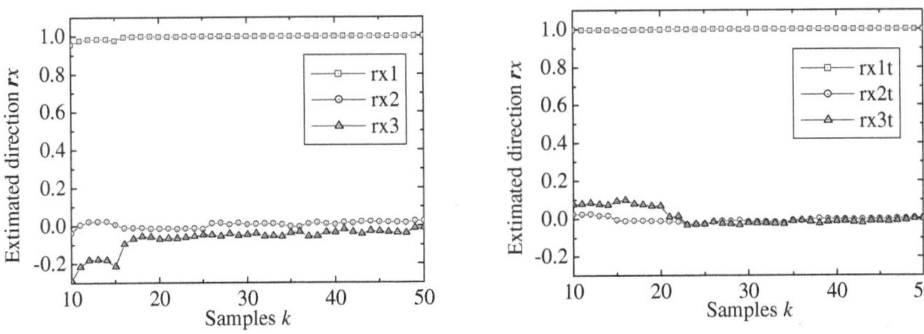

(a) Linear estimate (b) Typed estimate

Fig. 8. Estimated direction of contact line

5 Transition Detection of Contact Conditions

In Section 3 and 4, contact type did not change to simplify problem formulation. These cases enable equal treatment between past measured data and latest one. In actual practice, transition of contact type occurs during assembly tasks (Fig. 9). To

(a) Point contact (b) Line contact (c) Planar contact

Fig. 9. Transition of contact types

perform the assembly tasks by robots dexterously, we must detect the transition. However, past data may not satisfy current contact type, for example, past data is acquired at Fig. 9(a) and latest one is acquired at (c). Hence, weight coefficient of the latest data is enlarged, and the number of employed data for parameter estimation is limited.

This section replace (A4) with the following assumption.

(A4') Contact type changes during object manipulation

5.1 Calculation of Force and Moment Deviation

In a similar way of Section 4, to eliminate measurement data offset, we use force and moment deviation. The weighted data average $\boldsymbol{b}_0(k)$ and the deviation $\bar{\boldsymbol{b}}_{k-j}(k)$ are calculated as

$$\boldsymbol{b}_0(k) := \frac{1}{w_a}\sum_{j=0}^{n-1} w_j \boldsymbol{b}_{k-j}\,,\ w_a := \sum_{j=0}^{n-1} w_j \tag{84}$$

$$\bar{\boldsymbol{b}}_{k-j}(k) := \boldsymbol{b}_{k-j} - \boldsymbol{b}_0(k)\,,\ (j = 0,1,\cdots,n-1) \tag{85}$$

where k is the sample number, j relative number for past data (latest data is j=0), n the number of estimation samples, w_j weight, w_a total weight. Note that weighted average of deviation $\bar{\boldsymbol{b}}_{k-j}(k)$ is zero. Similarly, calculating $\bar{A}_{k-j}(k)$ and $\bar{\boldsymbol{n}}_{c,k-j}(k)$, (5) is transformed to

$$\bar{A}_{k-j}(k)\boldsymbol{x} + \bar{\boldsymbol{n}}_{c,k-j}(k) = \bar{\boldsymbol{b}}_{k-j}(k)\ ,\ (j = 0,1,2,\cdots,n-1) \tag{86}$$

5.2 Linear Estimate of Contact Conditions

We employ the following performance index.

$$J := \frac{1}{w_a}\sum_{j=0}^{n-1} w_j \left\| \bar{\boldsymbol{b}}_{k-j}(k) - \bar{A}_{k-j}(k)\boldsymbol{x} \right\|^2 \tag{87}$$

Minimizing (87) yields the following contact centroid:

$$\hat{\boldsymbol{x}}(k) := \left\{ \frac{1}{w_a}\sum_{j=0}^{n-1} w_j \bar{A}_{k-j}^T(k)\bar{A}_{k-j}(k) \right\}^{-1} \left\{ \frac{1}{w_a}\sum_{j=0}^{n-1} w_j \bar{A}_{k-j}^T(k)\bar{\boldsymbol{b}}_{k-j}(k) \right\} \tag{88}$$

Contact moment is estimated as

$$\hat{\bar{\boldsymbol{n}}}_{c,k-j}(k) := \bar{\boldsymbol{b}}_{k-j}(k) - \bar{A}_{k-j}(k)\hat{\boldsymbol{x}}(k)\,,\ (j = 0,1,\cdots,n-1) \tag{89}$$

We calculate the following symmetric matrix

$$N(k) := \frac{1}{w_a} \sum_{j=0}^{n-1} w_j \{\hat{\bar{\boldsymbol{n}}}_{c,k-j}(k)\}\{\hat{\bar{\boldsymbol{n}}}_{c,k-j}(k)\}^T \in \Re^{3\times 3} \tag{90}$$

Performing singular value decomposition, we have the following form.

$$N(k) = V(k)\mathrm{diag}[\lambda_1(k), \lambda_2(k), \lambda_3(k)]V(k)^T, \ \lambda_1 \geq \lambda_2 \geq \lambda_3 \geq 0 \tag{91}$$

where λ_i is eigenvalue and orthogonal matrix V is estimate of orientation ${}^cR_{c'}$. Contact type is judged by (57).

Section 4 corresponds to the case that $k = n$ and $w_i = 1$.

5.3 Transition Detection

From Section 3, the above identification method is usable when $k \geq 6$. However, estimated contact type is occasionally incorrect because measured data are contaminated with noise and n is limited. We check continuity of estimated contact type and revise contact type as follows:

$$\left.\begin{array}{ll} \text{if} & m(k) = m(k-1) = \cdots = m(k-n_t+1) \\ \text{then} & m_{rev}(k) = m(k) \\ \text{else} & m_{rev}(k) = m_{rev}(k-1) \end{array}\right\}, \tag{92}$$

where n_t is the number of data for transition detection, m_{rev} the revised contact type.

5.4 Re-identification Contact Conditions Considering Contact Type

Because contact moment $\bar{\boldsymbol{n}}_c$ is omitted in the index (87), we must derive more accurate estimate of contact parameters, so employ the following performance index (Typed estimate).

$$J_{type} := \frac{1}{w_a} \sum_{j=0}^{n-1} w_j \left\| \bar{\boldsymbol{b}}_{k-j}(k) - \{\bar{A}_{k-j}(k)\boldsymbol{x} + \bar{\boldsymbol{n}}_{c,k-j}(k)\} \right\|^2 \tag{93}$$

Typed estimate of contact conditions is derived in a similar way of Section 4.

5.5 Numerical Examples

Effectiveness of our proposed method is demonstrated through numerical examples. For actual practice, we exemplify the contact transition of Fig. 9. The cube shown in Fig. 10 is grasped by a robot hand. Its bottom makes contact with the environment, and contact type changes as shown in Fig. 11. Each transition occurs in 50 samples. Table 3 shows contact position, in which s and t are free parameters of contact direction. The first term of contact position is a contact centroid. In this numerical example, measured force and moment are generated with $s = t = 0$. Contact force and moment (f_c', n_c') are the same conditions as (79).

Gravitational effect is added to the measured force and moment, calculated by

$$\left.\begin{aligned} {}^o f_g &= w_g[\sin\phi_y, -\sin\phi_x\cos\phi_y, \cos\phi_x\cos\phi_y]^T \\ {}^o n_g &= {}^o r_g \times {}^o f_g \end{aligned}\right\} \tag{94}$$

where gravitational center is given by ${}^o r_g = [0,0,0.05]^T$ [m], the object mass given by $w_g = 10$[N]. The object incline (ϕ_x, ϕ_y) is given as Table 3. In this case, the object orientation changes with 0.004[rad] per one sample. Measurement noise in Σ_o is assigned with

$$\mathrm{Cov}[\varepsilon_{fo}] = \mathrm{diag}[0.040^2, 0.040^2, 0.100^2]^T [\mathrm{N}^2],$$
$$\mathrm{Cov}[\varepsilon_{no}] = \mathrm{diag}[0.002^2, 0.002^2, 0.003^2]^T [(\mathrm{Nm})^2]. \tag{95}$$

[Transition detection condition]
Number of data for parameter estimation: $n = 15$
Weight coefficient: $w_j = 0.9^j$
Threshold: $\theta = 5\times10^{-4}$
Number of data for transition detection: $n_t = 3$

If measured data are noiseless, four contact types are distinguishable when $k \geq 6$. If measured moment is contaminated, it is numerically assured that four contact types are distinguishable when $k \geq 20$. If measured force is also contaminated, sensing times k must be much increased more than 20 samples. But we use $n = 15$ in this simulation, to detect contact transition quickly.

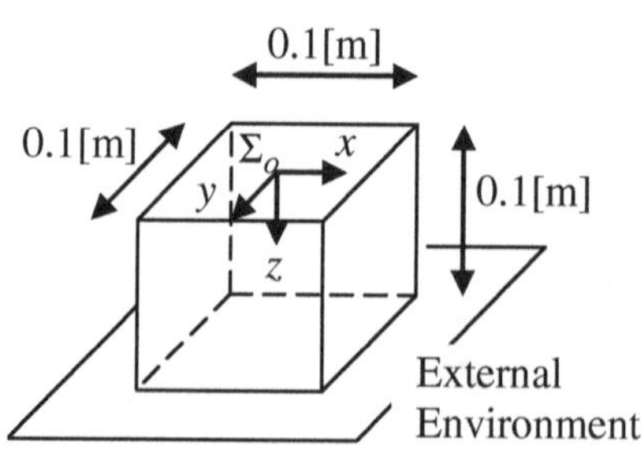

Fig. 10. Shape of the grasped object

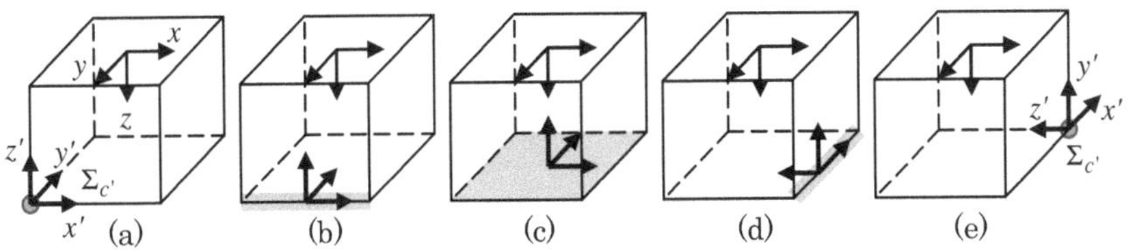

Fig. 11. Transition of the contact types (Gray color is contact area)

Table 3. Contact position between the grasped object and its environment, and orientation of the grasped object.

	Type	Contact Position [m]	ϕ_x [rad]	ϕ_y [rad]
(a)	Point	(-0.05, 0.05, 0.1)	0.2	from 0.2 to 0
(b)	Line	(0, 0.05, 0.1)+s(1,0,0)	from 0.2 to 0	0
(c)	Planar	(0, 0, 0.1)+s(1,0,0)+t(0,1,0)	0	0
(d)	Line	(0.05, 0, 0.1)+t(0,1,0)	0	from 0 to -0.2
(e)	Point	(0.05, -0.05, 0.1)	from 0 to -0.2	-0.2

Fig. 12(a) shows eigenvalues, (b) contact type, (c) contact position, (d) contact position in reidentification, (e) direction of contact line, and (f) direction of contact line in reidentification.

From Fig. 12, it is assured that contact transition is detected. By continuity check of contact transition, incorrect judgment around k=60, 170, 200 in Fig. 12(b) is revised. From Fig. 12(c) and (d), it is shown that contact position transits from (-0.05, 0.05, 0.1) to (0.05, -0.05, 0.1), and for line contact type, contact position is revised to more accurate value in reidentification. But x_1 in k=50-100 and x_3 in k=150-200 correspond to parameters s and t, and these are indeterminate. From Fig. 12(e) and (f), direction of contact line is estimated around (1,0,0) in case of Fig. 11(b), and around (0,1,0) in Fig. 11(d). These directions are revised in reidentification. Direction angle error is revised from 20 [deg] to 10 [deg]. The cause of angle error is as follows. (1) Contaminated data are used to detect contact transition. (2) The number of data for identification is small. If the active sensing data n is increased, more accurate estimates of contact direction are obtained, but transition of contact type is detected slowly. When the above direction error is $|\phi_e| \le 10$[deg] , the direction is located around

$$0.98 \le \cos\phi_e \le 1, \quad |\sin\phi_e| \le 0.17 \tag{96}$$

This direction is shown in Fig. 12(f). Estimated direction greatly changes around k=100 and k=215, because current contact type respectively already changed to planar and point contact types but judged contact type is line. This difference between current and judged contact types is affected by past data, so direction of contact line is indeterminate.

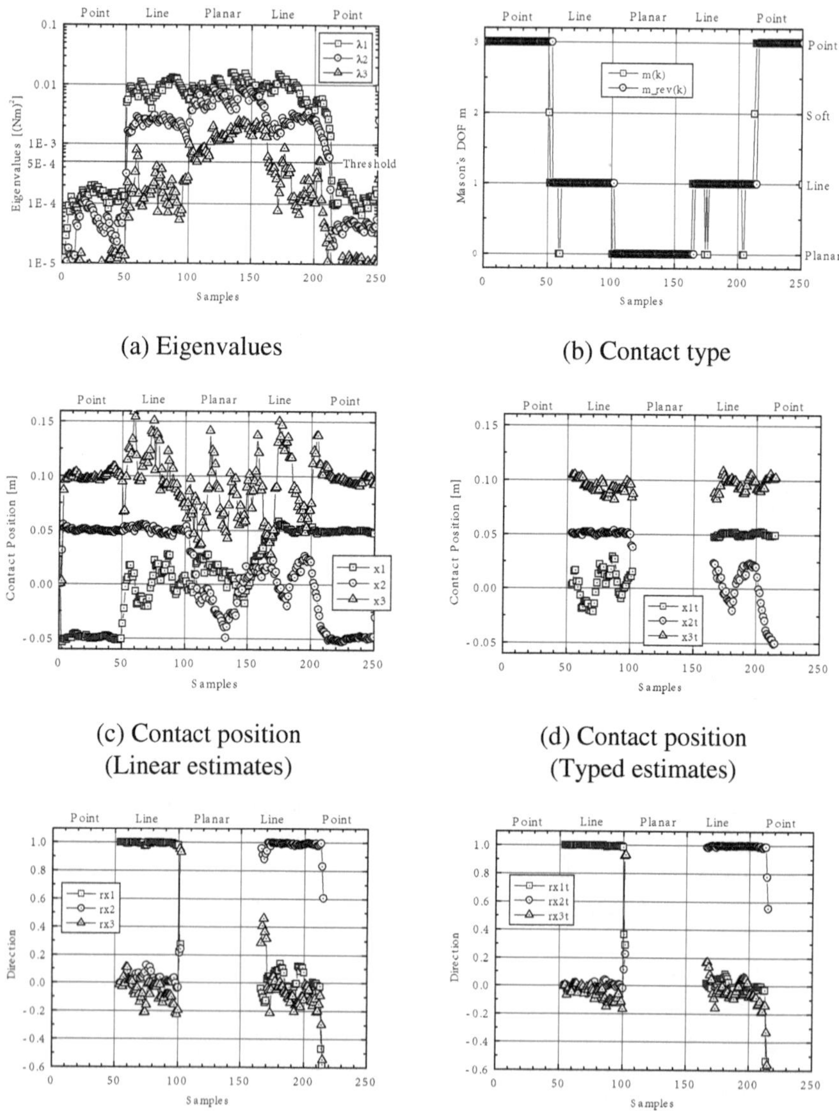

(a) Eigenvalues

(b) Contact type

(c) Contact position (Linear estimates)

(d) Contact position (Typed estimates)

(e) Direction of contact line (Linear estimates)

(f) Direction of contact line (Typed estimates)

Fig. 12. Estimated parameters of contact conditions in transition detection

Because data offset is eliminated by using data deviation, gravitational effect of object mass is reduced when change of object inclination is small. In this numerical example, smallest eigenvalue is bit enlarged, but this does not affect to contact type judgment.

6 Conclusions

This article has discussed the parameter identification problem of contact conditions between a grasped object and its environment. We have the following novel results: (1) It is assured that four contact types are distinguishable when the number of active force sensing times is at least 6. (2) Efficient algorithms distinguishing four contact types are proposed. Effectiveness of the algorithms is verified through numerical examples.

Our proposed methods have the following advantages: (1) The contact conditions between unknown shaped object and environment are identifiable. (2) Four contact types as point, soft-finger, line, and planar contact types are distinguishable. (3) If the type is judged as soft-finger contact type, the direction of contact normal is estimated. If the type is judged as line contact type, the direction of contact line is estimated. Moreover, (4) Contact transition is detectable.

As shown in Fig. 13, our proposed method is applicable to the identification of the contact conditions in a variety of tasks, such as the contact between a grasped object

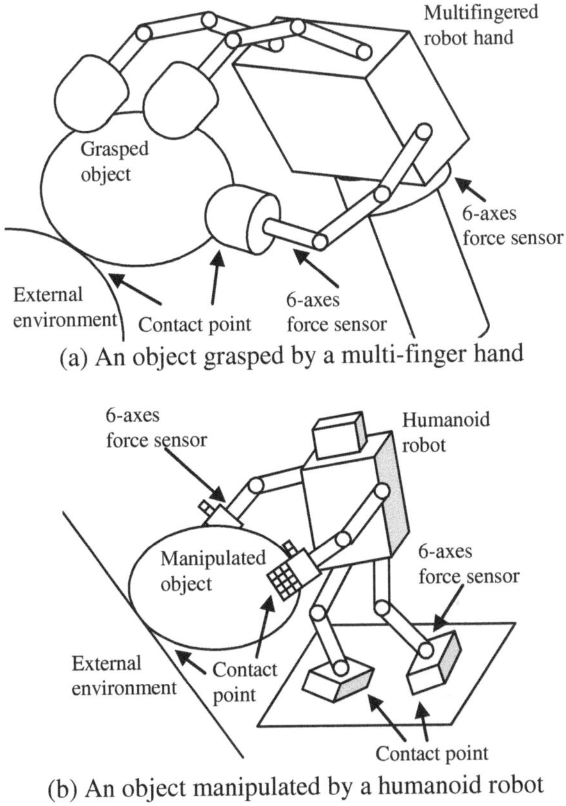

Fig. 13. Manipulation with contact

and an external environment, the contact between a grasped object and each finger of a robot hand, the contact between a manipulated object and each arm of a humanoid robot, the contact between a floor surface and each foot of a legged robot.

Acknowledgments

We would like to thank Dr. Tatsushi OOBA for useful discussion.

References

1. Shimokura, K., Muto, S.: A Study of a Munipulation System with a Detection Module for Contact-state- transition. Journal of the RSJ 12(6), 837–845 (1994) (in Japanese)
2. Takano, K., Nakai, T., Sasaki, K.: Control Level Analysis of Primitive Motion of Handling Rectangular Solid. Journal of the JSPE 65(7), 1051–1055 (1999) (in Japanese)
3. Nakai, T., Sasaki, K.: Signal Processing of Active Force Sensing on Detecting Contact State. Journal of the Horological Institute of Japan 45(1), 16–26 (2001) (in Japanese)
4. Skubic, M., Voltz, R.A.: Acquiring Robust Force-Based Assembly Skills from Human Demonstration. IEEE Trans. on R. and A. 16(6), 772–781 (2000)
5. Takamatsu, J., Ogawara, K., Kimura, H., Ikeuchi, K.: Understanding of Human Assembly Tasks for Robot Execution (Generation of Optimal Trajectory Based on Transitions of Contact Relations). Journal of the RSJ 22(6), 752–763 (2004) (in Japanese)
6. Shutter, J.D., Bruyninckx, H., Durte, S., Geeter, J.D., Kaptupitiya, J., Demey, S., Lefebvre, T.: Estimating First-Order Geometric Parameters and Monitoring Contact Transitions during Force-Controlled Compliant Motion. The Int. Journal of Robotics Research 18(12), 1161–1184 (1999)
7. Lefebvre, T., Bruyninckx, H., Schutter, J.D.: Polyhedral Contact Formation Identification for Autonomous Compliant Motion: Exact Nonlinear Bayesian Filtering. IEEE Trans. on R. and A. 21(1), 124–129 (2005)
8. Meeussen, W., Rutgeerts, J., Gadeyne, K., Bruyninckx, H., Schutter, J.D.: Contact State Segmentation Using Particle Filters for Programming by Human Demonstration in Compliant Motion Tasks. IEEE Trans. on R. 23(2), 218–231 (2007)
9. Kitagaki, K.: Groping Motion and Manipulation. Journal of the RSJ 11(8), 1138–1142 (1993) (in Japanese)
10. Ishikawa, M.: Active Sensing and Robot Hand. Journal of the RSJ 11(7), 6–10 (1993) (in Japanese)
11. Salisbury, J.K.: Interpretation of Contact Geometries from Force Measurements. In: Proc. of 1st Int. Symp. Robotics Research, pp. 565–577 (1983)
12. Tsujimura, T., Yabuta, T.: Object Detection by Tactile Sensing Method Employing Force/Torque Information. IEEE Trans. on R. and A. 5(4), 444–450 (1989)
13. Zhou, X., Shi, Q., Li, Z.: Contact Localization using Force/Torque Measurements. In: Proc. of the IEEE ICRA, vol. 2, pp. 1339–1344 (1996)
14. Nagata, K., Tsumura, M., Omata, T.: Development of a Fingertip-type 6D Force Sensor and Error Evaluation of Contact Point Sensing. Journal of the RSJ 14(8), 1221–1228 (1996) (in Japanese)
15. Nagase, K., Yoshinaga, K., Nakashima, A., Hayakawa, Y.: Estimation of Contact Point by Force Sensor with Measurement Noise. Journal of the RSJ 23(6), 85–91 (2005)

16. Bicchi, A.: Intrinsic Contact Sensing for Soft Fingers. In: Proc. of the IEEE ICRA, vol. 2, pp. 968–973 (1990)
17. Murakami, K., Hasegawa, T.: Novel Fingertip Equipped with Soft Skin and Hard Nail for Dexterous Multi-fingered Robotic Manipulation. In: Proc. of the IEEE ICRA, pp. 708–713 (2003)
18. Murakami, K., Hasegawa, T.: Tactile Sensing of Edge Direction of an Object using a Robotic Fingertip with Soft Skin. Journal of the RSJ 24(2), 240–247 (2006) (in Japanese)
19. Kitagaki, K., Ogasawara, T., Suehiro, T.: Contact State Detection by Force Sensing for Assembly Tasks. In: Proc. of the IEEE Int. Conf. on Multisensor Fusion and Integration for Intelligent Systems, pp. 366–370 (1994)
20. Mimura, N., Funahashi, Y.: Parameter Identification of Contact Conditions by Active Force Sensing. Transactions of the JSME, Series C 60(579), 3816–3821 (1994) (in Japanese)
21. Mimura, N., Funahashi, Y., Mouri, T.: An Algorithm for Identification of Contact Conditions. Transactions of the JSME, Series C 63(610), 2061–2068 (1997) (in Japanese)
22. Yamada, T., Mouri, T., Mimura, N., Funahashi, Y.: Identification of Contact Conditions from Contaminated Data. Transactions of the JSME, Series C 64(618), 584–589 (1998) (in Japanese)
23. Mouri, T., Yamada, T., Mimura, N., Funahashi, Y.: Identification of Contact Conditions from Contaminated Data of Contact Moment. Transactions of the JSME, Series C 66(648), 2685–2692 (2000) (in Japanese)
24. Yamada, T., Mouri, T., Shimosaka, K., Mimura, N., Funahashi, Y.: Parameter Identification and Transitions Detection of Contact Conditions of a Robot by Active Force Sensing. Transaction of the JSME, Series C 74(747), 2721–2728 (2008) (in Japanese)
25. Mason, M.T., Salisbury, J.K.: Robot Hands and the Mechanics of Manipulation. MIT Press, Cambridge (1985)
26. Horn, R.A., John, C.R.: Topics in Matrix Analysis, pp. 170–181. Cambridge University Press, Cambridge (1991)

Modelling User-Centric Pervasive Adaptive Systems – The REFLECT Ontology

Gerd Kock[1], Marko Ribarić[2], and Nikola Šerbedžija[1]

[1] Fraunhofer FIRST, Kekuléstr. 7, D 12489 Berlin, Germany
{gerd.kock,nikola.serbedzija}@first.fraunhofer.de
[2] Institute Mihailo Pupin, Volgina 15, 11060 Belgrade, Serbia
marko.ribaric@institutepupin.com

Abstract. Moving on from desktop computers, computing intelligence will be woven into the "fabric of everyday life". User-centric pervasive adaptive systems will deliver services adapted to ourselves and our context of use. Their practical development is a cross disciplinary endeavour requiring synergy of computer engineering, human science and practice. This work describes a novel *reflective approach* for development and deployment of adaptive systems. Special focus is on a *reflective ontology* which uses UML diagrams as graphical representations, and is employed for developing *reflective applications*. The semi-formal and semi-automatic development chain starts with UML elements, whereas the UML elements can be partially represented by XML, which in turn can be used to parameterize the Java implementation of the final system.

1 Introduction

Most of the present systems that deal with personal human experience (affective computing systems) are poorly engineered. As a consequence, maintenance and modification of affective computing applications becomes very difficult, if not even impossible. Furthermore, re-usability as a capability of re-deploying the same software structures in different application domains is not possible. This approach adds complexity to the current pervasive adaptive systems by introducing cognitive and physical experience and at the same time strives for a generic, flexible and re-usable solution.

One of the concrete answers to new challenges in building pervasive adaptive systems is the reflective framework (cf. [25]) aiming at developing methods and tools for developing such user-centric pervasive adaptive systems (in the following: REFLECTive systems). The reflective framework is a service- and component-oriented infrastructure implemented in Java using Eclipse and OSGi. A multidisciplinary team with computer engineering [18], human science [6] and praxis background [19] works together deploying a wide range of competences in a common endeavour. Often, the parties involved in such a common development do not understand each other well, as they are coming from different fields, have different educational background and sometimes even different way of thinking. These factors pose an extra burden to the already complex developing task.

The major goal of the Reflective ontology is to provide a common description of the concepts and structures needed for the design of reflective software and the development of the reflective applications. In this context the reflective system is a system that

N.T. Nguyen and E. Szczerbicki (Eds.): Intel. Sys. for Know. Management, SCI 252, pp. 307–329.
springerlink.com

is responsive to emotional cognitive and physical experiences of the users involved in a particular task or situation. Reflective ontology should also contribute to a common understanding of the problem domain within a cross-disciplinary team requiring synergy of computer engineering, human science and praxis in developing complex systems that seamlessly adapt to the users.

Here, we consider an ontology to be a structure of concepts or entities within a domain, organized by relationships [26]. We define it to be an explicit specification of an abstract, simplified view of a world to be represented, which specifies both the concepts inherent in this view as well as their interrelationships [14].

As proposed in [10], we have used UML [27] as the language for describing the REFLECT ontology. One reason is that this approach allows to base the discussion on easily understandable graphical representations[1], which will help in the future usage and elaboration of the ontology. Another important reason is, that by using UML the ontology can be related to the REFLECT software in a straightforward manner. Partially, it will be possible to (semi-)automatically generate XML from the given UML structures, which in turn can be used to parameterize the Java implementation of the REFLECT system.

Section 2 presents the global structure of the REFLECT ontology, relates it to the REFLECT software architecture, and points out its intended role in the development of applications. Section 3 details a specific part of the REFLECT ontology, and the case study in section 4 illustrates, how a REFLECT application can be modelled on the base of this ontology. Section 5 is devoted to implementation issues; it details the execution of REFLECT applications and points out XML based means, by which the UML elements of the ontology can be used in this process. Section 6 discusses the methodologies having been employed when developing the core ontology, and specifically reports about existing ontologies and their partial reuse. The section 7 concludes this chapter summarizing the achievements and pointing to the open problems for the further work.

2 Structure and Role of the Ontology

This chapter sketches REFLECT software structures, and then explains the overall organization of the REFLECT ontology and its role in the development of REFLECT applications.

2.1 REFLECT Software

There are two different appearances of the REFLECT software. One is the REFLECT framework that consists of the REFLECT infrastructure and a number of tools for developing and deploying reflective applications. It runs on top of OSGI and uses numerous Java –based tools. Another is the REFLECT application as a concrete application instance developed with the reflective framework.

The REFLECT development system consists of three layers (Figure 1(a)). The ***tangible layer*** includes basic services, which can be used to communicate with sensors and actuators. The ***reflective layer*** is the middleware of the REFLECT system,

[1] The UML diagrams used within this chapter are explained in footnotes.

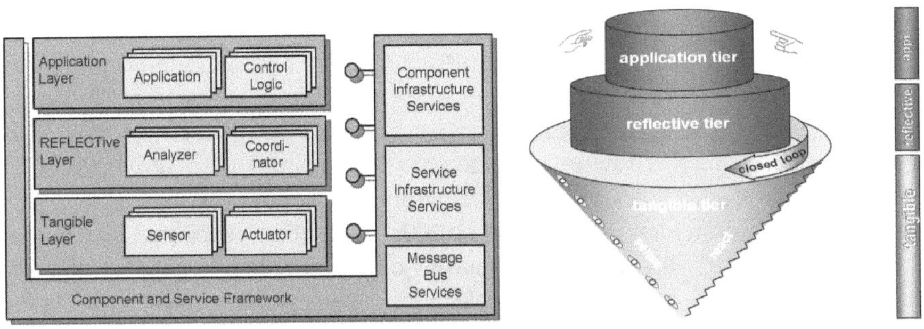

(a) **REFLECT Framework** (b) **REFLECT Application**

Fig. 1. REFLECT Software

comprising a service registry and other central components. And the ***application layer*** adheres to software components and tools for the development of applications and related configuration items[2].

Consequently, any REFLECT application contains components of each of these layers (Figure 1(b)). Such an application can be considered as a constantly adapting system, analysing sensor inputs, reflecting about the actual and former results of the analysis, and reacting according to a given application scheme via controlling actuating variables. Conceptually this means, that any REFLECT application is running in a ***closed loop***, reflecting the permanent cycle ***sense-analyse-react***. However, this does not imply that at any time the same loop is performed, it only means that perpetually sensors deliver inputs, leading to outputs for actuators, which in turn lead to modified inputs, etc. Actually, there will be applications, where a number of closed loops will run in parallel. Thus the Figure 1(b) represents a visual metaphor of a "spinning top" that performs fine-tuning and self-adaptation in an endless loop.

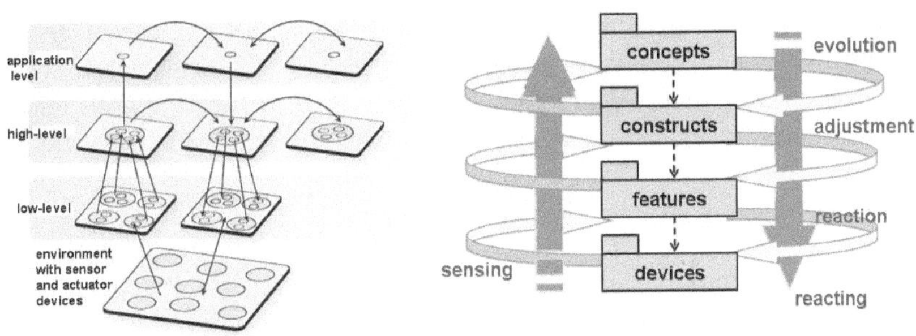

(a) **Conceptual view** (b) **Ontology view**

Fig. 2. REFLECT Viewpoints

[2] The goal of this section is not to give a comprehensive account of the REFLECT software, but its basic structures are explained, so that the role of the REFLECT ontology can be presented in a sensible was. Details about the REFLECT system architecture can be found in [20].

Figure 2(a) details this conceptual view. The basic idea is to hierarchically organize the sense-analyse-react cycle. At the bottom, there is the environment with sensor and actuator ***devices***. Above, the low-level software components perform the first and the last steps in any cycle; either ***features*** are extracted from sensors or are used to control actuators. Next, the high-level components deal with abstract ***constructs*** describing the user context, his psychological, cognitive or physical state, or high-level system goals. The software is organized such that these high-level components build on the low-level ones, but not directly on devices. And at the top application level, according to the ***concepts*** at hand, the effectively available user constructs (obtained from the connected sensors) are related to possible system goals (depending on the involved actuators). Again, the application level directly relies only on the high-level components.

Considering the diversity of system adaptations, one can distinguish between immediate ***reactions***, reflective ***adjustments***, and ***evolutionary*** adaptations (see Figure 2(b)). Roughly spoken, an immediate reaction takes place, if sensor features are directly used to control actuators, which specifically might be of interest in case of emergency. Reflective adjustments are reactions, which are based on the analysis of different sensor features, resulting in the availability of more abstract constructs. And evolutionary adaptations, last but not least, account for the long term ramifications of the adaptations mentioned above, and specifically perform "system evolution".

2.2 The Role of the REFLECT Ontology

The global structure of the ontology adheres to the conceptual view as depicted in Figure 2(b), i.e. the ontology is hierarchically organized into four modules being named ***devices***, ***features***, ***constructs*** and ***concepts***, and these modules incorporate those items or concepts, which make the ontology suited for discussing pervasive adaptive applications[3].

In the course of elaborating or specifying particular applications, the notions of the different modules have to be related to each other. During this activity, in spelling out the details of the applications under consideration, the items of the ontology will be enriched by attributes and operations. Sometimes, there will be the need for adding further notions. Consequently, the ontology outlined in section 3 is under permanent development. Because of its role to be used in UML specifications of REFLECT applications it constantly grows in depth and breadth.

The role of the REFLECT ontology as being the base for modelling REFLECT applications will in turn be mirrored in the REFLECT development system. Here, ontology entries can be used as a base to provide reusable software components.

In the next subsections, for each of the four levels introduced in Figure 2(b), we sketch the internal structure of the associated modules as well as the relations between

---

[3] The four modules are depicted by the UML package diagram standing for a collection of classes, which can be considered as a collection of notions. The dotted line expresses dependency: the ***features*** module depends on the ***devices*** module, the ***constructs*** module on the ***features*** module, and the ***concepts*** module on the ***constructs*** module.

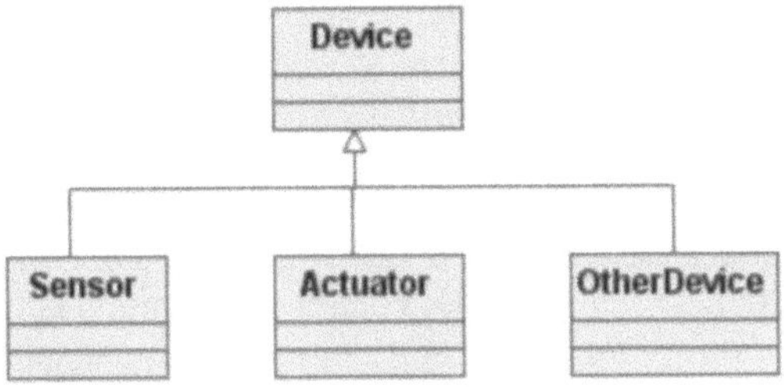

Fig. 3. Modelling Devices

close-by modules. In this, relations between close-by modules are illustrated by giving examples showing which items play together in the sensing or reacting stage of the sense-analyse-react cycle mentioned above.

2.3 Modelling Devices

The devices module contains entries for sensors, actuators, or other devices, like for example hard disks containing configuration settings. In Figure 3, the top level UML diagrams of the ***devices*** package are presented[4].

Sensor examples are ***Camera***, ***Thermometer*** or ***BloodPressureSensor***, ***Actuato***r examples are ***RoomLight*** or ***Radio***, a hard disk is an example for ***OtherDevice***.

2.4 Modelling Features

Features may be measurements like pulse or temperature stemming from sensors, actuating variables like tone or volume belonging to actuators, or facts like age or gender stemming from a hard disk. More complex examples for features are the mean heart rate or the mean temperature. In Figure 4, the top level UML diagrams of the ***features*** package are presented.

The left part expresses that each ***Feature*** is realized and/or implemented by one or more ***Device*** items, the middle part expresses that a ***Feature*** can be a ***UserFeature*** or

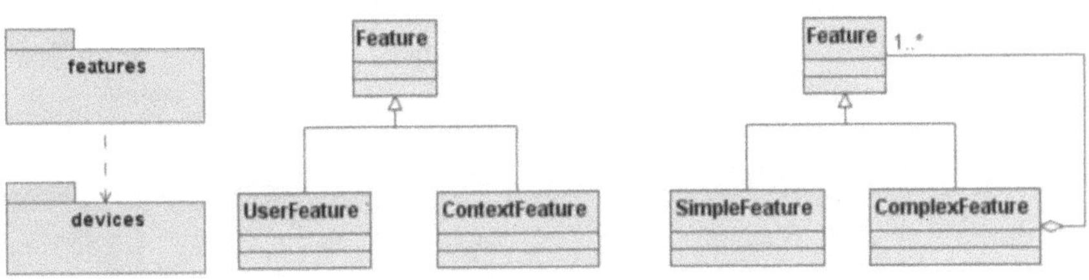

Fig. 4. Modelling Features

[4] For a single notion the UML class diagram is used. Beside the class name, a class diagram can also show attributes and operations belonging to a class, and later we will see examples for this. A line with a triangle at its head leads from a special notion (subclass) to a more general one (superclass): ***Sensor***, ***Actuator***, and ***OtherDevice*** are special forms of the superclass ***Device***.

a ***ContextFeature***, and the right part expresses that one has to distinguish between ***SimpleFeature*** and ***ComplexFeature***[5].

The term ***UserFeature*** is used, if the respective feature refers to the emotional, cognitive, or physical state of the user (e.g. ***Pulse*** or ***Movement***). The term ***ContextFeature*** refers to the user context (e.g. ***Time***). Note however that some features will be user and context features at the same time (e.g. ***Temperature***). Anyway, in many cases the final category depends on the application under consideration.

Notions like ***Pulse***, ***Movement***, or ***Temperature*** are simple features. Complex features are items being derived from one or more other (simple or complex) features, an example would be ***MeanTemperature***.

As noted above, features are used for sensing and reacting. Consequently, in modelling applications they will be related to sensor or actuator items, or to both. Examples for the first case would be ***FacialExpression*** and ***HeartRate***, and examples for the last case would be ***RoomTemperature*** and ***RoomLighting***. (A camera might be used to observe the facial expression, but there is no device to directly manipulate it. On the other hand, the room temperature can be measured by a thermometer and can be adjusted by a heater control.)

2.5 Modelling Constructs

Any REFLECT application considers the user's current context and state and adapts itself to it. As far as the user's state is concerned, one has to distinguish between the emotional state (e.g. annoyance) and cognitive engagement (e.g. high mental workload) of the user and physical conditions and actions (e.g. temperature and movement). Dealing with the system reaction to a given context and user state requires a notion of "goal", which is used with environment and user data to infer actions to be taken.

In Figure 5, the top level UML diagrams of the ***constructs*** package are presented. The left part expresses that each ***Construct*** item is an aggregate of one or more ***Feature*** items, and the right part says that one has to distinguish between ***UserContext***, ***EmotionalState***, ***CognitiveState***, ***PhysicalState***, and ***SystemGoal***.

Examples for emotional or cognitive states are ***Motivation***, ***MentalOverload***, ***Comfort***, ***Effort*** and ***Mood***.

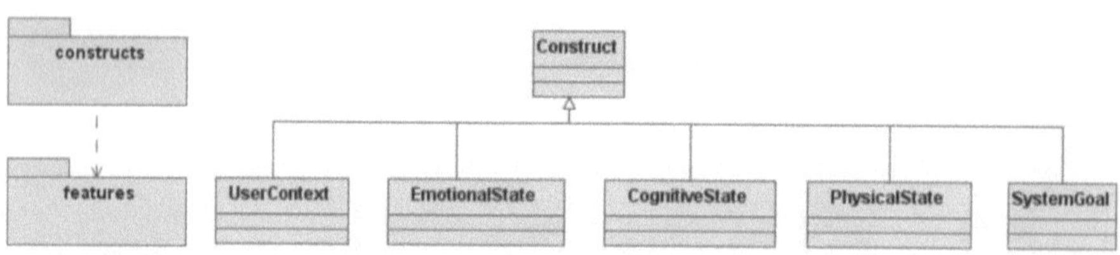

Fig. 5. Modelling Constructs

[5] A solid line with a diamond at its head expresses aggregation. The class pointed to by the diamond "contains" the source class. An additional annotation "1..*" at the source side means that it contains one ore more source class items: a ***ComplexFeature*** consist of one or more ***Feature*** items, which themselves may be of the ***SimpleFeature*** or the ***ComplexFeature*** type.

Again, in the process of modelling applications, notions from the constructs module are being related to notions from the feature module, to reflect the sensing or reacting stage respectively. As an example, consider the ***Mood*** construct. Determining the actual mood level belongs to the sensing stage and would be reflected in relating the mood notion for example to features like ***FacialExpression*** or ***HeartRate***. On the other hand, for the reacting stage, i.e. to influence the mood, ***SystemGoal*** items would build on features like ***RoomTemperature*** or ***RoomLight***, or on musical features.

2.6 Modelling Concepts

The idea of the concepts module is to provide a means for easy description of application scenarios. One example of a simple scenario is: "If the driver has been driving for a long time and he frequently changes his sitting positions then give him some visual/audio warning telling him to rest a bit". This module effectively connects available user constructs (obtained from the connected sensors) with possible system goals (depending on the involved actuators) and the application description. Development of this module is something to be undertaken in the future. In Figure 6, the top level UML diagram of the ***concepts*** package is presented.

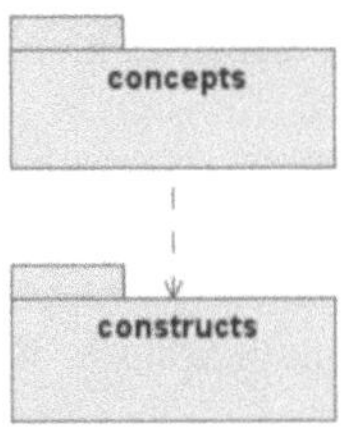

Fig. 6. Modelling Concepts

It expresses that each ***Concept*** is an aggregate of one or more ***Construct*** items.

3 The REFLECT Ontology

To present some technical details of the REFLECT ontology, this section illustrates the ***features*** module. More details about the ***devices***, ***constructs***, and ***concepts*** module can be found in [20].

3.1 The Features Module

The idea of the ***features*** module is to present ontology of both user and environmental or context features (Figure 4). Those features can be used as input values to some categorization system that would return high level concepts presented in ***constructs*** module based on those features (see the section 2.5).

As stated in section 2.4, one has to distinguish between simple and complex features, where complex features are items being derived from one or more other

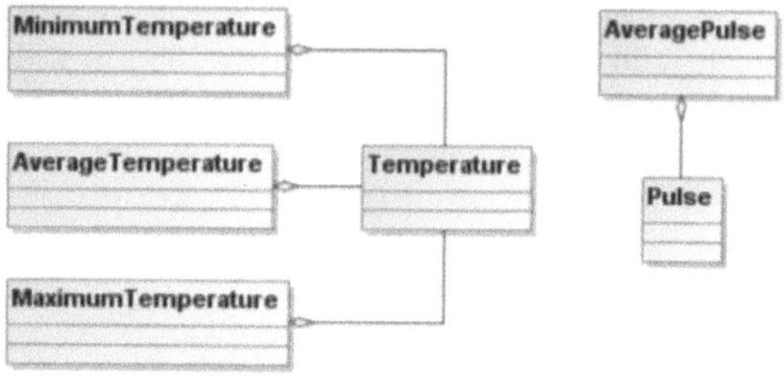

Fig. 7. Examples of complex features

(simple or complex) features. Examples of simple features are ***Pulse***, ***Age***, or ***Temperature***, and examples of complex features are ***AverageTemperature***, ***AveragePulse***, ***MinimumTemperature***, etc. (Figure 7)[6].

3.2 User Features

Figure 8 illustrates the full complexity of user features. A top level feature is further decomposed taking into account voice features, facial expressions and body postures, and numerous psycho-physiological measurements.

Firstly, as far as voice features are concerned, one has to take into consideration acoustic features of emotional expression, but also the fact that people express their feelings with the content they want to say, as well as through non-linguistic vocalizations like laughs, cries, sights, yawns. However, researchers found that linguistic messages are rather unreliable means to analyze human behaviour, and that it is very difficult to anticipate a person's word choice and the associated intent in affective expressions [24].

Secondly, concerning facial expressions, the work of Eckman and Friesen [5] needs to be taken into account, where authors developed the Facial Action Coding System (FACS), in which they defined 48 different action units (AUs) in relation to different muscle groups. Each emotion is described by a combination of AUs activity.

And thirdly, neuro-physiological features are often not convenient for emotion recognition in daily situations, as they require somewhat intrusive instruments for their measurement (when constructing taxonomy of neuro-physiological features the [7] document was considered).

To cope with the above mention problems numerous devices need to be included offering sometime redundant features which can lead into more precise diagnosis of users' psychological constructs. Especially in a given concrete situation with a predictable behaviour, the system can offer reliable diagnoses.

3.3 Context Features

The text that follows gives a few concrete examples of context features in concrete application settings.

---

[6] The rest of figures in this section do not show explicit distinction between simple and complex features, as their goal is to present grouping of features to either user or contextl features.

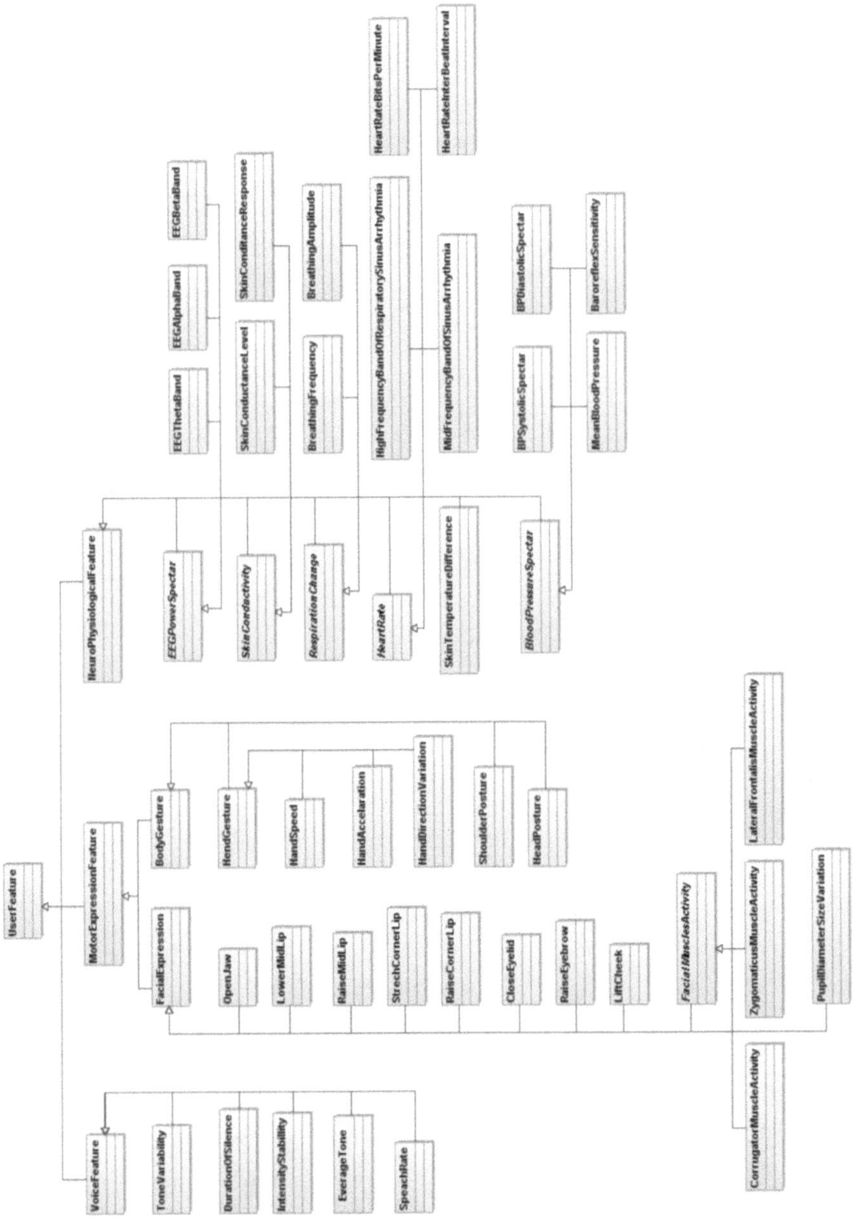

Fig. 8. Modelling User Features

Figure 9 illustrates the full complexity of context features. A top level feature is further decomposed taking into account multiple environment settings (music, weather, inside/outside features, etc.).

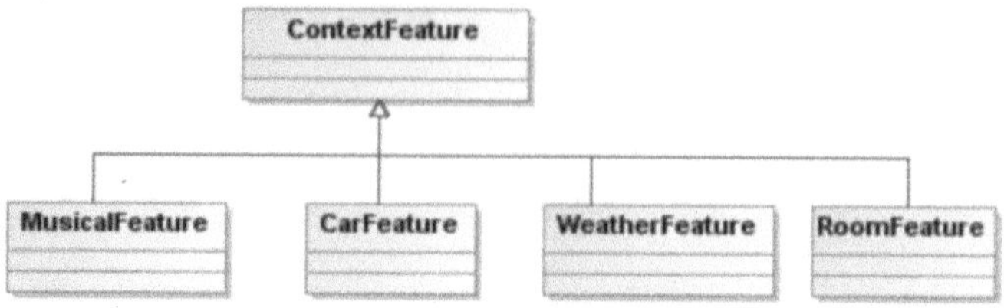

Fig. 9. Modelling context features

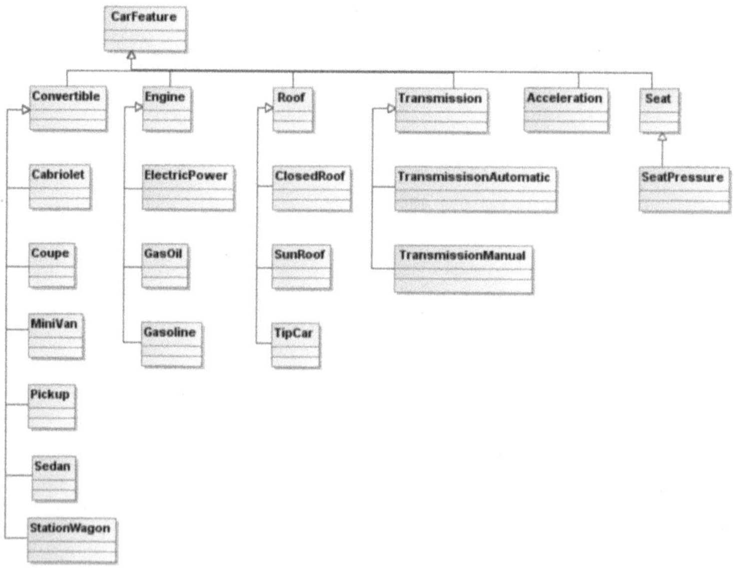

Fig. 10. Modelling car features.

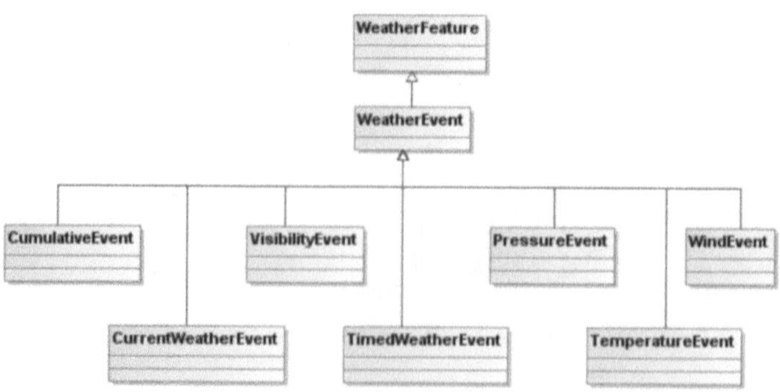

Fig. 11. Modelling weather features

Fig. 12. Modelling musical features

Figure 10 illustrates the full complexity of the vehicular environment. A top level feature is further decomposed taking into account vehicular ambient, engine etc.

Figure 11 illustrates the full complexity of weather features. A top level feature is further decomposed taking into account different elements that constitute the weather condition in a given moment.

Figure 12 illustrates the full complexity of music features. A top level feature is further decomposed taking into account a well known description of the music (taken from other sources).

3.4 User Context

Figure 13 illustrates the full complexity of the user context. The top level feature is further decomposed taking into account the aspects of personal, social, working and other situations that user may be presently engaged in.

In the process of constructing this module that models generic features a special care must be taken when dealing with the context (either environment, personal, task, social, or spatial-temporal context as shown in the Figure 13). In REFLECT system,

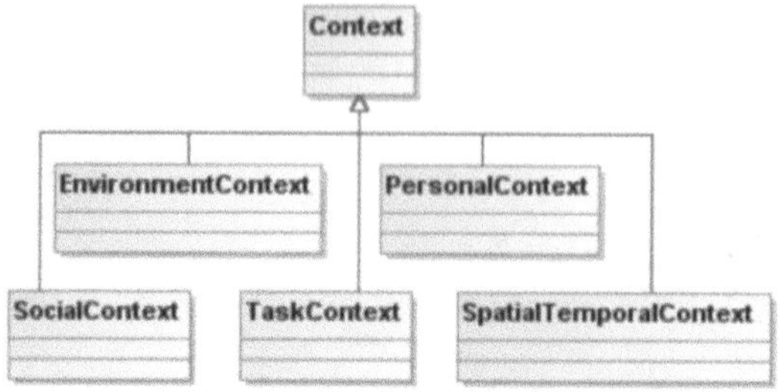

Fig. 13. Modelling context

there are multiple ways of obtaining context information, it can be derived from the previous context and history of user actions, or it can be gained interactively (e.g. in a car environment will the driver provide information to the system by the means of GUI application asking him whether he is driving to the work, or to the gym; or maybe, the system will derive that information without the driver's involvement from the GPS device).

4 A Case Study

In order to show how the ontology items presented in the last sections can be used to model an application, here we discuss the vehicle as a co-driver as introduced in [19]. That paper consists of an analysis part, where the fatigue and stress of a driver is derived from sensor input, and a related vehicular response part. For the sake of simplicity, here we only consider the analysis part.

The next section sketches the informal analysis from [19], and then we show how the ontology can be used to formalize this analysis.

4.1 Driving Scenario

Under the assumption that a driver has been driving for a long time, fatigue and stress will inevitably increase. An adaptive vehicular system would be able to detect changes in driver's psycho-physical state analysing three kinds of variations:

1. gesture/movements:
   a) augmented frequency of changes in sitting position using measures of pressure applied on the seat
   b) variations in head movements and eye blink using embedded CCD cameras

2. controls interactions:
   a) prolonged time and reaction (frequency and intensity at pedals)
   b) increased frequency of steering wheel corrections and grip force variations
   c) increased frequency of gear changes

3. psycho-physiological parameters:
   a) skin temperature variations
   b) augmented heart rate
   c) augmented skin conductivity

All this information is gathered by the adaptive-control system forming and maintaining the current driver's and driving conditions.

4.2 Modelling the Scenario

The topic of the informal analysis is to describe the evolution of the constructs ***Stress*** and ***Fatigue***. We consider the former as a special form of a ***PhysicalState***, and the latter as a form of ***EmotionalState***. Next, it is said that both constructs can be analysed by observing (1) gesture/movements, (2) controls interactions, and (3) psycho-physiological parameters. For these complex features we introduce the notions (1) ***DriverGesture***, (2) ***DriverInteraction***, and (3) ***DriverCondition***. Using the established notions, the overall formal modelling of the scenario is presented in Figures 14 and 15.

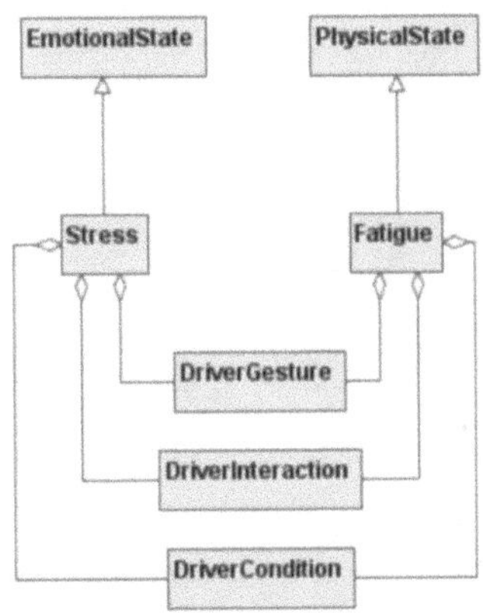

Fig. 14. Case Study (1)

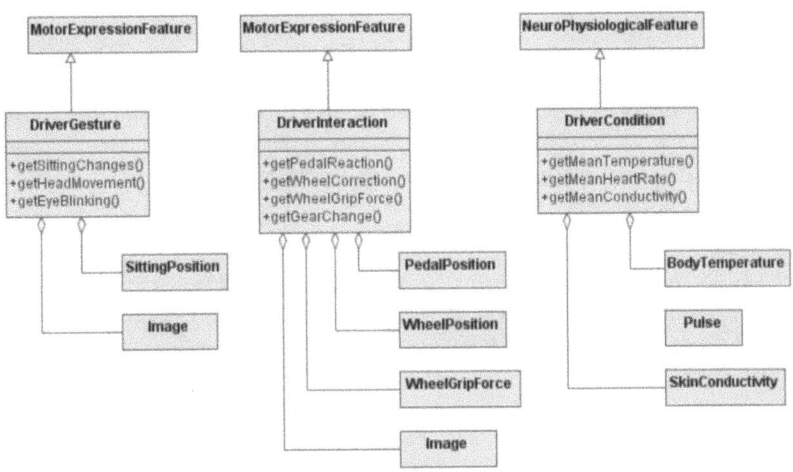

Fig. 15. Case Study (2)

Figure 14 expresses the top-level relations mentioned above, and Figure 15 says that the complex features ***DriverGesture*** and ***DriverInteraction*** are special forms of a ***MotorExpressionFeature***, where the ***DriverCondition*** is considered to be a ***Neuro-PhysiologicalFeature***. Moreover, the basic constituents of these complex features are detailed.

For example, it is said that the analysis of the ***DriverGesture*** is based on the features ***Image*** and ***SittingPosition***, and the mentioned ***getter***-operations reflect the subitems 1.a)[7] and 1.b)[8] of the last section. Similar statements hold for the other two complex features.

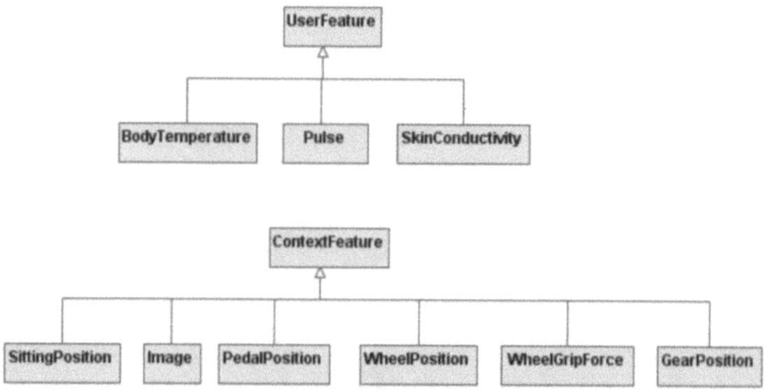

Fig. 16. Case Study (3)

Finally, let us consider Figure 16. Here, it is expressed that some of the basic constituents mentioned in Figure 15 are considered to be a ***UserFeature***, where others are considered to be a ***ContextFeature***.

5 Implementation Issues

This section elaborates implementation issues of the REFLECT system. As the focus here is on high-level design and the structural inter-dependency, only the course structure will be shown, indicating the data flow within the system components. A detailed software description is beyond the scope of this presentation.

5.1 REFLECT Design

The REFLECT system is designed to seamlessly collect information from humans and their surroundings (in this way forming a personal and contextual knowledge of a system), to adapt and react to their state, and finally to influence their state. This influence is achieved by impacting on a humans' environment, in a way that is according to the current context of a system (is environment a room, a car, a building, is it raining, is it a loud environment, etc.) and user's previous actions and preferences.

[7] "augmented frequency of changes in sitting position using measures of pressure …".

[8] "variations in head movements and eye blinking using embedded CCD cameras".

The REFLECT system is operating in the following way (see Figure 17):

- Each type of sensor that exists in the system (that is being used for gathering information about the environment) is described in the sensor ontology. This ontology offers, for each type of sensor, properties that match the output from this sensor's type (usually defined in a sensor specification).
- Data gathered by sensors then need to be processed in order to get features that can be used for identifying current emotional, physical and cognitive state of the user. This processing involves the use of complex computer algorithms for image processing, speech processing, voice analysis, etc. (e.g. algorithms for statistical signal processing). Today there are lots of available functional implementations in this area, and REFLECT system encourages their reusability and provides a mechanism for their easy integration.

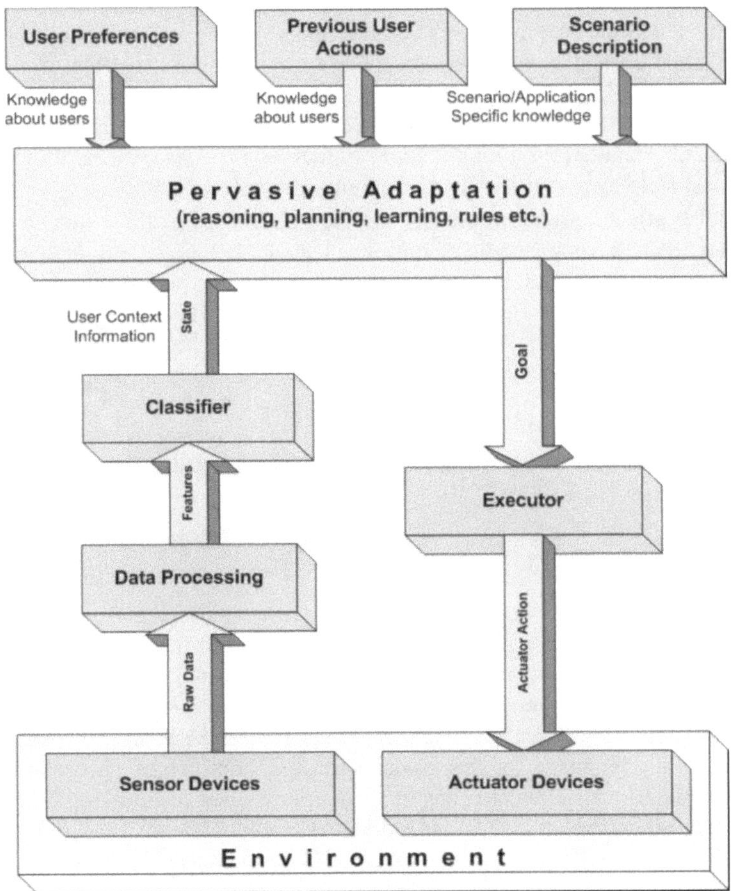

Fig. 17. High level view of the REFLECT system

- Each feature gained as an output of these processors is defined in the features ontology. Based on those features REFLECT system has to infer the state (emotional, physical and cognitive) of the user. There are lots of solutions that can be used for emotion recognition from available features – the majority of which fall into group of so called classifiers (machine learning methods for classification): e.g. support vector machines [1] neural networks [8] or rule-based classifiers [17]. REFLECT system does not offer its own classifier, but rather provides a way for easy integration of existing solutions.
- The output of classifier is a user's current state (each state is again defined in the state ontology). At this point REFLECT system contains information about the user's state, the user's preferences as well as history of previous user actions (system keeps records of previous choices a user made). It also knows which sensors and actuators are available. Based on this information system can decide on the future course of action, trying to adapt itself to the user needs. (Of course system needs to have a way for defining its initial behaviour that defines exact reaction in a certain situation. This initial behaviour defines system's course of action in a particular situation and with system's capability of learning it is able to continually enlarge "known" situations). REFLECT system relies on the use of rules for scenario description. These scenarios define goals (rules actions) that aim at changing the state of actuators on a high level (e.g. maintain the state of the user, reinforce emotional state by music and lightning, reinforce music, etc.)
- High level goals defined previously are then forwarded to the component called executor, that decomposes these high-level goals to the specific actuator actions (e.g. for a music player it changes a song within certain genre, for a lamp it changes a colour of the light, etc.). The executor knows what actuators are available in the system, and how concrete actuator needs to be changed in order to satisfy the given goal (because it keeps a track of the states of available actuators). The executor is aware of all the actions each actuator is able to perform, as all actuators are described in actuator ontology. This ontology gives classification of actuators and it defines each actuator through actions that it can perform (e.g. the "radio" concept has properties like playSong, stopSong, adjustVolume, etc.).

5.2 Presenting Ontologies

As shown, REFLECT architecture relies on the use of ontologies. There are many approaches for modelling ontologies: e.g. using frames and first order logic; using description logic; using UML; using the entity relationship model. Ontology modelling is an essential part of the Semantic Web, where OWL [11] is widely considered as the standard ontology modelling language. However (as proposed in [10]) we are using UML [27] as the language for describing REFLECT ontologies[9]. More specifically, UML profiles and their extension mechanism are used, for customizing UML models for particular domain: stereotypes, tagged values, and constraints [23]. For specifying additional constraints UML diagrams are enriched with OCL [15] expressions. One of

---

[9] We are also considering the use of OMG submission specification *Ontology Definition Metamodel* – ODM [16] as this specification presents a family of MOF metamodels, mappings between those metamodels, mappings to and from UML, as well as a set of UML profiles.

the benefits of using UML is that UML will ease the future elaboration of ontologies, as it allows basing the discussion on easily understandable graphical representations. Another important reason is, that by using UML, the ontology can be related to the REFLECT software design in a straightforward manner. It will be possible to (semi-)automatically generate XML like documents from the given UML diagrams, which in turn can be used to parameterize the intended Java implementation of the REFLECT system. This can be achieved by relying on the Eclipse Modeling Framework [28] and model-to-model transformations[10]. Also, we should mention a plugin for Protégé [29] that provides an import and export mechanism between the Protégé knowledge model and the UML diagram.

Relying our knowledge representation techniques on Semantic Web technologies, such as describing a context of our system within a RDF/RDFS document (e.g. that we obtained by serializing UML class diagram) gives us many advantages, some of which are: 1) reasoning could be applied in order to augment existing context information; 2) this document could be queried by other objects in the system to perceive overall context information. For example we can describe an environmental context with the following RDF file (this file is presented in the RDF/XML notation. Alternative formats for presenting RDF which are more human friendly are Notation3 [30], Turtle [31] and N-Triples [32]):

```
<?xml version="1.0"?>
<rdf:RDF
     xmlns:rdf="http://www.w3.org/1999/02/22-rdf-syntax-ns#"
     xmlns=" http://reflect.first.fraunhofer.de/example1#">
    <!-- … -->
   <rdf:Description rdf:about=
                       "http://reflect.first.fraunhofer.de/example1#radio1">
     <state rdf:datatype="http://www.w3.orq/2001/XMLSchema#boolean>1</state>
     <volume rdf:datatype="http://www.w3.orq/2001/XMLSchema#int">
            4
      </volume>
      <station rdf:resource="http://www.b92.net/radio"/>
     </rdf:Description>
     <rdf:Description rdf:about=
                           "http://reflect.first.fraunhofer.de/example1#tv1">
       <state rdf:datatype="http://www.w3.orq/2001/XMLSchema#boolean>1</state>
       <channel rdf:resource=" http://www.b92.net/tv"/>
     </rdf:Description>
   <!-- … -->
   </rdf:RDF>
```

While a user context can be described like:

```
<?xml version="1.0"?>
<rdf:RDF
     xmlns:rdf="http://www.w3.org/1999/02/22-rdf-syntax-ns#"
     xmlns=" http://reflect.first.fraunhofer.de/example2#">
    <!-- … -->
   <rdf:Description rdf:about=
                    "http://reflect.first.fraunhofer.de/example1#person1">
     <name rdf:datatype="http://www.w3.orq/2001/XMLSchema#string>
   firstName lastName</name>
     <age rdf:datatype="http://www.w3.orq/2001/XMLSchema#unsignedByte">
```

[10] Model transformation language like ATLAS Transformation Language (ATL) [2], as ATL has advantage of having a large repository of many different metamodels and developed transformations contributed by the community. One of the provided beneficial ATL use cases is "ODM Implementation (Bridging UML and OWL)" [3].

```
            20
      </age>
      <favouriteMusicGenre rdf:datatype="http://purl.org/ontology/mo#Genre">
         http://dbpedia.org/resource/Baroque_music </favouriteMusicGenre>
    </rdf:Description>
  <!-- … -->
  </rdf:RDF>
```

5.3 Defining REFLECT Scenarios

As stated in the section 2.6, the idea of the ***concepts*** module is to provide a means for easy description of application scenarios. One example of a simple scenario is: "If the driver has been driving for a long time and frequently changes the sitting position then re-shape the seat and suggest a coffee-break". Concepts module effectively connects available user constructs (obtained from the connected sensors) with possible system goals (depending on the involved actuators) and the application description. Development of this module is something to be undertaken in the future - here just a global solution is presented.

When one thinks of a way to define a behavioral description of REFLECT system (i.e. a way to define different scenarios), one needs to have in mind the reactive behavior of REFLECT system. This is the reason we propose the use of rules that have constructs like preconditions and postconditions, more specifically production rules, or reaction rules, for defining application scenarios. At this stage we are only interested in production rules as there are already available mature tools for working with productions rules: rule engines like Jess [33] and Drools [34]. Both rule engines define their own languages for presenting rules, plus they also offer XML-based versions of those languages. This is an option that is appropriate for us, as we saw XML is a predominant language in REFLECT system. It is easy to transform one XML format that we get as an output of a certain tool in REFLECT system (e.g. classifier) to the required format of a rule engine, by using model transformations or XSLT.

A major advantage of using rules is that the behaviour of REFLECT system can be modified dynamically according to the present context of the system, without the need to change other parts of a system. Also a rule based programming is declarative programming which is easier to understand then procedural programming.

To summarize, by presenting a context of REFLECT system (both environmental and user's context), and its behaviour in this way (i.e. with the use of taxonomies and rules) we get a system whose behaviour is not determined statically at its beginning. Rather, it is dynamically adapted to the new inputs, and it is able to react to those inputs in the appropriate manner.

6 Methodological Issues

When dealing with ontologies the idea is to take advantage of existing work that is being carried out in the area of Semantic Web. ("The Semantic Web is an evolving extension of the World Wide Web in which the semantics of information and services on the web is defined, making it possible for the web to understand and satisfy the requests of people and machines to use the web content" [35]). In the REFLECT project existing ontologies are reused, and only in the case where for certain domains no ontology is available, its creation is conducted explicitly. The ontological engineering process has

Table 1. REFLECT ontologies

(reengineered) Ontology	**Module**	**Information**
OntoSensor	Devices	OntoSensor includes defintions of concepts and properties adopted (in part) from SensorML, extensions to IEEE SUMO and references to ISO 19115. http://www.engr.memphis.edu/eece/cas/OntoSensor/OntoSensor
Music Ontology	Features	The Music Ontology Specification provides main concepts and properties of describing music (i.e. artists, albums, tracks, but also performances, arrangements, etc.) on the Semantic Web. http://musicontology.com/
Car Ontology	Features	http://sisinflab.poliba.it/colucci/files/webstart/Ontology_file/car.owl
Weather Ontology	Features	http://droz.dia.fi.upm.es/plugins/issues/weather/weather.owl
Emotion Ontology	Features/ Constructs	developed based on [13] and [22]
Medical Ontology	Features/ Constructs	Ontology Language for Medical and Biosciences Applications. http://www.medicalcomputing.net/owl/physiology.owl
Cognitive Ontology	Constructs	At the moment this ontology contains just four concepts, as proposed in [4]: interest, boredom, frustration, and puzzlement.

Remarks:

(1) Ontosensor ontology was chosen for the reason that it relies on the SensorML (OpenGIS Sensor Model Language Standard [36]): "SensorML provides standard models and an XML encoding for describing sensors and measurement processes".

(2) Regarding the weather ontology, ontology http://mnemosyne.umd.edu/~aelkiss/weather-ont.daml is also considered

(3) Regarding ontologies of emotions the work of Triezenberg [22] was considered as a good start. Also the HUMAINE [37] project was found to be extremely beneficial as this project offers many useful solutions regarding the recognition of emotional related states. E.g. the paper of Zeng et al [24], from the HUMAINE project, gives the detailed summary of affect recognition methods developed so far.

(4) Regarding the medical ontology, a simple solution was used – a solution that did not go in depth of human anatomy knowledge, as only concepts that are easily measured by the existing non intrusive methods are of interest to us. In the future projects like the digital human [38] and living human [39] will be considered in greater depth.

been based on methodologies and activities being defined in the literature (e.g. [9]), specifically those developed in the NeOn project [12] (e.g. specification of ontology, conceptualization, evaluation, formalization, implementation, maintenance, assessment and its use). Furthermore, an important role was played by a comparative survey [21] of some well recognized methodological approaches for ontology development (e.g. METHONTOLOGY, On-To-Knowledge, and DILIGENT methodologies).

For researching existing ontologies a number of specific search engines and repositories have been used:

- Watson (http://watson.kmi.open.ac.uk/WatsonWUI/)
- Swoogle (http://swoogle.umbc.edu/)
- Protege Ontology Library (http://protegewiki.stanford.edu/index.php/Protege_Ontology_Library)
- OntoSelect Ontology Library (http://olp.dfki.de/ontoselect/)

The chosen list of ontologies is presented in Table 1, where it is categorized according to the REFLECT ontology structure described earlier. The main criteria for the research was to find ontologies that imposed themselves as a standard in their domain, or otherwise covered the largest number of concepts for a certain domain. As ontology definition is an ongoing process, the list is liable to changes.

The following principles were applied during categorization of these ontologies: because the low level sensors concepts are assigned to the low level devices module, ontologies describing sensor devices are located to that module. The features module contains ontologies that cover features of both environmental context (we distinguish the following ontologies: music ontology, car ontology, weather ontology), and user context. As high level user's context and state (emotional state, cognitive engagement, physical conditions) is related to the constructs module, ontologies describing those states should be located at this module also.

However, because some of these ontologies do not fit nicely into our categorization (see Figure 2), some changes had to be made in their design, i.e. some reengineering had to be conducted - some concepts from existing ontologies were extracted and on such basis new concepts, relevant but not covered in original ontologies, were added. The process of reengineering existing domain ontologies requires that a set of activities (that concern the ontology development process) needs to be followed, which means that good ontological engineering practices have to be respected. This is why the NeOn methodology is proposed for building ontologies.

7 Conclusion

The ultimate goal of the approach described here is to supplement current smart systems with genuinely supportive behaviour in terms of doing what the users want, feel and desire in a seamless and personalized way. The novel approach introduces reflective technology that strives for simplicity offering a generic software/hardware solution for a wide range of application domains, ranging from vehicular, home ambient up to the embedded real-time systems.

Reflective software is structured in such a way to support various sensor and actuator devices (connected to possibly different processors) and dynamic (re-) configuration. That implies inherently parallel and distributed structures which are, according to the state-of-the-art in software technology, best achieved by service oriented computing and component-based structuring. The reflective software should be generic and applicable in a whole range of different application scenarios and event driven as it needs to react to the changing situation recognized and triggered by the sensor devices.

To follow the sound principles in software engineering, the reflective ontology has been developed that details concepts and structures that are involved in reflective systems. The ontology yields taxonomies which are used to implement low-level device interfaces and reflect services. The reflective ontology has been described in detail illustrating its elements, structures, relationships and the use. The ontology plays a central role in the whole process of reflective systems development as it captures generic features of the complex systems and allows for semi-formal description of the system components.

Reflective middleware is designed to be fully compatible with the reflective ontology. The tangible layer is responsible for the control of low level sensor/actuator devices. The second reflective layer performs analyses and evaluation offering an infrastructure for dynamic registry, event driven control, learning and reasoning. The application layer is responsible for scenario deployment through service/component composition and high level programming of the application logic.

The approach to design the ontology at first and then to design and to develop reflective software brought several benefits. Some of the advantages can be summarized in the following:

- abstract and generic approach to system design.
- separate development of low-level psycho-physiological measurement services (hidden in the tangible layer)
- separate development of the service and component oriented platform to support context-awareness
- separate development of high level adaptive and reflective components that combine kernel primitives, user profiles and application scenario (hidden in the application layer)
- early prototyping – allowing for different scenario probation and re-designing in an almost real framework
- easy interfacing to other existing taxonomies and ontologies of similar systems
- efficient final implementation of embedded control systems which requires only assembly of already tested modules from all three layers.

A comprehensive reflective ontology is under constant refinement and it allows for high-level programming and dynamic and flexible deployment of reflective system. Further work is oriented towards ontology extensions, both vertically and horizontally. Vertical refinement enriches the higher level tears: (1) reflective layer responsible for adaptivity and reflectivity achieved by composing lower level sensor/actuator services and (2) application layer responsible for user/application scenario profiles that are used to configure the application. Horizontal refinement is a constant task that involves enrichment of existing device descriptions. New sensors and actuators are to be added extending the capabilities of the tangible layer.

References

1. Ashraf, A.B., Lucey, S., Cohn, J.F., et al.: The painful face: pain expression recognition using active appearance models. In: Proc. of the ACM Int. Conf. on Multimodal Interfaces 2007 (ICMI 2007), pp. 9–14 (2007)
2. ATL (ATLAS Transformation Language), `http://www.eclipse.org/m2m/atl/`
3. ATL Use Case - ODM Implementation Bridging UML and OWL, `http://www.eclipse.org/m2m/atl/usecases/ODMImplementation/`
4. Chen, L.S.H.: Joint processing of audio-visual information for the recognition of emotional expressions in human-computer interaction. PhD Thesis, University of Illinois at Urbana-Champaign (2000)
5. Eckman, P., Friesen, W.V.: Facial Action Coding System: Investigator's Guide. Consulting Psychologists Press (1978)
6. Fairclough, S.H.: BCI and Physiological Computing for Computer Games: Differences, Similarities & Intuitive Control. In: CHI 2008, Florence, Italy (April 2008)
7. Fairclough, S.H., Dijksterhuis, C., et al.: Sensors: notes & specification. REFLECT WP3&4 deliverable (2008)
8. Fasel, B., Monay, F., et al.: Latent semantic analysis of facial action codes for automatic facial ex-pression recognition. In: Proc. of the 6th ACM SIGMM Int. workshop on Multimedia information retrieval, pp. 181–188 (2004)
9. Gómez-Pérez, A., Fernández-López, M., et al.: Ontological Engineering with examples from the areas of Knowledge Management, e-Commerce and the Semantic Web (2004)
10. Kogut, P., Cranefield, S., et al.: UML for Ontology Development. The Knowledge Engineering Review 17(1), 61–64 (2002)
11. McGuiness, D.L., van Harmelen, F.: OWL Web Ontology Language Overview. W3C Recommendation, February 10(2004)
12. NeOn - Lifecycle Support for Networked Ontologies, `http://www.neon-project.org/`
13. Obrenovic, Z., Garay, N., et al.: An Ontology for Description of Emotional Cues. In: Tao, J., Tan, T., Picard, R.W. (eds.) ACII 2005. LNCS, vol. 3784, pp. 505–512. Springer, Heidelberg (2005)
14. Obrenovic, Z., Starcevic, D., et al.: Using Ontologies in Design of Multimodal User Interfaces. In: Rauterberg, et al. (eds.) Human-Computer Interaction - INTERACT 2003, pp. 535–542. IOS Press & IFIP (2003)
15. OCL, Object Constraint Language. OMG Available Specification, Version 2.0, formal/06-05-01 (2006)
16. ODM, Ontology Definition Metamodel. In: OMG ODM 2005 (2005) (3rd Revised Submission)
17. Pantic, M., Patras, I.: Dynamics of facial expression: recognition of facial actions and their temporal segments form face profile image sequences. IEEE Transactions on Systems, Man and Cybernetics – Part B 36(2), 433–449 (2006)
18. Schroeder, A., van der Zwaag, M., Hammer, M.: A Middleware Architecture for Human-Centred Pervasive Adaptive Applications. In: 1st PerAda Workshop at SASO, Venice, Italy (2008)

19. Serbedzija, N.B., Calvosa, A.M., Ragnoni, A.: Vehicle as a Co-Driver. In: Proc. of the First Annual Int. Symposium on Vehicular Comp. Systems, ISVCS 2008, Dublin, Ireland, July 22 - 24 (2008)
20. Serbedzija, N.B., Kock, G., et al.: REFLECT Deliverable D1.1, First Year Report: Requirements and Design (2009)
21. Suárez-Figueroa, M.C., Dellschaft, K., et al.: NeOn Methodology for Building Contextualized Ontology Networks. NeOn Deliverable D5.4.1 (2008)
22. Triezenberg, K.: The ontology of emotions. PhD Theses, Center for Education and Research in Information Assurance and Security, Purdue University (2006)
23. Unified Modeling Language (OMG UML), Infrastructure, V2.1.2, http://www.omg.org/technology/documents/modeling_spec_catalog.htm#UML
24. Zeng, Z., Pantic, M., et al.: A Survey of Affect Recognition Methods: Audio, Visual, and Spontaneous Expressions (2007)
25. REFLECT Project, http://reflect.first.fraunhofer.de
26. Wikipedia entry, http://en.wiktionary.org/wiki/ontology
27. UML Resource Page, http://www.uml.org/
28. Eclipse Modeling Framework Project, http://www.eclipse.org/modeling/emf
29. The Protégé Ontology Editor and Knowledge Acquisition System, http://protege.stanford.edu
30. The Notation 3 Specification, http://www.w3.org/DesignIssues/Notation3
31. Turtle – Terse RDF Triple Language, http://www.w3.org/TeamSubmission/turtle
32. N-Triples Working Draft, http://www.w3.org/2001/sw/RDFCore/ntriples
33. Jess, the Rule Engine for the Java Platform, http://www.jessrules.com
34. Drools: Business Logic Integration Platform, http://www.jboss.org/drools
35. Wikipedia entry, http://en.wiktionary.org/wiki/Semantic_Web
36. Sensor Model Language, http://www.opengeospatial.org/standards/sensorml
37. The HUMAINE Portal, http://emotion-research.net
38. The Digital Human Open Source Software Consortium, http://www.fas.org/dh
39. The Living Human Digital Library, http://www.livinghuman.org

Author Index

FSC
www.fsc.org
MIX
Papier | Fördert
gute Waldnutzung
FSC® C083411

Kolibri 360 GmbH
Ferdinand-Jühlke-Straße 7
99095 Erfurt, Deutschland
produktsicherheit@kolibri360.de